Quasiconformal Mappings and Analysis

Springer Science+Business Media, LLC

Quasiconformal Mappings and Analysis

A Collection of Papers Honoring F.W. Gehring

Peter Duren Juha Heinonen Brad Osgood
Bruce Palka *Editors*

With 12 Illustrations

Springer

Peter Duren
Department of Mathematics
University of Michigan
4074 East Hall
Ann Arbor, MI 48109-1109
USA

Juha Heinonen
Department of Mathematics
University of Michigan
525 East University Street–East Hall
Ann Arbor, MI 48109-1109
USA

Brad Osgood
Department of Mathematics
Stanford University
Stanford, CA 94305-2125
USA

Bruce Palka
Department of Mathematics
University of Texas
Austin, TX 78712-1082
USA

Library of Congress Cataloging-in-Publication Data
Quasiconformal mappings and analysis : a collection of papers honoring F. W. Gehring / [edited by] Peter Duren . . . [et al.].
p. cm.
Includes bibliographical references.
ISBN 978-1-4612-6836-9 ISBN 978-1-4612-0605-7 (eBook)
DOI 10.1007/978-1-4612-0605-7
1. Quasiconformal mappings. 2. Mathematical analysis. I. Gehring, Frederick W. II. Duren, Peter L., 1935–
QA360.Q37 1997
515'.9–dc21 97-26432

Printed on acid-free paper.

Production managed by Anthony K. Guardiola; manufacturing supervised by Johanna Tschebull.
Photocomposed pages prepared from the editors' TeX files.

9 8 7 6 5 4 3 2 1

ISBN 978-1-4612-6836-9

Preface

In August 1995 an international symposium on "Quasiconformal Mappings and Analysis" was held in Ann Arbor on the occasion of Professor Frederick W. Gehring's 70th birthday and his impending retirement from the Mathematics Department at the University of Michigan. The concept of the symposium was to feature broad survey talks on a wide array of topics related to Gehring's basic research contributions in the field of quasiconformal mappings, emphasizing their relations to other parts of analysis. Principal speakers were Kari Astala, Albert Baernstein, Clifford Earle, Peter Jones, Irwin Kra, Olli Lehto, Gaven Martin, Dennis Sullivan, and Jussi Väisälä. Financial support was provided by the National Science Foundation, with additional grants from the University of Michigan and from the Institute for Mathematics and its Applications.

The symposium was a great success. The speakers rose to the occasion and presented excellent survey lectures. The present volume was conceived as a means for disseminating those expositions to a wider audience. Additional mathematicians, some of whom had not been able to attend the symposium, were invited to contribute similar articles. The result is a fitting tribute to Fred Gehring's pre-eminent role in developing the theory of quasiconformal mappings, through his own research and writings and lectures, and through his supervision of graduate students. The volume begins with descriptions of Gehring's mathematical career and an overview of his research achievements. Included are a complete (to date) bibliography of his publications and a list of his Ph.D. students. Next comes Olli Lehto's account of Gehring's interactions with the Finnish school of quasiconformal mappings. The remaining articles appear in alphabetical order.

The editors wish to acknowledge the efficient work of Candy Ellis, Ethel Rathbun, Lee Zukowski, and Kathleen Alcocer in coordinating the symposium at the University of Michigan. As this volume entered its final stages of preparation, the editors were especially fortunate to have the expert technical assistance of Margaret Combs at the University of Texas at Austin. The book owes much to her mastery of TEX typesetting and her tireless attention to detail. Ina Lindemann and the production staff of Springer-Verlag were also very helpful.

Finally, the editors take this opportunity to thank all of the writers for producing such a valuable collection of articles. Their hard work and obvious enthusiasm for the project is itself a fine tribute to Fred Gehring.

Peter Duren
Juha Heinonen
Brad Osgood
Bruce Palka

Contents

Preface vii

F.W. Gehring: A Biographical Sketch 1
Peter Duren

The Mathematics of F.W. Gehring 5
Juha Heinonen and Bruce Palka

Publications of F.W. Gehring 9

Ph.D. Students of F.W. Gehring 17

Fred Gehring and Finnish Mathematics 19
Olli Lehto

Planar Quasiconformal Mappings; Deformations and Interactions 33
Kari Astala

Local Minimality Results Related to the Bloch and Landau Constants 55
Albert Baernstein II and Jade P. Vinson

The Apollonian Metric of a Domain in $\mathbb{R}^n$ 91
A.F. Beardon

Conjugacy Invariants of Möbius Groups 109
C. Cao and P. L. Waterman

Extremal Quasiconformal Mappings in Plane Domains 141
Clifford J. Earle and Li Zhong

On the Asymptotic Behavior of Quasiconformal Mappings in Space 159
V.Ya. Gutlyanskiĭ, O. Martio, V.I. Ryazanov and M. Vuorinen

The Gehring Lemma 181
Tadeusz Iwaniec

Old and New on the Quasihyperbolic Metric 205
Pekka Koskela

Kleinian Groups; Eichler Cohomology and the Complex Theory 221
Irwin Kra

On the Geometry of Kleinian Groups 253
Gaven J. Martin

Old and New on the Schwarzian Derivative 275
Brad Osgood

The Gehring-Hayman Inequality in Conformal Mapping 309
Ch. Pommerenke and S. Rohde

Extremal Extensions from the Circle to the Disk 321
Edgar Reich

Construction of Quasiregular Mappings 337
Seppo Rickman

Mappings and Spaces 347
Stephen Semmes

Questions on Quasiconformal Maps in Space 369
Jussi Väisälä

List of Contributors 375

F.W. Gehring: A Biographical Sketch

Frederick William Gehring was born in Ann Arbor, Michigan on August 7, 1925. His father Carl Gehring was of German ancestry but the family had been established for many years in Cleveland, Ohio. Carl loved music and was an amateur composer. Although he started out studying engineering, he soon switched to journalism and later worked for the *Ann Arbor News* as state news editor and music critic. Fred's mother Hester Reed Gehring was the daughter of a physics professor at the University of Michigan (UM) who later became Dean of the Literary College. She and Carl met as undergraduates at UM. She went on to complete a Ph.D. in German and became a foreign language examiner for the UM Graduate School.

Fred attended school in Ann Arbor and graduated from University High School in 1943. He was admitted to M.I.T. to study physics or engineering, but there was a war on. He therefore enlisted instead in the U.S. Navy V-12 program, not knowing where he would be sent. By coincidence, the Navy sent him to Ann Arbor for a special year-round program in electrical engineering at UM. He graduated in 32 months with a double major in electrical engineering and mathematics. By then the war was over, but the Navy sent him to sea for 4 months.

Upon his return, he re-enrolled at UM and decided, at the suggestion of Ruel Churchill, to concentrate on mathematics. After a M.A. from UM in 1949, he went to Cambridge (England) on a Fulbright Scholarship and remained to get a Ph.D. in mathematics, taking courses from famous analysts such as J.E. Littlewood and A.S. Besicovitch while writing a thesis under the supervision of J.C. Burkill.

Fred and Lois at Kalle's christening in Helsinki

At Cambridge Fred met Lois Bigger, who had come from Iowa, also on a Fulbright, to continue her study of microbiology. Fred and Lois both got their Ph.D. degrees from Cambridge in 1952. Fred then went to Harvard as a Peirce Instructor, while Lois took a Research Fellowship at Yale. They were married in 1953 and recently celebrated their 43rd wedding anniversary.

In 1955, T.H. Hildebrandt managed to attract Fred back to the University of Michigan, where he has remained ever since. Then in 1958 came a decisive event in Fred's mathematical career. His colleague Jack Lohwater had spent the year 1956-57 in Helsinki, where he saw that Olli Lehto and others were working on the theory of normal functions, an area in which Fred was interested. Hoping to spend a year abroad, Fred applied for Fulbright, Guggenheim and NSF fellowships and was awarded all three. Lohwater put Gehring in contact with Lehto and Fred arranged to spend 1958-59 in Helsinki. But when he arrived, he found the analysts in Helsinki were now working on quasiconformal mappings. Fred knew little about this area, but he became interested in it and was able to collaborate with Finnish colleagues in developing some of the analytic properties of this class of mappings.

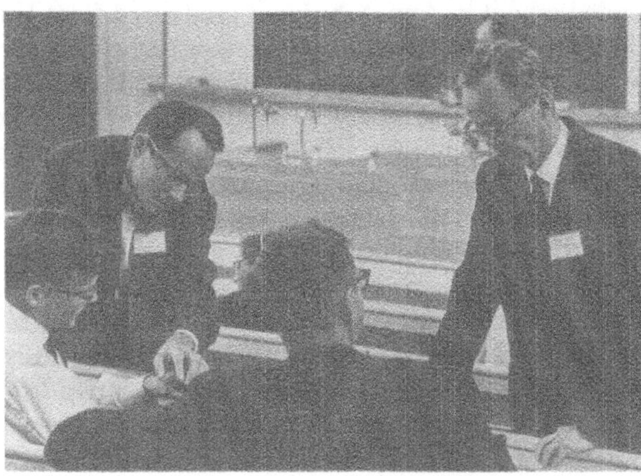

Gehring and Lehto in conversation with
Jussi Väisälä (left) and Kalevi Suominen

In 1959-60 Fred moved to Zürich where, stimulated by discussions with Albert Pfluger and a paper by C. Loewner, he began to develop the theory of quasiconformal mappings in higher dimensions. This theory included a proof of the Liouville theorem on conformal mappings in R^3 with no *a priori* differentiability hypotheses, in answer to a question raised earlier by Rolf Nevanlinna. In those two years abroad, Fred entered the developing field of quasiconformal mappings and made professional contacts, particularly with the Helsinki school, which would strongly influence the entire course of his career. (For a Finnish view, see Olli Lehto's article in this volume.)

In Helsinki the Gehrings learned to speak a little Finnish. Their son Kalle was born there and was named after Fred's father Carl and Kalle Virtanen. Their son Peter was born two years later in Ann Arbor.

Two Kalles: Virtanen and Gehring

On his return from Europe in 1960, Fred began training students in quasiconformal mappings, and his first student graduated from UM in 1963. By the time of his retirement in 1996, Fred had directed 28 Ph.D. students, including two who received degrees from other institutions. Almost all are currently employed in academic positions. As a measure of their loyalty and continuing scholarly activity, it may be reported that when the conference was held at UM in celebration of Fred's 70th birthday, 25 of his 28 students were in attendance.

During the 1960s, Ann Arbor became known as a center for research in quasiconformal mappings. Finnish colleagues, as well as mathematicians from other parts of the world, began coming to Ann Arbor as visitors, a tradition which would continue for many years. By the time of his retirement, Fred had sponsored visits by 42 foreign mathematicians at the University of Michigan.

At the University of Michigan, Fred served as Chairman of the Mathematics Department in 1973-75, 1977-80 and 1981-84. The University appointed him Collegiate (1984) and then Distinguished University Professor

Gehring, Olli Tammi et alia in the mathematical library in Helsinki

(1987), and honored him with a Distinguished Faculty Achievement Award (1981), the Henry Russel Lecturership (1990), and the Sokol Faculty Award (1994).

Fred was invited three times to address the International Congress of Mathematicians (1966, 1974, 1986). He was awarded honorary degrees from the Universities of Helsinki (1979) and Jyväskylä (1990), and the Norwegian University of Science and Technology (1997). He was elected to the American Academy of Arts and Sciences and the National Academy of Sciences in 1989. He also served 19 years on various committees of the American Mathematical Society, including 3 years on the Council, 4 years on the Executive Committee, and 10 years on the Board of Trustees.

Peter Duren

The Mathematics of F.W. Gehring

By widespread consensus, the modern theory of quasiconformal mappings had its birth in a 1954 *Journal D'Analyse* article of Lars Ahlfors that presented the first systematic treatment of various definitions for quasiconformal mappings in the correct generality; i.e., with no a priori smoothness imposed on the mappings. Before the Ahlfors paper, authors had largely elected to side-step the smoothness issue by considering quasiconformality only within the category of diffeomorphism. From the seeds planted by Ahlfors in that paper have developed three main ways of introducing quasiconformal mappings in the setting of Euclidean n-space, giving rise to the so-called metric, analytic and geometric definitions. (We refer the reader to Väisälä's article in this volume for further details on this subject.) The metric definition, which indeed makes sense for homeomorphisms between arbitrary metric spaces, stipulates uniform control over the infinitesimal distortion of spheres. It is arguably the most natural of the three definitions — and usually the easiest to verify in practice — but would appear, on the surface, to be less restrictive than either the analytic or the geometric definition. It was an open problem throughout most of the 1950s whether a homeomorphism between plane domains that satisfied the conditions of the metric definition would meet the requirements of the other two, seemingly stronger, definitions. In his first paper dealing with quasiconformal mappings [13], Gehring settled this question in the affirmative. The ingenious methods he pioneered in that paper remained for many years the only route to deriving regularity properties of quasiconformal mappings starting from minimal smoothness assumptions. They played a crucial part, for example, in Mostow's work on the rigidity of hyperbolic space forms. More importantly as far as the present narrative is concerned, [13] was the public announcement of Fred's baptism into the quasiconformal persuasion.

Foundational papers on three-dimensional quasiconformal mappings followed quickly on the heels of [13]. Among the most fundamental, and most frequently cited, of Fred's papers are [20] and [22]. Article [20] contains the deep results about the symmetrization of condensers and related extremal functions that number among their many significant implications the sharp local Hölder estimate for quasiconformal mappings. In [22] Gehring established the celebrated theorem that 1-quasiconformal mappings in space are restrictions of Möbius transformations, thus extending the classical Liouville theorem to mappings which are not a priori smooth. Another of

Fred's early papers on higher dimensional quasiconformal mappings that merits special comment is [26], which generalizes the Carathéodory convergence theorem to the province of quasiconformal mappings in space. This paper highlights two of the most important features of the class of K-quasiconformal mappings: its wealth of normal families and the fact that the limit of a sequence from this class, if a homeomorphism, is itself K-quasiconformal.

By the mid-1960s, the foundations of the theory of quasiconformal mappings in Euclidean spaces had been firmly established, thanks in large part to the efforts of Gehring, Väisälä and the very active Russian school. It had been manifest from the outset that the higher dimensional theory would be quite different from its plane counterpart, a beautiful illustration of which was provided by Gehring and Väisälä in their substantial *Acta* article [29]. In this paper, the authors conducted an elegant study of the mapping problem in three-space — a problem on which, it might be added, only piecemeal advances have been made in the time since the publication of [29]. (By the "mapping problem" is meant the problem of deciding whether two domains in Euclidean n-space can be mapped one to the other by a quasiconformal homeomorphism.) For instance, we learned from [29] that a Jordan domain in three-space whose boundary is smooth except for an outward-directed "spike" can be mapped quasiconformally to a ball, whereas a domain of similar description in which the spike points inward cannot be so mapped. This paper also introduced the technique of quasiconformally "bubbling-up towers" of arbitrary height with mappings of fixed dilatation, from its inception a mainstay in the construction of higher dimensional quasiconformal mappings with exotic boundary behavior. It was further demonstrated in [29] that the complement in three-space of the quasiconformal image of a ball has a property we now call "linear local connectivity," a simple but important concept which had surfaced for the first time in [27]. Subsequent works by Gehring on the mapping problem include [33], [49] (with Palka), where the quasihyperbolic metric and quasiconformal groups made their initial appearance, and [67] (with Martio). Similar in spirit is the lesser known but very interesting paper [35], which characterizes those quasiconformal images of round balls in three-space whose complements are contained in topological two-spheres.

In a 1966 article [30], Gehring and Reich attacked the area distortion problem for quasiconformal mappings in the plane, giving new estimates for the distortion exponent. The connection between this problem and the determination of the L^p-norm of the Beurling transform was also spelled out in the paper, thus providing great impetus for further research. Only recently did Astala provide a complete solution to the area distortion problem, an account of which can be found in his article in this volume.

There is little doubt as to which of Fred Gehring's papers is best known and has had the greatest impact on analysis — surely, it is the 1973 *Acta* paper [43] on the higher integrability of the Jacobian of a quasiconformal mapping. Not only did this paper generalize the highly acclaimed theorem of Boyarskii from the plane to higher dimensions, it also ushered into harmonic analysis a striking new notion of "self-improvement," which was later found to have a direct bearing on other significant concepts in harmonic analysis, notably functions of bounded mean oscillation and Muckenhoupt's A_∞-weights. Iwaniec's contribution to the book at hand sheds further light on this groundbreaking work. Let it just be mentioned here in passing that the influence of [43] extends far beyond quasiconformal mappings, harmonic analysis and partial differential equations. The fact that the Jacobian of a quasiconformal mapping belongs to a higher L^p-space than might be expected was, for instance, of the utmost importance in the work Donaldson and Sullivan did on four-manifolds and quasiconformal gauge theory in the late 1980s.

Before Fred's interests moved in the middle and late 1970s toward more classical problems on functions of one complex variable, he wrote a remarkable paper on lower dimensional absolute continuity properties of quasiconformal mappings [46]. This article had a lovely companion piece in [45], which studied the relationship between Hausdorff measure and the linking of sets in Euclidean space. Today, over twenty years later, questions of how topology can affect measures are important in both Riemannian and synthetic geometry. (See the essay on this theme written for inclusion here by Semmes.)

A perusal of Fred Gehring's list of publications for the decade 1977-87 leads to an inescapable conclusion: much of Fred's mathematical life during those years was spent wandering around in circles — quasiconformal circles, that is. The appearance in 1977 of a paper on univalent functions and Schwarzian derivatives [52] not only marked Fred's reentry, after some years' absence, into the published realm of good old-fashioned geometric function theory, but also set the dominant tone for his research over the next ten years. In [52] Gehring demonstrated that a generalization of the Nehari univalence criterion which Ahlfors had earlier shown to be valid in quasidisks actually characterized such domains. (For a more detailed discussion of this and related items consult the article by Osgood in the present volume.) It was followed by an impressive series of papers, some solo efforts and others collaborative undertakings, on a wide-ranging set of topics in geometric function theory (e.g., the Schwarzian derivative, the hyperbolic metric, the universal Teichmüller space). A unifying thread that ran through most of these papers was, to quote the master [72], "the ubiquitous quasidisk." The crown jewel amidst this assortment of gems is, by universal accord, the exquisite 1978 *Acta* paper [53], where Fred answered in the negative a long-standing question of Lipman Bers by proving that the universal Teichmüller space $T(1)$ is not dense in the space U of Schwarzian derivatives of functions which are analytic and univalent in the lower half-plane. The nature of the boundary of $T(1)$ in U is a fascinating subject whose continuing investigation was launched by the Gehring paper.

The year 1987 witnessed a dramatic shift in the overall direction of Fred Gehring's research. The change, attributable in no small measure to the chemistry of a productive collaboration with Gaven Martin, saw Fred, the consummate analyst, transformed into a dynamicist, a hyperbolic geometer, a three-manifold topologist and even — Saints, preserve us! — a dabbler in the algebraic arts! The Gehring & Martin tandem made its debut in 1987 with noteworthy papers on discrete quasiconformal groups [71] and discrete convergence groups [73]. (The latter article, which was somewhat overlooked at the time of its publication, has watched the value of its stock soar of late because of the prominent role convergence groups played in the settling of the Seifert conjecture for three-manifolds, a feat accomplished independently by Casson and Gabai.) What began as a steady, but tranquil stream of intriguing ideas soon developed into a veritable torrent of exciting results about Kleinian groups and hyperbolic orbifolds, some of which are surveyed in Martin's contribution to this book. The event that really caused the floodgates to open was the discovery of an unexpected relationship between the geometry of Kleinian groups and the iteration of certain complex polynomials [78]. Furthermore, although the bonanza of high quality information that has already been retrieved as a result of this insight is impressive, the strike appears at the present time nowhere near to being played out. Whatever the ultimate yield of this endeavor might prove to be, one thing is already perfectly clear: item No.100 on the Gehring publication list will by no means be his last word on this subject — or others. The beat, as they say, goes on.

Juha Heinonen
Bruce Palka

Publications of F.W. Gehring

1. *Images of convergent sequences in sets*, J. London Math. Soc. **26** (1951), 249–256.

2. *A scale of integrals from Lebesgue's to Denjoy's*, Quart. J. Math. Oxford Ser. **4** (1953), 210–220 (with J.C. Burkill).

3. *A study of α-variation*, Trans. Amer. Math. Soc. **76** (1954), 420–443.

4. *A note on a paper by L.C. Young*, Pacific J. Math. **5** (1955), 67–72.

5. *On the Dirichlet problem*, Michigan Math. J. **3** (1956), 201.

6. *On the radial order of subharmonic functions*, J. Math. Soc. Japan **9** (1957), 77–79.

7. *The Fatou theorem and its converse*, Trans. Amer. Math. Soc. **85** (1957), 106–121.

8. *The Fatou theorem for functions harmonic in a half-space*, Proc. London Math. Soc. **8** (1958), 149–160.

9. *The asymptotic values for analytic functions with bounded characteristic*, Quart. J. Math. Oxford Ser. **9** (1958), 282–289.

10. *On solutions of the equation of heat conduction*, Michigan Math. J. **5** (1958), 191–202.

11. *On the Lindelöf theorem*, Math. Nachr. **19** (1958), 165–170 (with A.J. Lohwater).

12. *On the total differentiability of functions of a complex variable*, Ann. Acad. Sci. Fenn. **272** (1959), 3–9 (with O. Lehto).

13. *The definitions and exceptional sets for quasiconformal mappings*, Ann. Acad. Sci. Fenn. **281** (1960), 3–28.

14. *Harmonic functions and Tauberian theorems*, Proc. London Math. Soc. **10** (1960), 88–106.

15. *The boundary behavior and uniqueness of solutions of the heat equation*, Trans. Amer. Math. Soc. **94** (1960), 337–364.

16. *The transformations which preserve the harmonic functions*, Ann. Acad. Sci. Fenn. **293** (1960), 3–13 (with H. Haahti).

17. *Rings and quasiconformal mappings in space*, Proc. Nat. Acad. Sci. U.S.A. **47** (1961), 98–105.

18. *On the geometric definition for quasiconformal mappings*, Comment. Math. Helv. **36** (1961), 19–32 (with J. Väisälä).

19. *A remark on the moduli of rings*, Comment. Math. Helv. **36** (1961), 42–46.

20. *Symmetrization of rings in space*, Trans. Amer. Math. Soc. **101** (1961), 499–519.

21. *Extremal length definitions for the conformal capacity of rings in space*, Michigan Math. J. **9** (1962), 137–150.

22. *Rings and quasiconformal mappings in space*, Trans. Amer. Math. Soc. **103** (1962), 353–393.

23. *An inequality in the theory of conformal mapping*, J. Math. Pures Appl. **41** (1962), 353–361 (with W.K. Hayman).

24. *A distortion theorem for functions univalent in an annulus*, Ann. Acad. Sci. Fenn. **325** (1963), 3–16 (with G. af Hällström).

25. *Quasiconformal mappings in space*, Bull. Amer. Math. Soc. **69** (1963), 146–164.

26. *The Carathéodory convergence theorem for quasiconformal mappings in space*, Ann. Acad. Sci. Fenn. **336/11** (1963–64), 3–21.

27. *Extension of quasiconformal mappings in three space*, J. Analyse Math. **14** (1965), 171–182.

28. *Angles and quasiconformal mappings*, Proc. London Math. Soc. **14A** (1965), 1–21 (with S.B. Agard).

29. *The coefficients of quasiconformality of domains in space*, Acta Math. **114** (1965), 1–70 (with J. Väisälä).

30. *Area distortion under quasiconformal mappings*, Ann. Acad. Sci. Fenn. **338** (1966), 3–15 (with E. Reich).

31. *Coefficients of quasiconformality of domains in three space*, in *Contemporary Problems in Theory of Analytic Functions (Erevan, 1965)*, pp. 83–88; Izdat. "Nauka", Moscow, 1966.

32. *Definitions for a class of plane quasiconformal mappings*, Nagoya Math. J. **29** (1967), 175–184.

33. *Extension theorems for quasiconformal mappings in n-space*, J. Analyse Math. **19** (1967), 149–169.

34. *Extension theorems for quasiconformal mappings in n-space*, Proceedings of the International Congress of Mathematicians (Moscow, 1966), pp. 313–318, Izdat. "Mir", Moscow, 1968.

35. *Quasiconformal mappings of slit domains in three space*, J. Math. Mech. **18** (1969), 689–703.

36. *Quasiconformal mappings which hold the real axis pointwise fixed*, in *Mathematical Essays dedicated to A.J. MacIntyre*, Ohio University Press, 1970, pp. 145–148.

37. *Extremal mappings between tori*, in *Certain Problems in Mathematics and Mechanics (on the Occasion of the Seventieth Birthday of M.A. Lavrent'ev)*, Izdat. "Nauka", Leningrad, 1970, pp. 146–152 (Russian); English transl. in Amer. Math. Soc. Transl. **104** (1976), 129–134.

38. *Inequalities for condensers, hyperbolic capacity, and extremal lengths*, Michigan Math. J. **18** (1971), 1–20.

39. *Lipschitz mappings and the p-capacity of rings in n-space*, in *Advances in the Theory of Riemann Surfaces*, Annals of Math. Studies No.66, Princeton Univ. Press, 1971, pp. 175–193.

40. *Quasiconformal mappings of a cylinder*, Proc. Romanian-Finnish Seminar on Teichmüller Spaces and Quasiconformal Mappings (Brasov, 1969), Publ. House of Acad. of Socialist Republic of Romania, Bucharest, 1971, pp. 171–186 (with F. Huckemann).

41. *Dilatations of quasiconformal boundary correspondences*, Duke Math. J. **39** (1972), 89–95.

42. *Hausdorff dimension and quasiconformal mappings*, J. London Math. Soc. **6** (1973), 504–512. (with J. Väisälä).

43. *The L^p-integrability of the partial derivatives of a quasiconformal mapping*, Acta Math. **130** (1973), 265–277.

44. *Quasiconformal mappings and Lebesgue density*, in *Discontinuous Groups and Riemann Surfaces (Proc. Conf. Univ. Maryland, 1973)*, Annals of Math. Studies, no.79, Princeton Univ. Press, 1974, pp. 171–179 (with J.C. Kelly).

45. *The Hausdorff measure of sets which link in Euclidean space*, in *Contributions to Analysis: A collection of papers dedicated to Lipman Bers*, Academic Press, New York, 1974, pp. 159–167.

46. *Lower dimensional absolute continuity properties of quasiconformal mappings*, Math. Proc. Cambridge Philos. Soc. **78** (1975), 81–93.

47. *Some metric properties of quasiconformal mappings*, Proceedings of the International Congress of Mathematicians (Vancouver, B.C., 1974), Vol.2, Canad. Math. Congress, Montreal, 1975, pp. 203–206.

48. *Absolute continuity properties of quasiconformal mappings*, Symposia Mathematica **18** (Convegno sulle Transformazioni Quasiconformi e Questioni Connesse, INDAM, Rome, 1974), Academic Press, London, 1976, pp. 551–559.

49. *Quasiconformally homogeneous domains*, J. Analyse Math. **30** (1976), 172–199 (with B.P. Palka).

50. *A remark on domains quasiconformally equivalent to a ball*, Ann. Acad. Sci. Fenn. **2** (1976), 147–155.

51. *Quasiconformal mappings*, in *Complex Analysis and its Applications (Lectures, Internat. Sem., Trieste, 1975)*, Vol. II, International Atomic Energy Agency, Vienna, 1976, pp. 213–268.

52. *Univalent functions and the Schwarzian derivative*, Comment. Math. Helv. **52** (1977), 561–572.

53. *Spirals and the universal Teichmüller space*, Acta Math. **141** (1978), 99–113.

54. *Remarks on the universal Teichmüller space*, Enseign. Math. **24** (1978), 173–178

55. *Some problems in complex analysis*, Proceedings of the First Finnish-Polish Summer School in Complex Analysis (Podlesice, 1977), Part II, pp. 61–64; Univ. Łódź, Łódź, 1978.

56. *Schwarzian derivatives, the Poincaré metric and the kernel function*, Comment. Math. Helv. **55** (1980), 50–64 (with A.F. Beardon).

57. *Uniform domains and the quasihyperbolic metric*, J. Analyse Math. **36** (1979), 50–74 (with B.G. Osgood).

58. *Analytic functions satisfying Hölder conditions on the boundary*, J. Approx. Theory **35** (1982), 243–249 (with W.K. Hayman and A. Hinkkanen).

59. *Characteristic Properties of Quasidisks*, Les Presses de l'Université de Montréal, Montréal, Que. (1982), 97 pp.

60. *Injectivity of local quasi-isometries*, Comment. Math. Helv. **57** (1982), 202–220.

61. *Quasidisks and the Hardy-Littlewood property*, Complex Variables Theory Appl. **2** (1983), 67–78 (with O. Martio).

62. *Conformally invariant length sums*, Indiana Univ. Math. J. **32** (1983), 809–829 (with J.B. Garnett and P.W. Jones).

63. *Injectivity criteria and the quasidisk*, Complex Variables Theory Appl. **3** (1984), 45–54 (with K. Astala).

64. *On the Nehari univalence criterion and quasicircles*, Comment. Math. Helv. **59** (1984), 226–242 (with Ch. Pommerenke).

65. *Polynomial approximation in quasidisks*, in *Differential Geometry and Complex Analysis* (H.E. Rauch memorial volume), Springer-Verlag, 1984, pp. 75–86 (with J.M. Anderson and A. Hinkkanen).

66. *Quasiconformal analogues of theorems of Koebe and Hardy-Littlewood*, Michigan Math. J. **32** (1985), 99–107 (with K. Astala).

67. *Lipschitz classes and quasiconformal mappings*, Ann. Acad. Sci. Fenn. **10** (1985), 203–219 (with O. Martio).

68. *Quasiextremal distance domains and extensions of quasiconformal mappings*, J. Analyse Math. **45** (1985), 181–206 (with O. Martio).

69. *Extension of quasi-isometric embeddings of Jordan curves*, Complex Variables Theory Appl. **5** (1986), 245–263.

70. *Injectivity, the BMO norm and the universal Teichmüller space*, J. Analyse Math. **46** (1986), 16–57 (with K. Astala).

71. *Discrete quasiconformal groups I*, Proc. London Math. Soc. **55** (1987), 331–358 (with G.J. Martin).

72. *Uniform domains and the ubiquitous quasidisk*, Jahresber. Deutsch. Math.-Verein. **89** (1987), 88–103.

73. *Discrete convergence groups*, in *Complex Analysis, I (College Park, Md., 1985-86)*, Lecture Notes in Math. vol. 1275 (Springer-Verlag, 1987), pp. 158–167 (with G.J. Martin).

74. *Remarks on uniform and quasiconformal extension domains*, Complex Variables Theory Appl. **9** (1987), 175–188 (with K. Hag).

75. *Topics in quasiconformal mappings*, Proceedings of the International Congress of Mathematicians (Berkeley, Calif., 1986); pp. 62–80, Amer. Math. Soc., Providence, R.I., 1987.

76. *Quasiconformal groups and the conical limit set*, in *Holomorphic Functions and Moduli, Vol.II (Berkeley, Calif., 1986)*; Math. Sci. Res. Inst. Publ., 11 (Springer-Verlag, 1988), pp. 59–67 (with J.B. Garnett and P.W. Jones).

77. *The matrix and chordal norms of Möbius transformations*, in *Complex Analysis: Articles dedicated to Albert Pfluger*, Birkhäuser Verlag, 1988, pp. 51–59 (with G.J. Martin).

78. *Iteration theory and inequalities for Kleinian groups*, Bull. Amer. Math. Soc. **21** (1989), 57–63 (with G.J. Martin).

79. *Stability and extremality in Jørgensen's inequality*, Complex Variables Theory Appl. **12** (1989), 277–282 (with G.J. Martin).

80. *Circular distortion of curves and quasicircles*, Ann. Acad. Sci. Fenn. **14** (1989), 381–390 (with Ch. Pommerenke).

81. *Quasihyperbolic geodesics in John domains*, Math. Scand. **65** (1989), 75–92 (with K. Hag and O. Martio).

82. *Crickets, zippers and the Bers universal Teichmüller space*, Proc. Amer. Math. Soc. **110** (1990), 675–687 (with K. Astala).

83. *Inequalities for Möbius transformations and discrete groups*, J. Reine Angew. Math. **418** (1991), 31–76 (with G.J. Martin).

84. *Some universal constraints for discrete Möbius groups* in *Paul Halmos: Celebrating 50 years of mathematics*, Springer-Verlag, 1991, pp. 205–220 (with G.J. Martin).

85. *Axial distances in discrete Möbius groups*, Proc. Nat. Acad. Sci. U.S.A. **89** (1992), 1999–2001 (with G.J. Martin).

86. *6-torsion and hyperbolic volume*, Proc. Amer. Math. Soc. **117** (1993), 727–735 (with G.J. Martin).

87. *The iterated commutator in a Fuchsian group*, Complex Variables Theory Appl. **21** (1993), 207–218 (with G.J. Martin).

88. *On the minimal volume hyperbolic 3-orbifold*, Math. Research Letters **1** (1994), 107–114 (with G.J. Martin).

89. *Commutators, collars and the geometry of Möbius groups*, J. Analyse Math. **63** (1994), 175–219 (with G.J. Martin).

90. *Chebyshev polynomials and discrete groups*, Proceedings of the Conference on Complex Analysis (Tianjin, 1992), International Press, Cambridge, Mass., 1994, pp. 114–125 (with G.J. Martin).

91. *Holomorphic motions, Schottky's theorem and an inequality for discrete groups*, in *Computational Methods and Function Theory*, World Scientific Publishing Co., 1995, pp. 173–181 (with G.J. Martin).

92. *On the Margulis constant for Kleinian groups, I*, Ann. Acad. Sci. Fenn. **21** (1996), 439–462 (with G.J. Martin).

93. *On the discreteness of the free product of finite cyclic groups*, Mitteilungen Math. Seminar Giessen **228** (1996), 9–15 (with C. Maclachlan and G.J. Martin).

94. *Lattice constants and a lemma of Zagier*, Contemporary Math. (1997), to appear (with C. Cao and G.J. Martin).

95. *Arithmeticity, discreteness and volume*, Trans. Amer. Math. Soc. (1997), to appear (with C. Maclachlan, G.J. Martin and A.W. Reid).

96. *Geodesics in hyperbolic 3-folds*, Michigan Math. J. **44** (1997), to appear (with G.J. Martin).

97. *Hyperbolic axes and elliptic fixed points in a Fuchsian group*, Complex Variables Theory Appl. (1997), to appear (with G.J. Martin).

98. *The volume of hyperbolic 3-folds with p-torsion, $p \geq 6$*, submitted (with G.J. Martin).

99. *Precisely invariant collars and the volume of hyperbolic 3-folds*, submitted (with G.J. Martin).

100. *The spectrum of elliptic axial distances in Kleinian groups*, submitted (with T.H. Marshall and G.J. Martin).

Ph.D. Students of F.W. Gehring

Name	Degree	Current Position
1. J.A. Kelingos	1963	1936-1996, requiescat in pace
2. S.B. Agard	1965	University of Minnesota
3. G.D. Anderson	1965	Michigan State University
4. S.T. Bagby	1966*	Indiana University
5. D.E. Wilson	1967†	Wabash College
6. M.F. Walker	1969	University of Capetown
7. M.K. Vamanamurthy	1969	University of Auckland
8. J.R. Quine	1971	Florida State University
9. D.K. Blevins	1972	Connecticut College
10. K.J.B. Hag	1972	University of Trondheim-NTH
11. J. Hesse	1972	Computer consultant in Minneapolis
12. B.P. Palka	1972	University of Texas at Austin
13. B.G. Osgood	1980	Stanford University
14. Y. Krishnangkura	1982	Srinakarinviot University
15. B.A.B. Flinn	1983	National Security Agency
16. D.A. Herron	1984	University of Cincinnati
17. G.J. Martin	1985	University of Auckland
18. C.A. Nolder	1985	Florida State University
19. W-D. Hwang	1986	National Taiwan University
20. D.T. Calvis	1988	Baldwin-Wallace College
21. S.G. Staples	1988	Texas Christian University
22. M. Ghamsari	1990	University of Cincinnati
23. N.R. Soderborg	1991	Ford Motor Company
24. S. Yang	1991	Emory University
25. K-W. Kim	1991	Pusan Women's University
26. L. Miller-Van Wieren	1994	Pennsylvania State University
27. C. Cao	1996	Boston College
28. N. Langmeyer	1996	National Security Agency

* Degree received from Harvard University.

† Degree received from Kansas State University.

All other degrees awarded by the University of Michigan.

Fred Gehring and Finnish Mathematics

Olli Lehto

1 Complex analysis comes to Finland

The title "Fred Gehring and Finnish mathematics" for this lecture was proposed to me by the organizers of this Conference. The topic is unusual, but unusually important, too, is the interplay between Fred Gehring and Finnish mathematics. By chance, it so happened that Fred's entry onto the Finnish mathematical scene enjoyed perfect timing, with the result that the impacts in both directions, Fred's on Finnish mathematics and that of Finnish mathematics on Fred, were exceptionally strong. In order to tell the whole story, which could also be entitled "How Fred became a quasi-person," let me go back all the way to the time when serious mathematical research began in Finland.

In the mid-nineteenth century, Finland had only one university, in Helsinki, and there was only one chair in mathematics. In 1858, the 29-year old Lorenz Lindelöf was appointed to this position. I mention his name here because there is a continuous connection from him to Fred Gehring. Lorenz Lindelöf contributed indirectly to the creation of the Finnish function-theoretic tradition through his son Ernst, through his foreign contacts, and through the decisive role he played in having Gösta Mittag-Leffler appointed as his successor.

Lorenz Lindelöf realized that it was mandatory for the remote Finland to be in close contact with some world mathematical center. For Finnish scientists of the day it was customary to seek wisdom from German sources. But Lindelöf was a Francophile who, as a young professor, had studied and done research in Paris six consecutive years, so it was easier for his son Ernst to gain access to the inner circles of the Parisian mathematical community at the end of the century. The choice of Paris was to be decisive for the development of Finnish mathematics.

Lorenz Lindelöf won international recognition for his research, but rather soon he turned to politics. Before leaving the University he fought vigorously in order to have Mittag-Leffler named as his successor. The Faculty agreed that, of all the candidates, Mittag-Leffler was scientifically the best qualified, but the case was clouded by a political language problem. In the

1870s the major teaching language at the University was Swedish. However, there was a strong movement at the time to make the University more Finnish, a language which Mittag-Leffler, who was a Swede, did not know. Ultimately, by the narrow margin of 12 to 10, the Faculty voted to recommend Mittag-Leffler, and he was appointed.

Mittag-Leffler's great achievement in Finnish mathematics was the introduction of complex analysis into Finland. Mittag-Leffler had spent some years in Berlin, where he had become a disciple and admirer of Weierstrass. He was offered a position there but, fortunately for Finland, preferred Helsinki. Mittag-Leffler was the first to give lectures on the theory of analytic functions at the University of Helsinki, and the first Finnish doctoral theses in function theory were done under his guidance.

Mittag-Leffler is linked to Fred Gehring through another connection independent of their shared love for complex analysis. While in Helsinki, Mittag-Leffler married the daughter of a Finnish general. Thanks to the wife's inherited wealth, the Mittag-Leffler Institute in Djursholm could be established, a place which Fred and many others of us know well.

2 The Finnish school of function theory

Mittag-Leffler left Finland after being invited to become professor at the newly established University of Stockholm. But the interest he had awakened in complex analysis survived his departure from Helsinki. His successor, Professor Edvard Neovius, who was an uncle of Rolf Nevanlinna and a first cousin of Ernst Lindelöf, was a geometer. He was a close personal friend of Hermann Amandus Schwarz, who visited Finland several times and kept mathematicians in Helsinki informed about the latest developments in complex analysis.

After Neovius, Ernst Lindelöf dominated Finnish mathematics for a long time. The Lindelöf family's close ties with Mittag-Leffler may have had an influence on Ernst Lindelöf's interest in complex analysis. Lorenz Lindelöf in all probability encouraged his son's interest in Paris, where Ernst twice spent year-long sabbaticals at the end of the century. Parisian mathematics ultimately determined the main direction Finnish mathematical research would take for the next hundred years.

In the 1890s, Paris was a center where an exceptionally high number of elite mathematicians were actively engaged in research in analysis. The leading figure was Henri Poincaré, but on the scene were Picard, Hadamard, and Painlevé, who were slightly older than Lindelöf, as well as Borel and others of roughly Lindelöf's age, i.e., between 25 and 30. Complex analysis was high Paris fashion then, especially the theory of entire functions. Lindelöf became the first non-French mathematician to make significant contributions to the value distribution theory of entire functions.

Despite the renown he gained for his outstanding research, Lindelöf was, above all, an educator. In the 1910s and early 1920s, a great number of high-class doctoral theses were produced under his guidance at the University of Helsinki. Internationally, the best known of his function-theoretic students were Felix Iversen, P.J. Myrberg, Frithiof Nevanlinna and Rolf Nevanlinna.

G. Mittag-Leffler

E. Lindelöf

Thanks to Lindelöf and his students, in the 1920s Helsinki became a stronghold of complex analysis. In particular, it was during this period that Rolf Nevanlinna developed his theory on the value distribution of meromorphic functions, known today under the name Nevanlinna theory. At the end of the decade the Helsinki group was further strengthened by the addition of Lars Ahlfors. At the age of 21, he settled the Denjoy conjecture concerning the relationship between the order and the number of asymptotic values of an entire function. This made him famous overnight. At the International Congress of Mathematicians in Oslo in 1936, Ahlfors was awarded one of the first two Fields Medals.

The connection between Fred Gehring and Finnish mathematics by way of Lorenz and Ernst Lindelöf and Mittag-Leffler may strike some as a bit far-fetched. By contrast, Rolf Nevanlinna and Lars Ahlfors have both steered events in such a way as to exert a direct influence on Fred's Finnish contacts and to be instrumental in his transformation into a quasi-conformalist.

They took their first steps, still tentative, as early as the 1930s. Ahlfors noticed that a class of diffeomorphisms which Grötzsch had originally introduced as a generalization of the class of conformal mappings not only provided an efficient tool in function theory but actually played an intrinsic role in it. In a paper of 1935, Ahlfors christened these functions "quasiconformal mappings" [1].

At the time, Ahlfors and Nevanlinna saw their research interests turning in the direction of Riemann surfaces. Nevanlinna first studied them in connection with his deficiency relation but later regarded them as objects worthy of attention in their own right. He spent the academic year 1936-37 at the University of Göttingen as a visiting professor, lecturing about Riemann surfaces. In his class was a young algebraist, Oswald Teichmüller, a student of Helmut Hasse. After Nevanlinna's visit Teichmüller became a function-theorist. It is tempting to see here a connection between Finnish mathematics and Teichmüller theory.

R. Nevanlinna

L. Ahlfors

3 The war and its aftermath

At the outbreak of the Second World War, complex analysis was flourishing at the University of Helsinki. Rolf Nevanlinna, Lars Ahlfors and P.J. Myrberg were the three full professors, and Ernst Lindelöf, who had retired in 1938, was the president of the Finnish Mathematical Society. However, this setting was soon to collapse almost entirely.

The war years had a devastating effect on Finnish mathematics. Ahlfors left the country in 1944. Lindelöf died two years later, and that same year Nevanlinna, who had been Rector of the University since 1941, accepted a position in Switzerland. At the Mathematics Department in Helsinki all now depended on Myrberg. He had his contacts with Germany and France, but the center of gravity of mathematical research was shifting to the other side of the Atlantic and English was becoming the dominant mathematical language. Moreover, Myrberg soon became an administrator. Although keeping up his research activities, he left the Mathematics Department, where contact with current mathematics was seriously weakened. Almost none of the teachers was engaged in significant research.

Fortunately, Nevanlinna spent part of his time in Finland, where doctoral education in mathematics was almost entirely in his hands. Nevanlinna gradually lost interest in complex analysis, which he abandoned entirely in the early 1950s, having completed his monograph *Uniformisierung*. After that time he no longer gave function-theoretic topics to his students.

In the mid-1950s, the Finnish tradition in complex analysis was seriously imperiled. But it was so strong that it did not perish.

4 Quasiconformal mappings come to Finland

Sometime around 1950, Nevanlinna had suggested to his doctoral student Yrjö Juve that he study classical questions related to Denjoy's problem and that, following ideas first developed by Teichmüller, he use quasiconformal mappings as a tool. Juve was one of the four assistants in the Helsinki Mathematics Department, which at that time was very small in every respect. There was but one room to house all the teachers and the library! Thus the few members of the staff were in near and frequent contact with each other. K.I. (Kalle) Virtanen and I often spoke with Juve about problems related to his dissertation [2] and, in the process, became interested in quasiconformal mappings. In 1953 I heard a lot about these mappings from Hans Künzi in Zürich and from Hans Wittich in Karlsruhe. Wittich had known Teichmüller and been with him in Nevanlinna's class in Göttingen. Both Künzi and Wittich used quasiconformal mappings as a tool in classical complex analysis, specifically in the Nevanlinna theory.

Probably decisive for the growth of interest in quasiconformal mappings in Helsinki was the paper of Ahlfors which appeared in the Journal d'Analyse in 1954 [3]. In this article Ahlfors showed that Teichmüller's results, which to some extent had been shrouded in mystery because of their partly heuristic proofs, could be put on a firm basis. Ahlfors discarded the requirement that a quasiconformal mapping be a diffeomorphism, calling a sense-preserving homeomorphism quasiconformal if it left the modules of quadrilaterals quasi-invariant. As far as I know, the notion and the name "Teichmüller space" appeared for the first time in this paper of Ahlfors.

We in Helsinki always followed Ahlfors's doings with special attention, but found this work to be of exceptional importance. At that time Virtanen and I were studying normal meromorphic functions. After some thought on the matter it struck me that some of our results could be extended so as to apply to quasi-meromorphic functions. I suggested to Virtanen that we write a paper about such a generalization in order to learn about quasiconformal mappings in a concrete context. Virtanen rejected the idea with contempt: such generalization with no new ideas was beneath the dignity of a senior mathematician. Still, he agreed that it would not be so stupid to learn more about quasiconformal mappings. They were increasing in importance, and we somehow felt that they might offer a way to restore to function theory in Helsinki some of its former vitality.

As a first step towards generating wider interest in quasiconformal mappings in Helsinki, Virtanen gave an introductory course into their theory in 1957-58. We also agreed to hold a seminar lasting the whole academic year

1958-59, for the purpose of organizing and presenting the existing knowledge about these mappings. Following Ahlfors's 1954 paper, the subject had undergone rapid progress in several different directions. But the results were scattered in the literature, and there was no uniform presentation of the main features of the theory.

In the early summer of 1958 I spent some time in Zürich, with Albert Pfluger, who was highly interested in quasiconformal mappings, and Künzi, who was writing a monograph about them for Springer Verlag's Ergebnisse-series. One evening, shortly before I returned to Finland, a serious discussion about the state of quasiconformal mappings took place at Pfluger's home. Nevanlinna was present, amongst many others. The sentiment was expressed that a yellow volume treating quasiconformal mappings should now be written for Springer's "Grundlehren der Mathematischen Wissenschaften." Before the evening was over, I had consented to write such a book, under the condition that Virtanen would agree to be co-author. That evening I realized the power of the Finnish tradition. My well-founded argument that, as a complete novice in the area of quasiconformal mappings, I was not up to the job was dismissed with remarks like "we are convinced that in Helsinki you know complex analysis."

Knowing Virtanen, I thought that the real risk of being obligated to write the monograph was small. To my astonishment, however, Virtanen agreed to undertake the project. The book cast our planned seminar in a new light. We realized that we were facing quite a challenge, and felt the particular need to have somebody around to consult with when inevitable problems related to real analysis arose.

5 Gehring becomes a quasiconformalist

At this same time Fred Gehring, unaware of the developments and plans in Helsinki, was making preparations to spend an academic year in Finland. Two persons influenced his decision to come to Helsinki. In 1953 Rolf Nevanlinna attended a Conference on Analytic Functions in Ann Arbor. There Nevanlinna made the acquaintance of Jack Lohwater, a colleague and friend of Fred. The result of their meeting was that Lohwater spent the academic year 1955-56 as a Fulbright-Hays grantee at the University of Helsinki. He had some trying experiences, partly self-inflicted, during that exceptionally cold winter, but evidently liked it enough in Helsinki to give Fred a positive recommendation.

The other pro-Helsinki advice came from Ahlfors. After earning his Master's degree at the University of Michigan, Fred pursued his doctoral studies in England, at the University of Cambridge. His thesis, completed in 1952 under the supervision of Professor Burkill, dealt with a topic in real analysis. After graduating from Cambridge, Fred spent three years at Harvard

University, where he got to know Ahlfors and became increasingly interested in complex analysis. He had noticed the work that Virtanen and I had done on normal meromorphic functions and, encouraged by Ahlfors, made the decision to spend an academic year in Helsinki.

Already in the early fifties, even without a permanent position in our Department of Mathematics, I had become the unofficial liaison with foreign visitors, corresponding with them and taking care of their arrangements. I do not remember, however, having had much contact with Fred before his arrival in Helsinki. He had obtained his funding from America, a Senior Fulbright-Hays Fellowship and a Guggenheim Fellowship, and with an efficiency characteristic of Fred, he must have arranged the practical details of his stay with the United States Information Service in Helsinki. A couple of weeks before Fred's arrival in Helsinki, I met him for the first time. Ahlfors brought us together at a party held in Edinburgh during the 1958 International Congress of Mathematicians. We were both accompanied by our wives. Because of the huge crowd in the reception room, we were standing close to each other and could hardly make a move. The noise level was such that it hindered all reasonable conversation. For a short while, our paths separated: Eva and I flew back home, whereas the Gehrings took a boat from London to Helsinki.

Fred's arrival at the Department of Mathematics of the University of Helsinki in early September 1958 is the highlight of this narration. I remember it well, because Fred himself has given an account of it so many times. With the years the story has improved a bit, because the essential points have been brought forth more clearly. In describing it now again, I am forced to stick to the truth to the best of my ability, since Fred himself is present in the audience.

According to Fred, he met Kalle Virtanen and me together. We asked him why on earth he had come to Helsinki, and for a whole year at that. What was he going to do here? "Learn about meromorphic functions, of course." Oh, meromorphic functions, we are no longer interested in them. There was a short pause. Fred then asked what we were interested in. Quasiconformal mappings, how would that be? "They are fine, very fine. What are they?" Do not worry about that. We shall start a seminar next week, and there you will learn about them. By the way, the seminar will be in Finnish.

In converting Fred Gehring to a quasiconformalist on that September day, some 37 years ago, two Finns had a marked influence on the future course of complex analysis. A lesser person might have been daunted by the prospect of having to learn about the fundamentals of the theory of quasiconformal mappings and the Finnish language during an extended weekend, but not Fred. Very soon he played the role of a third teacher in the seminar, speaking an odd mixture of English and Finnish. By the time the seminar was over, the English part had been very much pushed to the side by his Finnish. It was clear from the outset that Fred had decided to learn Finnish, and to learn it well. At that time, the University was launching its first "Finnish for foreigners" courses. Fred was a student in such a course, but soon became a planner and adviser also. He was able to impress his teachers with tricky model sentences displaying the subtleties of Finnish grammar with mathematical precision. These examples were actually constructed for him during lunch and coffee breaks by his mathematical colleagues, who were infected by Fred's determination to penetrate the mysteries of the Finnish language. When the first text-book related to the course "Finnish for foreigners" was published, Fred Gehring's contribution was duly noted in the preface.

Returning to mathematics, I must say that Fred certainly contributed to the atmosphere of the seminar by his indefatigable enthusiasm. For Virtanen and me, as well as for other participants, he often provided the expertise in real analysis that was needed. Time may guild past memories, but I remember this quasiconformal seminar in 1958-59 as the best teaching achievement during my academic career. It marked the beginning of a new era in mathematical research in Finland in the sense that, from that period on, quasiconformal mappings have played and continue to play a very dominant role in Finnish mathematical life. Our tradition in function theory was saved. Fred Gehring deserves great credit for his part in the salvage operation. And his contribution to Finnish mathematics was by no means restricted to the remarkable period 1958-59, as you will soon hear.

A good part of the program of the seminar was inspired by the paper of Ahlfors. As related above, the quasiconformal mappings he used were a priori just homeomorphisms. (Such a definition had been suggested in a paper by Pfluger as early as 1951, but Pfluger himself later spoke emphatically of Ahlfors as the father of the definition.) Classical quasiconformal mappings were assumed to be continuously differentiable and were characterized by their local behavior. It was thus natural to investigate the question of how much the new homeomorphic quasiconformal mappings deviated from the classical ones. Ahlfors himself did not touch such problems, but an intensive effort to clarify the properties of quasiconformal homeomorphisms began more or less simultaneously in different parts of the world. Kurt Strebel, who had attended lectures by Ahlfors at Stanford, proved in 1955 that a quasiconformal mapping is absolutely continuous on lines [4]. This result implies easily that a quasiconformal mapping has partial derivatives almost everywhere. A little later, in a posthumous paper which came out in 1957, Akira Mori settled the outstanding problem in the proper framework by proving that quasiconformal mappings are differentiable almost everywhere [5]. It followed that they behave locally like the Grötzsch mappings at almost all points.

Mori's proof of almost everywhere differentiability was quite short. It was based on a relatively easy distortion theorem for quasiconformal mappings and the classical Rademacher-Stepanov differentiability theorem. The use of the Rademacher-Stepanov theorem intrigued us in Helsinki, since all known proofs of it were long and complicated. We studied Saks's mono-

F.W. Gehring at the 100th Anniversary
of the Finnish Mathematical Society

graph, *Theory of the Integral*, but still remained in the dark about why a quasiconformal mapping is almost everywhere differentiable.

In my attempts to prove differentiability without resorting to the Rademacher-Stepanov theorem, I noticed that quasiconformality was not an essential condition. A mapping of a plane domain into the plane is differentiable a.e. if it is continuous and open and has finite partial derivatives a.e. But my proof of this result still depended on the use of the omnipresent Rademacher-Stepanov theorem. I told Fred about my progress in the wrong direction. Having thought about the same problem, he immediately grasped my ideas and, in a relatively short time, was able to advance in the desired direction. With obvious satisfaction he announced that he now had a "very simple" proof, one which was essentially based on the use of the maximum principle and avoided Rademacher-Stepanov completely. We decided to write a joint paper, which appeared in 1959 [6]. It was the first in a long series of papers Fred has published with Finnish mathematicians. According to the list "Publications of F.W. Gehring," as of February 1994 there were 16 such joint papers, with 7 different Finnish co-authors.

Rapidly acquiring knowledge of quasiconformal mappings, Fred launched his own research in this field. Virtanen and I ran the seminar and did our research with the demands of the Springer-monograph foremost in our minds. Fred's first paper on quasiconformal mappings, "The definitions and exceptional sets for quasiconformal mappings," was of direct use to this project. It appeared in the Finnish Annales in 1960 [7]. But Fred evidently also wanted to move away from the beaten track. At an early stage, he turned his interest to higher-dimensional quasiconformal mappings. While the two-dimensional theory had developed rapidly in the late 1950s, the theory of n-dimensional mappings with $n > 2$ was still in its infancy. In no time, Fred established himself as one of the leading figures in the field. Although mappings in higher dimensions were not part of the work Virtanen and I were doing, once again Fred's research was not without a Finnish connection.

6 The Gehring connection

Already in the course of the seminar, and in subsequent years, Fred Gehring became well-acquainted with Finnish mathematicians. The first with whom he had close scientific cooperation was Jussi Väisälä. Väisälä had also become interested in space quasiconformal mappings, and soon the two joined forces. In 1965, Gehring and Väisälä published in Acta Mathematica a 70-page joint paper "The coefficients of quasiconformality of domains in space" [8]. Long in preparation under the working name "Monster," it contained a wealth of information about the properties of higher-dimensional quasiconformal mappings. There are two other joint Gehring-Väisälä papers, but I think that the span of time during which they appeared, 1961-1973, speaks more eloquently than does their mere number to the close and long-standing mathematical contacts between the two.

Analysis is Fred Gehring' s field of mathematics. From his early production on, however, he liked to sprinkle it with geometry. For example, in the theory of plane quasiconformal mappings, quasicircles and quasidiscs (i.e., images of circles and discs under quasiconformal mappings of the plane) have always fascinated him. Quasicircles were introduced by Pfluger in 1960, and three years later Ahlfors gave them a simple geometric characterization. From the general theory it follows immediately that a quasicircle has 2-dimensional Lebesgue measure zero. On the other hand, Pfluger proved that such a curve can be non-rectifiable. In a joint paper of 1973, Gehring and Väisälä showed that a quasicircle can be much "worse" than that. Using 2-dimensional Cantor sets and appealing to the criterion of Ahlfors, they constructed quasicircles whose Hausdorff dimension was any prescribed number less than 2 [9].

A few years later, Fred returned to the topic of quasidiscs. He called a simply-connected domain A "Schwarzian" if there is a positive constant a, such that the following is true: whenever f is a meromorphic function in A with a Schwarzian derivative S_f whose hyperbolic sup-norm $\sup |S_f(z)|/h(z)^2$ (h is the Poincaré density for A) is not greater than a, then f is univalent. By a theorem of Ahlfors, a quasidisc is Schwarzian. In 1977, relying on geometric arguments, Fred proved the interesting converse: Every Schwarzian domain is a quasidisc [10].

This theorem was one more proof of the intrinsic role of quasiconformal mappings in the theory of analytic functions. Fred gave the result a beautiful interpretation in the language of Teichmüller theory: In the Bers imbedding, the universal Teichmüller space is precisely the interior of the space U consisting of the Schwarzian derivatives of univalent functions. Shortly thereafter Fred completed this theorem by disproving a famous and long outstanding conjecture of Bers: The closure of the universal Teichmüller space is a proper subset of the space U [11]. Bers had promised a prize (cognac) to the one who proved that the conjecture was true. Fred was under the impression that the prize was for any solution of the problem. This may have saved him from heading too long in the wrong direction... .

In the late 1970s, both Fred and I were both interested in Schwarzian derivatives and Teichmüller spaces. This brought us mathematically close to each other again, and soon Fred was to have a decisive influence on my mathematical work. I was already shifting towards administration, first in the International Mathematical Union and then at my University. Perhaps wanting to save me from this fate, Fred urged that I write a monograph on Teichmüller spaces from the point of view of classical complex analysis. Without his encouragement I would not have undertaken the task. Even then Fred's intervention occurred at the last moment. I had just begun the work in 1982, when I was elected Secretary of the International Mathematical Union and the following year Rector of the University of Helsinki. I was thus fully engaged with administrative duties. But I could not disappoint Fred, so with a few years' delay my book appeared in Springer-Verlag's series "Graduate Texts in Mathematics," of which Fred and Paul Halmos were the Editors [12].

In the 1980s, Fred found two new Finnish mathematicians with whom to work. In studying problems related to quasidiscs, uniform domains, etc., i.e., problems with a geometric flavour, he collaborated with Olli Martio. For problems dealing with BMO-functions, universal Teichmüller space and the like, often characterized by the use of hard real analysis, his working partner was Kari Astala. In both cases, cooperation was intense and undeniably fruitful. Within six years Fred had four joint papers with each of them.

The problem of determining the exact degree of distortion of area under a 2-dimensional K-quasiconformal mapping or, what is more or less the same, the local integrability of the partial derivatives of such a mapping, arose in the mid-1950s. It was easy to prove that the derivatives were locally in L_2. But in 1955 Bojarski proved more: the derivatives are actually locally L_p-integrable for $2 < p < 2 + c$, where c is a positive constant which depends only on K. From Bojarski's reasoning a lower bound was obtained for c in terms of the L_p-norm of the 2-dimensional Hilbert transformation, but this bound was implicit. A simple example led to the conjecture that $c = 2/(K - 1)$.

If I remember correctly, there was already some discussion of all this in the 1958-59 seminar. Important progress was made by Gehring and Edgar Reich in 1966, with methods and results that were restricted to the 2-dimensional case [13]. In higher dimensions the problem remained completely intractable for a long time. But in 1973, Fred Gehring extended Bojarski's theorem to all dimensions greater than or equal to 2. Not only was the result remarkable, but equally so was the methodology: a straightforward, careful comparison of the L_p-norms for different values of p, a kind of reverse Hölder inequality [14]. Fred himself called his proof "quite elementary."

Twenty years later Astala proved that the old 2-dimensional conjecture $c = 2/(K - 1)$ was correct [15]. In his solution, which brought him the Salem Prize, Astala introduced a new way of looking at the problem and new technical machinery. In addition, he utilized the results of Gehring and Reich. I also see here an implicit Gehring connection in the sense that, having worked with Fred, Astala had attended the "right school."

This aspect, Fred Gehring's non-measurable influence on Finnish mathematics, deserves special attention. Since his first visit to Finland in 1958-59, our Michigan connection has worked without interruption. Thanks to Fred, one Finnish mathematician after another has been given the opportunity to spend a good amount of time at the Mathematics Department in Ann Arbor. I am convinced that these visits have contributed very much to the advancement of Finnish mathematics. The visitors have learned much and brought their acquired wisdom back to Finland. But there has been more than the direct mathematical gain. Living in a small country, we in Finland are dependent on international contacts to an exceptionally high degree. Establishing them is not so easy, it requires connections and funds, language problems have to be overcome, etc. The example of Lorenz Lindelöf from the mid-1850s and several others show that there have been mathematicians in Finland willing and able to cultivate international relations. However, for a very long time, international mathematical contacts depended on very few individuals. This is not the case any longer. Now there is a large number of Finnish mathematicians who are accustomed to live abroad, who have developed good networks of contacts, who have personal friends in foreign countries. This development is not due exclusively to Michigan, but this connection, more precisely the Gehring connection, has been a great and important contributor to its advancement.

7 Gehring and International Congresses of Mathematicians

Today we know that when Fred Gehring became a convert to quasiconformalism in Finland, a great service was done for mathematics, or more exactly, for the area of quasiconformal mappings. This statement can be rigorously proved. Achievements of mathematicians the world over are constantly being weighed and ranked in conjunction with the setting up of programs for the International Congresses of Mathematicians (ICM) held every fourth year. Since 1962, the process of selecting speakers has followed well defined rules. Decisions are made by an international Committee under the advisement of international panels, one for each of the sections into which the Committee has divided the mathematical program.

Fred's ICM record is impressive, to say the least. He was invited to speak for the first time at the 1966 International Congress in Moscow. The title of his lecture was "Extension theorems for quasiconformal mappings in n-space." Only eight years later, at the 1974 Congress in Vancouver, he was again an invited speaker. This time he spoke about "Some metric properties of quasiconformal mappings." These two lectures fell under the heading of the section that included Complex Analysis. The competition for speaking slots is heavy, since the average number of speakers in this section is about six. But even more recognition for his work was to come. Top honors befell him at the 1986 International Congress, which convened in Berkeley. There he was invited to deliver one of the 16 plenary lectures, for which the competition is between mathematicians in all fields. Fred's address was entitled "Topics in quasiconformal mappings."

Who knows what the future may hold, as years seem to have no negative effect on Fred!

References

[1] L. Ahlfors, Zur Theorie der Ueberlagerungsflächen, *Acta Math.*, **65** (1935).

[2] Y. Juve, Ueber gewisse Verzerrungseigenschaften konformer und quasikonformer Abbildungen, *Ann. Acad. Sci. Fenn. A* I, **174** (1954).

[3] L. Ahlfors, On quasiconformal mappings, *J. Analyse Math.*, **3** (1954).

[4] K. Strebel, On the maximal dilation of quasiconformal mappings, *Proc. Amer. Math. Soc.*, **6** (1955).

[5] A. Mori, On quasi-conformality and pseudo-analyticity, *Trans. Amer. Math. Soc.*, **84** (1957).

[6] F.W. Gehring and O. Lehto, On the total differentiability of functions of a complex variable, *Ann. Acad. Sci. Fenn. A* I, **272** (1959).

[7] F.W. Gehring, The definitions and exceptional sets for quasiconformal mappings, *Ann. Acad. Sci. Fenn.*, **281** (1960).

[8] F.W. Gehring and J. Väisälä, Hausdorff dimension and quasiconformal mappings, *J. London Math. Soc.* **6** (1973), no.2.

[9] F.W. Gehring and J. Väisälä, The coefficients of quasiconformality of domains in space, *Acta Math.* **114** (1965).

[10] F.W. Gehring, Univalent functions and the Schwarzian derivative, *Comment. Math. Helv.* **52** (1977).

[11] F.W. Gehring, Spirals and the universal Teichmüller space, *Acta Math.*, **141** (1978).

[12] O. Lehto, Univalent Functions and Teichmüller Spaces, in *Graduate Texts in Mathematics* vol.109, Springer Verlag, 1986.

[13] F.W. Gehring and E. Reich, Area distortion under quasiconformal mappings, *Ann. Acad. Sci. Fenn. A* I, **388** (1966).

[14] F.W. Gehring, The L_p-integrability of the partial derivatives of a quasiconformal mapping, *Acta Math.*, **130** (1973).
[15] K. Astala, Area distortion of quasiconformal mappings, *Acta Math.*, **173** (1994), no.1.

Department of Mathematics
University of Helsinki
P.O. Box 4
Hallituskatu 15
SF-00014 Helsinki
Finland

Planar Quasiconformal Mappings; Deformations and Interactions

Kari Astala

1 Introduction

The theory of quasiconformal mappings divides traditionally into two branches, the mappings in the plane and the case of higher dimensions. Basically, this is not due to the history of the topic but rather since planar quasiconformal mappings admit flexible methods (so far) not available in space. In this expository paper we wish to describe some recent trends and activities in quasiconformal theory peculiar to the plane. It is obvious, though, that not all topics can be covered no matter which point of view is taken; many important advances and connections must necessarily be bypassed. Therefore we concentrate on a specific theme, a property that singles out the difference between mappings in plane and in space: Planar quasiconformal mappings admit an effective deformation theory; any such mapping can be deformed to the identity, within the family of all quasiconformal mappings.

Our goal is to discuss different aspects of these parametric representations of planar quasiconformal mappings, with a few occasional words of the background of the topics, to see how and why these mappings can be deformed, and above all which are the results and consequences obtained, which are the questions still open. In particular, in many respects it is the deformation properties which are the "cause" of the many connections and applications of quasiconformality. Thus we try to give a particular attention to discuss how this field of ideas interacts with other parts of mathematical analysis.

It will be no surprise to see the strong influence of Gehring also within these contexts.

By the analytic definition a homeomorphism $f : \Omega \to \Omega'$ between planar domains Ω and Ω' is called *K-quasiconformal* if it is contained in the Sobolev class $W^{1,2}_{\mathrm{loc}}(\Omega)$ and its directional derivatives satisfy

$$\max_{\alpha} |\partial_\alpha f(x)| \le K \min_{\alpha} |\partial_\alpha f(x)| \qquad \text{a.e. } x \in \Omega.$$

A corresponding definition works of course for higher dimensional mappings as well, but what is special to the mappings in plane is the well known Beltrami differential equation, that the above inequality is equivalent to

$$\bar{\partial} f(z) = \mu(z)\, \partial f(z) \,, \tag{1.1}$$

where μ is the complex dilatation or the Beltrami coefficient of f with $\|\mu\|_\infty = \frac{K-1}{K+1} < 1$. This is a first order linear differential equation, solvable by singular integrals.

Note also that although the recent work of Donaldson and Sullivan [12], developed also by Iwaniec and Martin [20], found a counterpart of (1.1) in even dimensional spaces this machinery still lacks the required properties for finding parametric representations in space. However, in the planar case (1.1) always does admit quasiconformal solutions: by the *measurable Riemann mapping theorem* for any $\mu \in L^\infty(\mathbf{C})$ with $\|\mu\|_\infty < 1$ we have a homeomorphic $W^{1,2}_{\mathrm{loc}}$-solution of (1.1), see [2] [9] [21], unique up to postcomposing with a Möbius transformation.

This result is the starting point of the deformation theory for planar quasiconformal mappings. It turns out that using (1.1) we can actually construct *two* natural and similar but in many respect different ways of parametrizing the mappings.

To be more precise, the basic idea for the first deformation is that we may embedd our mapping f to a flow $\{f_t\}_{t\in\mathbf{R}}$ of a (nonautonomous) vector field v. In fact, by careful choices obtain a geodesic in the Teichmüller metric,

$$\log K(f_t \circ f_s{}^{-1}) = |t-s|, \quad f_0 = id.$$

For this approach, to control the flow we must find the infinitesimal generating vector field v for which

$$\partial_t f_t(x) = v(t, f_t(x)), \; f_{t_0} = f. \tag{1.2}$$

Thus the questions here are how to find v, what are its properties and, especially, how to characterize vector fields generating quasiconformal flows.

The second approach arises in connection with deformations of complex analytic structures, where the holomorphicity of the parametrization is an essential feature required. Then one must abandon the semigroup setting and more directly use the properties of the equation (1.1). Namely, if $\mu \in L^\infty(\mathbf{C})$ has a compact support, let $f = f^\mu$ be the solution of (1.1) with $f(z) = z + O(|z|^{-1})$. Then,

$$\text{if } \mu \text{ varies holomorphically, so does } f^\mu(z), \tag{1.3}$$

thus giving us holomorphic ways of deforming f to the identity. Indeed, as is well known this holomorphic dependence follows with the help of the classical Beurling operator, the two dimensional counterpart of the Hilbert transform,

$$S\omega(z) = -\frac{1}{\pi} \int_{\mathbf{C}} \frac{\omega(\zeta)\, dm(\zeta)}{(\zeta - z)^2} .$$

It defines an isometry in $L^2(\mathbf{C})$, and moreover, what makes it useful for the Beltrami equation is that S transforms $\overline{\partial}$-derivatives to ∂-derivatives,

$$S(\overline{\partial} u) = \partial u, \quad \forall\, u \in W^{1,2}(\mathbf{C}) . \tag{1.4}$$

Using this identity for $u(z) = f(z) - z$ and combining with the equation $\overline{\partial} f(z) = \mu(z)\partial f(z)$ we have

$$\partial f(z) = 1 + (I - S\mu)^{-1} S(\mu) . \tag{1.5}$$

Since as an operator on $L^2(\mathbf{C})$, the norm of $g \mapsto S(\mu g)$ is at most $\|\mu\|_\infty \|S\|_{L^2}$
< 1, the expression is well defined and shows that $\partial f(z)$ depends holomorphically on μ. Since by the Cauchy formula

$$f(z) = z - \frac{1}{\pi} \int_{\mathbf{C}} \frac{\mu \partial f(\zeta)\, dm(\zeta)}{(\zeta - z)} ,$$

we have the same conclusion for $f(z)$.

2 Flows of quasiconformal mappings

The parametric representation in terms of flows of quasiconformal mappings was originally derived by Shah Dao-Shing [32]. However Gehring and Reich [17] were the first who systematically developed and utilized the method. Their goal was to find a new approach to quasiconformal distortion problems. We return to these applications below after first analysing the basic properties and some consequences of their construction.

To build up the flow, let us agree first on the notation f_μ for the mapping with complex dilatation μ, normalized so that 0, 1 and ∞ are fixed. To embed a given K-quasiconformal mapping f to a continuous flow, we may assume that $f = f_\mu$ and define then a further family of dilatations by

$$\mu_t(z) = \frac{\mu(z)}{|\mu(z)|} \tanh\left(\frac{t}{T} \tanh^{-1} |\mu(z)|\right) , \qquad T = \log K .$$

According to the measurable Riemann mapping theorem for each $t \in \mathbf{R}$ there exists a unique normalized mapping $f_t = f_{\mu_t}$. Geometrically, $\mu_t(z)$ is chosen to be the point on the radius through $\mu(z)$ such that in the unit disk the hyperbolic distance from 0 to $\mu_t(z)$ is equal to t/T times the distance between 0 and $\mu(z)$. From the analytic view, or by the transformation formulas of the complex dilatation for a composition of two quasiconformal mappings [21], the $\mu_t(z)$'s are constructed precisely so that we have

$$\log K(f_t \circ f_s^{-1}) = \frac{1}{T}|t-s|.$$

We have made the above choice $\frac{1}{T}$ for the speed of the flow for later purposes, c.f. (2.6), (2.7).

Sometimes it is also convenient to have a finite version of the flow. That is, considering only a discrete set of parameter values and noticing that $K(f_{\mu_t}) \leq e^t$ we have the *factorization theorem*: For any $\epsilon > 0$, all quasiconformal mappings in $\mathbf{C}$ can be written as a composition of a finite number of $(1+\epsilon)$-quasiconformal mappings.

The generating vector field can next be given simply as

$$v(z,t) = \partial_t f_t \circ f_t^{-1}(z)\,. \tag{2.6}$$

In other words, the equation $\partial_t f_t(x) = v(t, f_t(x))$, $f_T = f$ is satisfied directly by the definition. Less obvious from here is to see any specific properties and, in particular, any explicit characterizations of vector fields that generate quasiconformal flows. However, by a clever combination of (1.1) and (1.2) Gehring and Reich found two interesting identities

$$\partial_{\bar{z}} v \circ f_t = \frac{1}{2}\frac{\mu(z)}{|\mu(z)|}\frac{\partial_z f_t}{\overline{\partial_z f_t}} \tag{2.7}$$

and

$$\partial_t J_{f_t} = 2\,Re[\partial v \circ f_t] J_{f_t} \tag{2.8}$$

for the vector field v and the Jacobian determinant J_{f_t} of f_t. The first of these answer immediately our question:

Corollary 2.1 *The vector field v generating the above flow of normalized quasiconformal mappings $\{f_t\}_{t\in\mathbf{R}}$ satisfies*

$$\|\bar{\partial} v\|_\infty \leq \frac{1}{2}\,.$$

Indeed, as shown later by Reimann [28] and Semenov [33] the corollary admits a direct converse.

Theorem 2.2 *Suppose we have a continuous vector field $v \in W^{1,1}_{\mathrm{loc}}(\mathbf{C})$ such that $v = O(|z|\log|z|)$ as $|z| \to \infty$ and*

$$\|\bar{\partial} v\|_\infty \leq C < \infty\,.$$

Then the flow $\{g_t\}_{t\in\mathbf{R}}$ of v consists of quasiconformal mappings with

$$K(g_t) \leq e^{Ct}\,.$$

The exact value of the constant C counts, of course, only for the speed of the flow.

Remarkably this direction holds in any dimensions ([28]),([33]). We only need to reinterpret the boundedness of $|\overline{\partial}v(z)|$ in terms of the differential operator

$$\Sigma V(x) = \frac{1}{2}[DV(x)^t + DV(x)] - \frac{1}{n}(TrDV(x))I \ , \qquad x \in \mathbf{R}^n \ ,$$

in the case of vector fields in $\mathbf{R}^n$. As shown by Ahlfors [3], in two dimensions $|\Sigma V(z)| = |\overline{\partial}V(z)|$.

To indicate the usefulness of this approach let us recall another closely related argument of Reimann [30]: Suppose we have a quasiconformal mapping f on the sphere $\mathbf{S}^2$ which defines an isomorphisms between two groups of Möbius transformations G and Γ, i.e., $f^{-1} \circ \Gamma \circ f = G$. Since the Möbius groups themselves extend canonically to the unit ball $\mathbf{B}^3$ we may ask if there is a way to extend the isomorphism as well.

For this [30] shows first that the generating vector field v constructed above from the (two dimensional) mapping f is G-invariant in an appropriate sense. Moreover, in all dimensions an Möbius-equivariant extension of *vector fields* is easily found by a Poisson-type integraloperator. After a calculation, it turned out that a similar representation holds also for the differential operator $\Sigma V(x)$. Therefore, both the equivariance and the condition $\|\Sigma V\|_\infty \leq C < \infty$ remain valid under the vector field extension. Thus integrating back the function, i.e. by using the three-dimensional version of Theorem 2.2 and the fact that G-invariance of vector fields is inherited by their flows, we see that also the quasiconformal mapping f has a G-equivariant extension.

The result is of importance e.g., in the study of hyperbolic three-manifolds, see [38]. The equivariant extension from one to two dimensions is due to Tukia [40], but these two lower dimensional cases are the only situations where such general group-invariant extensions are known to exist.

Next, returning back to the work [17], does the latter identity (2.8) have similar consequences? Following Reimann [29] let us first assume that μ is compactly supported. Then we may interpret the identity in the form

$$\partial_t \log J_{f_t} = 2\,Re[S(\overline{\partial}v) \circ f_t] \ . \tag{2.9}$$

This leads us to the function space BMO, consisting locally integrable functions u for which

$$\|u\|_* = \mathrm{Sup}_B \left\{ \frac{1}{|B|} \int\limits_B |u - u_B| dx \right\} .$$

Here the supremum is taken over all disks $B \subset \mathbf{C}$, $|B|$ denotes the area of B and u_B the mean value of u over B. As is well known the Beurling operator takes bounded functions to elements in BMO. Furthermore, $\|u \circ f\|_* \leq C(K)\|u\|_*$ for any BMO function and K-quasiconformal mapping f [31]. Therefore Corollary 2.1 shows that the right hand side of (2.9) are BMO functions with norms bounded by constants depending only on t. In other words, recalling the relation $T = \log K$ between the parameters, we have

$$\| \log J_f \|_* \leq C_1(K) \log K \to 0 \quad \text{as} \quad K \to 1 \ .$$

Lastly, by a limiting argument this holds for any K-quasiconformal mapping f on $\mathbf{C}$.

It is interesting to note that from here we may obtain a proof for the inverse Hölder inequality for the Jacobian J_f. By the John-Nirenberg inequality, when the norm $\| \log J_f \|_*$ is small enough, we have

$$\left(\frac{1}{|B|} \int_B J_f^p \, dx \right)^{1/p} \leq C \left(\frac{1}{|B|} \int_B J_f \, dx \right)$$

for some $1 < p < \infty$ and constant C depending only on K. But using the factorization theorem this generalizes to any K-quasiconformal mapping f in $\mathbf{C}$.

A more direct approach would, of course, use the representation (1.5) and the boundedness of the Beurling transform on the L^p-spaces, $1 < p < \infty$. The inverse Hölder inequality, as well as the finiteness of $\| \log J_f \|_*$, are known to hold also in higher dimensions by the work of Gehring [18] and Reimann [31], respectively, but the proofs there lie much deeper.

Finally let us return to the original setup and background of the work of Gehring and Reich in [17]. In the 50's Ahlfors [1] and Mori [26] had established that K-quasiconformal mappings are locally $1/K$-Hölder continuous. The function

$$f_0(z) = z|z|^{\frac{1}{K}-1} \tag{2.10}$$

shows that those estimates optimal. Around the same time Bojarski [9] had proved with (1.5) that under quasiconformal mappings also the area of sets is distorted in a Hölder continuous manner. The precise or best possible exponent, however, remained open. The same example (2.10) indicated that $1/K$ would again be the correct bound.

To study the distortion of area Gehring and Reich reinterpreted (2.8) in the following manner: Suppose $E \subset \frac{1}{2}\Delta$, where $\Delta = \{z : |z| < 1\}$ is the open unit disk. If we are given a quasiconformal mapping f with the corresponding flow $\{f_t\}$ and generating vector field v, let $A(t) = |f_t E|$. Then we have from (2.8), using the symmetry of the Beurling transform that

$$\frac{d}{dt} A(t) = 2 \, Re \int_\Delta \partial_{\bar{z}} v \, S(\chi_{f_t E}) dx + c(t) A(t) \ , \tag{2.11}$$

where c is bounded in t. Using the identity (2.7) and estimates for Beurling transform we get an upper bound for the right hand side and thus a differential inequality for $A(t)$. When that is solved one has the estimate $|f(E)| \leq M_K|E|^{K^{-\alpha}}$. Indeed, here α is the smallest number for which the inequality

$$\int_{\Delta} |S(\chi_F)|dx \leq \alpha|F|\log\left(\frac{b}{|F|}\right) \tag{2.12}$$

holds for some constant b and all subsets F of the unit disk.

Remarkably, Gehring and Reich showed also the converse! That is, with a control on M_K the correct bound in the distortion exponent was equivalent to proving (2.12) with $\alpha = 1$; an argument was then shown for the inequality with a constant $\alpha \leq 40$.

A closer look shows that actually further information can be squeezed from this argument. Namely, suppose the quasiconformal mapping f has the special property that it is conformal outside the set E, $\mu|_{E^c} \equiv 0$. Then, cf. (2.7), $\overline{\partial}v$ vanishes outside f_tE and the integral term in (2.11) can be estimated from above simply by

$$\int_{f_tE} |S(\chi_{f_tE})|\,dx \leq |f_tE|^{1/2}(\int |S(\chi_{f_tE})|^2\,dx)^{1/2} = |f_tE| = A(t)\ , \tag{2.13}$$

where in the first equality we have used the fact that the Beurling transform S is an isometry on L^2. Consequently the method shows that

$$|f(E)| \leq C(K)|E| \tag{2.14}$$

for all quasiconformal mappings f, fixing 0, 1 and ∞, which are conformal outside the set E.

As we shall see, to understand the complementary case, when f is conformal *in* E, we need to apply the second method of deforming quasiconformal mappings, i.e., the holomorphic deformations.

3 Holomorphic motions

The very geometric theory of quasiconformal mappings has been closely connected to holomorphic deformations of different setups ever since its early days when Teichmüller used these mappings to understand complex structures on Riemann surfaces. Recently, after the pioneering work of Sullivan, quasiconformal mappings have become a basic tool in complex dynamics. However, it was not apparent at all why these quite differently looking topics should be related. In fact it turns out that understanding complex dynamical systems has clarified and explained the role of quasiconformality in general. Therefore we start by recalling the beautiful explanation in terms of holomorphic motions.

Definition 3.1 *A function $\Phi : \Delta \times A \to \overline{\mathbf{C}}$ is called a holomorphic motion of a set $A \subset \overline{\mathbf{C}}$ if*

(i) for any fixed $a \in A$, the map $\lambda \mapsto \Phi(\lambda, a)$ is holomorphic in Δ

(ii) for any fixed $\lambda \in \Delta$, the map $a \mapsto \Phi_\lambda(a) = \Phi(\lambda, a)$ is an injection, and

(iii) the mapping Φ_0 is the identity on A.

Note in particular that no assumptions are made on the set A or on the a'priori continuity of Φ_λ; this will obtained later as consequence. This flexibility gives us a wide variety of examples. The typical ones come of course from complex dynamical systems.

Example 3.2 *Holomorphic deformations of parameters of a dynamical system:*

a) Let Γ be Kleinian group, a discrete group of Möbius transformations with limit set $L(\Gamma) \neq \overline{\mathbf{C}}$. We assume Γ has no elliptic elements. Then changing the coefficients of the group elements in a holomorphic manner gives a holomorphic motion of the limit set, as long as the new groups Γ_λ obtained are discrete, the natural identification $(gh)_\lambda = g_\lambda h_\lambda$ remains an isomorphism and no accidental parabolics are formed.

b) Let R be a rational function which is hyperbolic, i.e., $|(R^n)'(z)| \geq ck^n$ with $k > 1$, uniformly on the Julia set $J(R)$, $n \in \mathbf{N}$. Then a (small) holomorphic change of the coefficients of R produces a family of rational functions with Julia sets moving holomorphically.

Below we give a proof of b) in an important special case. The proof of a) is similar.

From the first section we obtain another type of motion.

Example 3.3 *Holomorphic deformations of a quasiconformal mappings:*

If we use the notation f^μ for the quasiconformal mapping which has (compactly supported) complex dilatation μ and is normalized by $f(z) = z + O(|z|^{-1})$ at ∞, consider the mapping

$$\lambda \mapsto f^{\lambda\mu}(z) \,, \qquad \lambda \in \Delta \,.$$

Then $\Phi(\lambda, z) = f^{\lambda\mu}(z)$ is a holomorphic motion.

The crucial point here is that these seemingly unrelated examples are actually just different aspects of one and the same phenomenon. This was realized first by Mañé, Sad and Sullivan [23] in their study of stability and geometric rigidity in the complex dynamical systems of rational functions. We formulate their result in the following form

Theorem 3.4 *Let $\Phi : \Delta \times A \to \overline{\mathbf{C}}$ be a holomorphic motion of a set $A \subset \overline{\mathbf{C}}$. Then*

(i) Φ is jointly continuous in $\Delta \times A$,

(ii) Φ extends to a holomorphic motion of the closure $\overline{A}$ and

(iii) if $A = \mathbf{C}$, then the mappings Φ_λ are quasiconformal with $K(\Phi_\lambda) \le \frac{1+|\lambda|}{1-|\lambda|}$.

Proof. By normalizing with a Möbius transformation we may assume that all Φ_λ's fix ∞. Consider then the different points $x, y, z \in \mathbf{C}$. The function

$$h_\lambda = \frac{\Phi_\lambda(x) - \Phi_\lambda(y)}{\Phi_\lambda(x) - \Phi_\lambda(z)}$$

is holomorphic in the unit disk and omits the three values $0, 1$ and ∞. Since $\overline{\mathbf{C}} \setminus \{0, 1, \infty\}$ admits a hyperbolic metric ρ and since holomorphic mappings do not increase hyperbolic distances [22],

$$\rho(h(\lambda), h(0)) \le \rho(\lambda, 0) = \log \frac{1 + |\lambda|}{1 - |\lambda|} .$$

On the other hand, in the hyperbolic metric $\overline{\mathbf{C}} \setminus \{0, 1, \infty\}$ is complete, or $\rho(z, w) \to \infty$ when z is fixed and $w \to 0$. This means that we can interpret the inequality in the form

$$\left| \frac{\Phi_\lambda(x) - \Phi_\lambda(y)}{\Phi_\lambda(x) - \Phi_\lambda(z)} \right| \le \eta \left(\left| \frac{x - y}{x - z} \right| \right) \tag{3.15}$$

for some continuous strictly increasing function $\eta = \eta_\lambda$ with $\eta_\lambda(0) = 0$.

The first two conclusions follow now from this uniform estimate and the compactness properties of analytic functions.

As to the third, (3.15) says (by definition) that each Φ_λ is *quasisymmetric*, a property equivalent to quasiconformality for mappings of $\overline{\mathbf{C}}$ [22]. Therefore the Φ_λ's have complex dilatations μ_{Φ_λ} and these depend holomorphically on λ. Here $\|\mu_{\Phi_\lambda}\|_\infty < 1$ and $\mu_{\Phi_0} = \mu_{id} = 0$. Hence the last claim follows from Schwarz lemma (for mappings with values in L^∞). □

It is of course natural to ask if the motion extends not only to the closure $\overline{A}$ but to a motion of the whole complex plane. After a number of partial answers the picture was completed by Slodkowski who, using methods from several complex variables, proved the following *generalized λ-lemma* .

Theorem 3.5 *Any holomorphic motion Φ of any set $A \subset \overline{\mathbf{C}}$ extends to a holomorphic motion of $\overline{\mathbf{C}}$.*

Combining Example 3.3 with Theorems 3.4 and 3.5 we see that quasiconformal mappings, with a choice of a holomorphic deformation to identity, are precisely the same as holomorphic motions,

Planar quasiconformal mappings $\equiv$ Holomorphic motions.

In particular, this explains and clarifies the role of quasiconformality in complex dynamics or even complex analysis in general. These results show that whenever we have (any) structure defined in terms of the complex plane and we perturb the parameters of the structure in a holomorphic manner, then quasiconformal mappings and phenomena will necessary appear.

As a concrete example of the use of these ideas let us recall the proof of 3.2 in the case of quadratic polynomials.

Example 3.6 *Let $P_c = z^2 + c$. Then, if c and γ are in the same (hyperbolic) component of the interior of the Mandelbrot set, their Julia sets are quasiconformally equivalent.*

Proof: The Julia set of $P_c = z^2 + c$ is well known to be the closure of the repelling periodic points of P_c [16]. But periodic points are (for fixed c) precisely the zeroes of the functions $F(c,z) = P_c^n(z) - z$. Since, by hyperbolicity,

$$\partial_z F(c,z) = (P_c^n)'(z) - 1 \neq 0, \quad z \in J(P_c)$$

the implicit function theorem shows that each periodic point can be continued holomorphically to neighbouring parameter values. This can be done as long as the polynomial stays hyperbolic, i.e., c stays in the same component of the interior of the Mandelbrot set. Furthermore, since the solutions of the implicit function theorem are unique, different periodic points will not "collide" and we obtain a canonical holomorphic motion of the repelling periodic points. Using 3.4 *(ii)* that extends to an (equivariant) motion of the Julia set. Lastly, by Theorems 3.4 and 3.5, an extension of the motion gives quasiconformal equivalence of the Julia sets. □

Remark. With a different argument the assumption of the hyperbolicity of the component can be dropped.

In dynamical settings like iteration of rational functions or actions of discrete groups or semigroups of Möbius transforms, the holomorphic motions of the limits sets are naturally invariant under the dynamics, cf. Examples 3.2 and 3.6. It would be often desirable to ask more, whether the Slodkowski extension can be chosen to be equivariant in the whole complex plane $\overline{\mathbb{C}}$. Unfortunately , this is not possible in general. For instance in our Example 3.6 every hyperbolic component of the Mandelbrot set contains precisely one parameter value $c = c_0$ such that $P(z) = z^2 + c$ has a periodic critical point. On the other hand, periodicity of the critical point is invariant under global conjugacies and, in particular, under equivariant motions.

However, in the case of group structures one can require global equivariance. This was shown by Earle, Kra and Krushkal [14] and Slodkowski [35]. Their basic idea is as follows: Let $\{\Gamma_\lambda\}_{\lambda \in \Delta}$ be a holomorphic deformation of a Kleinian group Γ, $\Gamma = \Gamma_0$, in the sense of Example 3.2. Then there exists a holomorphic motion Φ of $\overline{\mathbf{C}}$ conjugating the groups,

$$\Phi_\lambda \circ \Gamma = \Gamma_\lambda \circ \Phi_\lambda \quad \text{on } \overline{\mathbf{C}} . \tag{3.16}$$

Namely, in this situation the fixed points x of the elements of Γ are easily seen to move holomorphically; hence by definition $\Phi_\lambda(\gamma(x)) = \gamma_\lambda(\Phi_\lambda(x))$. According to 3.4 this defines a Γ-invariant motion of the limit set $L(\Gamma)$, which is the closure of the fixed points. Next, choose a point $z \notin L(\Gamma)$ and use Slodkowski's result to extend Φ to a motion of $\{z\} \cup L(\Gamma)$. Here any choice will do, but then the group structure defines the motion of the orbit $\Gamma(z)$ by

$$\Phi_\lambda(\gamma(z)) = \gamma_\lambda \circ \Phi_\lambda(z) , \qquad \gamma \in \Gamma .$$

Since $z \notin L(\Gamma)$ it is not a fixed point for any $\gamma \in \Gamma$ and the extension is well defined. Similarly, if we had $\Phi_\lambda\gamma(z) = \Phi_\lambda g(z)$ for different $\gamma, g \in \Gamma$, then $(g^{-1}\gamma)_\lambda\Phi_\lambda(z) = \Phi_\lambda(z)$ which is not possible as $\Phi_\lambda(z) \notin L(\Gamma_\lambda)$. By the invariance of the corresponding limit sets we see, in fact, that Φ is a holomorphic motion of $\Gamma(z) \cup L(\Gamma)$.

Choosing next a point $w \notin \Gamma(z) \cup L(\Gamma)$ and continuing in this manner will, with a compactness argument, produce a holomorphic motion satisfying (3.16).

Note that the complex dilatations of the above group-invariant motions can be easily characterized. Writing $\mu_\lambda = \mu_{\Phi_\lambda}$, we have from (3.16) and from the transformation rules of the complex dilatations that

$$\mu_\lambda(\gamma(z)) = \mu_\lambda(z)\frac{\overline{\gamma'(z)}}{\gamma'(z)} \qquad \forall\, \gamma \in \Gamma . \tag{3.17}$$

The condition is also sufficient: By the measurable Riemann mapping theorem and the uniqueness of solutions of (1.1) up to Möbius transforms, the μ's satisfying (3.17) yield Γ-invariant holomorphic motions. This means that the space of all Γ-equivariant holomorphic motions is essentially equal to the space consisting of holomorphic functions on Δ with values in the open unit ball of L^∞, vanishing at $\lambda = 0$ and satisfying (3.17).

In this picture the complex deformations of quasiconformal mappings are tied to the Teichmüller theory in an especially elegant manner: Let Γ be a finitely generated Fuchsian group such that $L(\Gamma) = \mathbf{S}^1$. Then classically any finite type Riemann surface R_0 can be represented as a quotient space $R_0 = \Delta^*/\Gamma$ of such a group, where $\Delta^* = \overline{\mathbf{C}} \setminus \Delta$. Considering now Γ-invariant holomorphic motions (3.16) for which Φ_λ is *conformal in* Δ^* for all λ, these produce holomorphically varying Riemann surfaces. Indeed, here $L(\Gamma_\lambda)$ is a quasicircle, $\overline{\mathbf{C}} \setminus L(\Gamma_\lambda) = \Phi_\lambda(\Delta) \cup \Phi_\lambda(\Delta^*)$ and the two quotient spaces $R_0 \simeq \Phi_\lambda(\Delta^*)/\Gamma_\lambda$, $R_\lambda = \Phi_\lambda(\Delta)/\Gamma_\lambda$ are then quasiconformally equivalent Riemann surfaces, or different complex structures on a surface of finite topological type. Expressing this also in the language of Bers' [8], the deformed group Γ_λ simultaneously uniformizes both surfaces R_0 and R_λ.

However, to be precise the complex dilatations (3.17) determine the quasiconformal mappings up to a Möbius transformation; hence they determine the group Γ_λ only up to a Möbius conjugate. To avoid trivial deformations, Möbius conjugates of Γ, we should normalize our motions properly, say, by requiring that three different points remain fixed. An alternative route around this problem was introduced by Bers, cf. [8], by using the Schwarzian derivatives

$$S_f = \left(\frac{f''(z)}{f'(z)}\right)' - \frac{1}{2}\left(\frac{f''(z)}{f'(z)}\right)^2$$

of locally conformal maps. Since $S_{\gamma \circ f} = S_f$, $S_{f \circ g} = S_f(g(z))g'(z)^2$ for all Möbius transforms γ, g and here S_f determines f up to the choice of γ, this quantity has the correct flexible normalizing properties. That is, defining

$$S(\Gamma) = \{S_f : f \text{ is conformal in } \Delta^* \text{ with } S_{f \circ g} = S_f \, \forall g \in \Gamma\} ,$$

each element in $S(\Gamma)$ corresponds to a unique conjugacy class of Möbius groups isomorphic to Γ. An additional feature of this *Bers' model* is that it gives for these conjugacy classes a natural topology and complex structure: The norm $\|S_f\| = \text{Sup}_{\Delta^*}(|z|^2 - 1)^2|S_f(z)| \leq 6$ for all $S_f \in S(\Gamma)$ [8]. Therefore $S(\Gamma)$ is a bounded subset of the (finite dimensional) complex Banach space $E(\Gamma)$ consisting of functions ϕ holomorphic in Δ^* with $\phi = (\phi \circ \gamma)\gamma'^2$ for all $\gamma \in \Gamma$.

Finally, in this way one can define the Teichmüller space by

$$T(\Gamma) = \{S_f \in S(\Gamma) : f \text{ has a quasiconformal extension to } \overline{\mathbf{C}}\} . \quad (3.18)$$

This Bers' model for Teichmüller space has a number of fascinating properties. For instance, $T(\Gamma)$ is an open set in $E(\Gamma)$, homeomorphic to a ball. Furthermore, in the language of holomorphic deformations, using the argument of [14], [35] one quickly sees that in this complex structure the holomorphic disks in $T(\Gamma)$ through the origin are precisely the normalized Γ-invariant holomorphic motions Φ for which Φ_λ are conformal in Δ^*.

For a number of different questions it would be useful if the conjugacy classes of Möbius groups determined by $S_f \in S(\Gamma)$ could all be approximated by holomorphic deformations of Γ, i.e., by elements in $T(\Gamma)$. This led Bers to state the following

Conjecture 3.7 *For all finitely generated Fuchsian groups of the first kind ($L(\Gamma) = \mathbf{S}^1$) we have*

$$S(\Gamma) = \overline{T(\Gamma)} .$$

The above notions $E(\Gamma), S(\Gamma)$ and $T(\Gamma)$ are, of course, well defined for all Fuchsian groups and especially for the trivial group $\Gamma = \{\mathbf{1}\}$; only the finite dimensionality is lost in case of groups not of the first kind. Since $T(\{1\})$ contains all other Teichmüller spaces, it is often called the *universal Teichmüller space*. A possible way for Bers' problem could have been to look for a similar property in the universal space, that all conformal mappings could be approximated by elements in $T(\{1\})$. Thus Bers stated also the stronger

Conjecture 3.8 $S(\{1\}) = \overline{T(\{1\})}$.

However, Gehring [19] found beautiful counterexamples to this latter conjecture by using domains with spiral boundaries.

Theorem 3.9 *Let* $\alpha > 0$ *and*

$$\Omega = \overline{\mathbf{C}} \setminus \{\pm e^{(1+i\alpha)\log t} : 0 < t \leq 1\} .$$

Then for each conformal $f : \Delta^* \to \Omega$, $S_f \notin \overline{T(\{1\})}$.

For other, later examples see [4], [39].

Very recently Minsky [24] has succeeded in finding a positive answer to Bers' Conjecture 3.7 in the case where Δ^*/Γ is a punctured torus. The general case, however, as well as the generalized conjecture due to Thurston that any finitely generated Kleinian group is the limit of geometrically finite groups, still remain as one of main open questions in the theory of Kleinian groups.

4 Quasiconformal tools in complex dynamics

In the previous section we saw how closely the quasiconformal mappings are related to different dynamical systems. Once this was understood, the geometric quasiconformal methods became a basic and standard tool in the area, in particular in the recent studies of iteration of rational functions. This field and these applications are now in an quickly expanding phase and thus no general picture can be given here. But to have an idea of the method we describe two important early uses of quasiconformal deformations in dynamics of rational functions. In many respects, these still underlie many of the current works and applications.

As the first we consider the problem of wandering domains of rational functions. Put more precisely, the question is whether the Fatou set of a rational function could have a component U not periodic or preperiodic, a component U such that

$$R^n(U) \cap R^m(U) = \emptyset \text{ for all } n, m \in \mathbf{N} . \tag{4.19}$$

Already Fatou conjectured that this was not possible for any rational function. Combining parametric representations with dynamical aspects Sullivan [36] was finally able to give a proof. Among the important corollaries we have, for example, a complete classification of the dynamical properties of rational functions on their Fatou sets.

Theorem 4.1 *Every component of the Fatou set of a rational function is either periodic or preperiodic.*

Proof: We describe here only the idea behind Sullivan's proof. Namely, if a rational function R had a wandering component, a component U satisfying (4.19), then we could build a family of complex dilatations μ as follows. Choose first a small disk B inside U and define μ in U by taking any function from $L^\infty(B)$ with norm strictly less than one and requiring μ to be zero in $U \setminus B$.

Next, we can define μ in the components $R^m(U)$, $m \in \mathbf{Z}$, by the action of R,

$$\mu(R(z)) = \mu(z)\frac{\overline{R'(z)}}{R'(z)} . \tag{4.20}$$

According to (4.19) this extension of μ is well defined in $\bigcup_m R^m(U)$. Lastly, let us set μ equal to zero elsewhere.

By the measurable Riemann mapping theorem we find a quasiconformal mapping f_μ with complex dilatation μ. From the invariance properties (4.20) we deduce that actually f_μ conjugates R to another rational function R_μ. The space of all possible Beltrami coefficients μ above is clearly infinite dimensional and thus the same must be true for the set of R_μ's. (In fact, to show that not all of the above deformations of R are trivial requires further work, see e.g., [16] or [36])

We have now reached a contradiction since the space of all rational functions of a given degree is obviously finite dimensional. □

Renormalizing methods are nowadays central tools in many different branches of modern dynamical systems. In the area of complex dynamics their use is also possible, by a quasiconformal argument due to Douady and Hubbard: They introduced the important and very useful notion *polynomial like mappings* and proved their basic properties in [13]. According to Douady and Hubbard, a complex analytic mapping $g : U \mapsto V$ between topological disks $U, V \subset \mathbf{C}$ is a (degree d-) polynomial like mapping if firstly U is compactly contained in V, $\overline{U} \subset V$, and secondly, g is proper map of degree d. As a typical example, if P is a hyperbolic polynomial and A is an attractive k-periodic component of the Fatou set $F(P)$, choose a small enough neighbourhood U of A. Then setting $V = P^k(U)$ gives the polynomial like mapping $g = P^k|_U : U \mapsto V$. Note that usually the degree of g is much lower than $deg(P^k)$.

What makes the polynomial like mappings so useful is the Straightening theorem of Douady and Hubbard:

Theorem 4.2 *If $g : U \mapsto V$ is a polynomial like mapping of degree d, then*

$$f \circ g = P \circ f$$

for some quasiconformal mapping f and some polynomial P with $\deg(P) = d$.

Moreover, the mapping f can chosen to be conformal in the interior of the filled-in Julia set $\bigcap_n g^{-n}V$ of g.

Remark. With help of the last consequence one can often identify the new polynomial P.

Idea of proof: Without loss of generality we may assume that V is the disk $B(0, R)$ for some $R > 1$. The mapping g is now defined only in U but we may set $g(z) = z^d$ for $|z| \geq R$. A simple argument shows that we can then extend g to a quasiregular function on $V \setminus \overline{U}$. Then g is quasiregular in the whole plane $\mathbf{C}$ and in fact by the construction, the dilatations $K(g^n) \leq K_{abs} < \infty$ are uniformly bounded in $n \in \mathbf{N}$.

This means that the values of the corresponding complex dilatations $\mu_{g^n}(z)$ are uniformly bounded in the hyperbolic geometry of the unit disk Δ. Hence we can define a new dilatation μ by choosing for each $z \in \Delta$ the value $\mu(z)$ to be the (hyperbolic) barycenter of the set

$$\{\mu_{g^n}(z) : n \in \mathbf{N}\} .$$

Let f be the quasiconformal mapping (fixing ∞) with complex dilatation μ. Then again using the transformation properties of the dilatation we see that $f \circ g = P \circ f$ for some holomorphic function P. But the topological properties of the extended g show that P must be a polynomial of degree d. □

Among the consequences of the straightening theorem we have, for instance, the self-similarity properties of the Mandelbrot set.

Finally, it is interesting to note that one of the main open problems in complex dynamics is actually equivalent to a question on quasiconformal mappings. Namely, inside all rational maps the properties of hyperbolic ones are the easiest to understand. The conjecture, due to Fatou, says that actually any rational map can be approximated in the parameter space by hyperbolic ones. For quadratic polynomials $P(z) = z^2 + c$ this is equivalent to

Conjecture 4.3 *The Julia set of a quadratic polynomial $P(z) = z^2 + c$ does not support an invariant linefield, i.e., a dilatation μ with*

$$\mu(P(z)) = \mu(z)\frac{\overline{P'(z)}}{P'(z)} \quad \text{a.e. } z \in J(P) .$$

5 Holomorphic motions revisited

In the last topic we return to the distortion properties of quasiconformal mappings. We have seen how strongly quasiconformal methods can be used to understand geometric phenomena in different complex dynamical systems. Here we wish to show that also a converse direction is possible, in the sense that ideas from dynamical systems can be helpful in the study of quasiconformal mappings themselves.

Following the work [5] of the author we will consider the problem, left open from the second section, of finding the precise quasiconformal area distortion estimates. In particular, we describe a proof of the estimate

$$|f(E)| \le M|E|^{1/K} \tag{5.21}$$

valid now for all properly normalized K-quasiconformal mappings f and all subsets $E \subset \mathbf{C}$.

We shall assume that f is conformal near ∞ and use the normalisation $f(z) = z + O(|z|^{-1})$ for $z \to \infty$. Since quasiconformal mappings preserve the class of sets of zero area [21] by the Vitali covering theorem it is enough to prove the inequality when E is a finite union of nonintersecting disks $B_i = B(z_i, r_i) \subset \Delta$, $1 \le i \le n$. Furthermore, using the measurable Riemann mapping theorem we can factorize f as a composition $f = h \circ g$, where g is conformal in E and h conformal outside $g(E)$. There the inequality (2.14) can be applied to h. This means that without loss of generality we may assume that f *is conformal in* E.

To study this complementary case we use a holomorphic argument, or the holomorphic deformations

$$\lambda \mapsto f_\lambda(z) \equiv f^{\lambda\mu}(z)\,, \qquad \lambda \in \Delta\,,\ \|\mu\|_\infty = 1\,.$$

Here $f_0 = id$ and the coefficient μ is chosen so that $f_\lambda = f$ when $\lambda = \frac{K-1}{K+1}$. In particular,

$$K(f_\lambda) \equiv \frac{1+|\lambda|}{1-|\lambda|}\,.$$

This family of quasiconformal mappings gives evidently a holomorphic motion. However, using the classical Koebe distortion theorem of conformal mappings we may replace it by a simpler one. In fact, as f and hence each f_λ is conformal in E, by Koebe's theorem

$$B_i(\lambda) \equiv B\Big(f_\lambda(z_i), \frac{r_i}{4}|f'_\lambda(z_i)|\Big) \subset f_\lambda B(z_i, r_i)\,, \qquad 1 \le i \le n\,, \tag{5.22}$$

and so the disks $B_i(\lambda)$ are disjoint. Moreover $B_i(\lambda) = F_{i,\lambda}B_i(0)$, where

$$F_{i,\lambda}(z) = f'_\lambda(z_i)\,(z - z_i) + f_\lambda(z_i)\,.$$

In other words, the $F_{i,\lambda}$'s are similarities with coefficients depending holomorphically on λ. Defining

$$F(\lambda, z) = F_{i,\lambda}(z)\,, \qquad z \in B_i(0)\,,\quad 1 \le i \le n\,,$$

we have a holomorphic motion F of the union of the disks $B_i(0) = B(z_i, r_i/4)$. Another way to interpret this special motion is to consider it as a holomorphic family of disjoint disks

$$\{B_i(\lambda)\}_{i=1}^n \,.$$

Without loss of generality we may assume that all disks are compactly contained in the unit disk Δ. By the basic quasiconformal distortion theorems the estimate (5.21) is now reduced, cf. (5.22), to

$$\sum_{i=1}^n |B_i(\lambda)| \leq C\Big(\sum_{i=1}^n |B_i(0)|\Big)^{\frac{1-|\lambda|}{1+|\lambda|}} . \tag{5.23}$$

In view of Slodkowski's extension theorem, this simpler looking problem is actually equivalent to (5.21).

To see how to approach (5.23) we iterate the configuration: For each i choose a similarity ϕ_i mapping $B_i(0)$ onto Δ and define a map

$$g_\lambda|_{B_i(\lambda)} = \phi_i \circ F_{i,\lambda}^{-1} \,, \qquad 1 \leq i \leq n \,,$$

so that $g_\lambda^{-1}\Delta = \bigcup_i B_i(\lambda) \subset \Delta$, i.e., g_λ is a generalized polynomial like mapping. Taking consecutive inverse images by g_λ we obtain Cantor sets

$$J_\lambda = \bigcap_{n=0}^{\infty} g_\lambda^{-n}\Delta$$

moving holomorphically with λ. Note that the geometric information is now provided by the shifts g_λ.

Finally, on these Cantor sets we dress up (5.23) to a dynamical formulation. As we shall see this will give the advantage that the required estimates are then transparent. Indeed, by the so called thermodynamical formalism, cf. [10], for each Hölder continuous function ψ on J_λ there is a unique g_λ-invariant probability measure for which the supremum

$$P(\psi) = \sup \{\, h_\nu(g) + \int_{J_\lambda} \psi \, d\nu \; : \; \nu \text{ is } g-\text{invariant}\} \tag{5.24}$$

is attained, here $h_\nu(g_\lambda)$ is the entropy [10]. For example, Hausdorff dimension of J_λ is the unique s for which $P(-s \,\log|g_\lambda'|) = 0$. More generally,

$$P(-t \,\log|g_\lambda'|) = \log\Big(\sum_{i=1}^n rad(B_i(\lambda))^t\Big) = h_\nu(g) - t\int_{J_\lambda} \log|g_\lambda'| \, d\nu \tag{5.25}$$

For us $t = 2$ is the interesting choice.

Clearly $\lambda \mapsto \log|g'_\lambda|$ is harmonic. If instead of choosing always the maximizing measure $\nu = \nu_\lambda$ in (5.25) we fix $\lambda = \lambda_0$ and change ν_{λ_0} by the motion (call these measures $\hat{\nu}_\lambda$) then the whole last expression in (5.25) remains harmonic (entropy is an isomorphism invariant). But $P(-2\,\log|g'_\lambda|) \le 0$ for all $\lambda \in \Delta$. Thus using Harnack's inequality in the first estimate and (5.24) in the latter gives

$$P(-2\,\log|g'_\lambda|) \le \frac{1+|\lambda|}{1-|\lambda|}(h_{\hat{\nu}_0}(g_0) - 2\int_{J_\lambda} \log|g'_0|\,d\hat{\nu}_0) \le \frac{1+|\lambda|}{1-|\lambda|} P(-2\log|g'_0|)\,.$$

This is equivalent to (5.23) with $C = 1$. Thus we have proved

Theorem 5.1 *For each $K-$quasiconformal mapping f of $\mathbf{C}$ fixing $0, 1$ and ∞ we have*

$$|f(E)| \le M_K |E|^{1/K}\,, \qquad E \subset \mathbf{C}\,,$$

where M_K depends only on K.

One can also give a proof along these lines without using the dynamical formalism as shown later by Eremenko and Hamilton [15].

The area distortion theorem has a number of corollaries on the properties of quasiconformal mappings. For instance, by Chebyshev's inequality $t|\{z \in \Delta : J_f \ge t\}| \le |f\{z \in \Delta : J_f \ge t\}|$ which now yields

Corollary 5.2 *A $K-$quasiconformal mapping is in the Sobolev class $W^{1,p}_{\mathrm{loc}}$ for all $p < \frac{2K}{K-1}$.*

The function of (2.10) shows that the estimate is optimal, i.e., it does not hold any more for $p = \frac{2K}{K-1}$. Likewise, we have the optimal control of distortion of Hausdorff dimension.

Corollary 5.3 *If f is K-quasiconformal, then for any set E*

$$\frac{1}{K}\Big(\frac{1}{\dim(E)} - \frac{1}{2}\Big) \le \frac{1}{\dim(fE)} - \frac{1}{2} \le K\Big(\frac{1}{\dim(E)} - \frac{1}{2}\Big)\,. \tag{5.26}$$

A refinement of the proof [6] shows also

Theorem 5.4 *If E has Hausdorff $\frac{2}{K+1}$-measure zero, then E is removable for all K-quasiregular functions.*

Note that for any $t > \frac{2}{K+1}$ there are sets E of dimension t which are not removable for some K-quasiregular mappings.

Lastly, as shown by Gehring and Reich, Theorem 5.1 has a number of consequences for the Beurling transform. In particular, the approach of Eremenko and Hamilton shows that

$$\int_{\Delta\setminus E} |S(\chi_E)|\,dx \le |E| \log\left(\frac{\pi}{|E|}\right) \tag{5.27}$$

where we have the equality when $E = B(0,r), r < 1$. Other implications of Theorem 5.1 include

$$\int_{\Delta} |Sv|\, dm \leq \int_{\Delta} |v| \log\Big(1 + \frac{|v|}{|v|_{\Delta}}\Big)\, dm + c\log\Big(\int_{\Delta} |v| \log\Big(1 + \frac{|v|}{|v|_{\Delta}}\Big)\, dm\Big) \tag{5.28}$$

for all $v \in L\log L(\Delta)$, here $|v|_{\Delta} = \frac{1}{\pi}\int_{\Delta} |v|\, dm$ is the integral mean of $|v|$. This estimate can also be shown to be optimal, up to the correct value of c. In addition, for bounded functions, for $|\omega(z)| \leq \chi_{\Delta}(z)$ *a.e.*, similar estimates show

$$|\{z \in \Delta : |\Re\, S\omega(z)| > t\}| \leq 2\alpha\pi e^{-t}. \tag{5.29}$$

However, in spite of all above sharp estimates the important question of the precise value of the L^p-norm of the Beurling transform remains open. It has been conjectured that $\|S\|_p = \max\{p-1, 1/(p-1)\}$. The best estimate so far, 4 times the conjectured value, is due to Banuelos and Wang [7]. Their approach is based on probabilistic methods.

On the other hand, Šverák has pointed out that the problem of knowing the norm $\|S\|_p$ is closely related to a question of a quite different nature, the Morrey problem in the calculus of variations. Indeed, there one is interested in the variational integrals

$$I(u) = \int_{\Omega} F(Du(x))\, dx\ ,$$

where Ω is a domain in $\mathbf{R}^m$ and the vector valued function $u : \Omega \to \mathbf{R}^n$ is sufficiently smooth. The function F is assumed to be continuous in the space of all $n \times m$ matrices. A basic question is to determine for which functions F the integral is lower semicontinuous. According to Morrey [25] that is the case (with respect to the w^*-convergence in $W^{1,\infty}$) if and only if F is *quasiconvex*, i.e., for any constant matrix $A \in M^{n\times m}$ and any smooth compactly supported function $u : \Omega \to \mathbf{R}^n$ we have the inequality

$$F(A) \leq \frac{1}{|\Omega|}\int_{\Omega} F(A + Du(x))\, dx\ . \tag{5.30}$$

In practice this condition is hard to verify. A more concrete notion is the *rank-one convexity*, that for any matrix $A \in M^{n\times m}$ and each rank-one matrix B, $t \mapsto F(A+tB)$ is convex. For smooth F this is equivalent to the ellipticity of the corresponding Euler-Lagrange equations.

Quasiconvexity of F always implies that it is rank-one convex. If either $m = 1$ or $n = 1$ then the converse is also true. Morrey conjectured in the early 50's that this remains true for all dimensions. However, recently Šverák [37] showed that if $m \geq 2$ and $n \geq 3$ (or vice versa) then Morrey's conjecture is false.

What is not decided is the 2×2 case, a question related to planar quasiconformality or the Beurling transform: Given a matrix $A \in M^{2\times 2}$ let us denote by A_+ the conformal and by A_- the anticonformal part of A and by $\|A\|$ the norm $\|A\| = Tr(AA^T)^{1/2}$. Let also $p \geq 2$. Then using an argument due to Burkholder [11] it can be shown that the function

$$F_p(A) = ((p-1)\|A_-\| - \|A_+\|)(\|A_-\| + \|A_+\|)^{p-1}$$

is rank-one convex. Moreover, for a constant $c = c_p > 0$

$$F_p(A) \leq c((p-1)^p\|A_-\|^p - \|A_+\|^p)\ .$$

Since for a differentiable complex valued function u in the plane $|\partial u(z)| = \|Du(z)_+\|$ and $|\bar{\partial} u(z)| = \|Du(z)_-\|$, we may interpret this inequality in terms of the Beurling transform, cf. (1.4). A positive answer to Morrey's conjecture, that in two dimensions rank-one convexity implies quasiconvexity, yields the optimal value of the L^p-norm of the Beurling transform. Or as well, if $\|S\|_p > \max\{p-1, 1/(p-1)\}$ then F_p fails (5.30) at $A = 0$ and we have a counterexample for the remaining case in the Morrey problem!

In conclusion the above areas and topics show a perhaps surprising variety of applications and relations of quasiconformal mappings in the plane. We should, however, emphasize that we still have been able to cover here only a part of quasiconformal connections to other fields, i.e., those most directly connected to different parametric representations. For instance, the mappings have an important role in many questions of partial differential equations and their applications. Such connections arouse already in the 50's in the work of Bers on fluid mechanics. Very recently, a completely different application was realized by Nesi [27], to homogenization or G-closure problems, where the results of the last section were applied. These and many other examples show the naturality and flexibility of the notion of quasiconformality.

In many respects the applications considered here depended on finding efficient parametric representations for planar quasiconformal mappings. What we have seen makes it clear how fruitful similar methods would be for quasiconformal mappings in space as well. Unfortunately though, the existence of their deformations or factorizations remains open.

References

[1] Ahlfors, L., On quasiconformal mappings. *J. Analyse Math.*, 3 (1954), 1–58.

[2] Ahlfors, L. & Bers L., Riemann's mapping theorem for variable metrics. *Ann. Math.*, 72 (1960), 385–404.

[3] Ahlfors, L., Quasiconformal deformations and mappings in $\mathbf{R}^n$. *J. Analyse Math.* 30 (1976), 74–97.

[4] Astala, K., Selfsimilar zippers. *Holomorphic functions and moduli*, Vol. I (Berkeley, CA, 1986), 61–73, *Math. Sci. Res. Inst. Publ.*, 10, Springer, New York-Berlin, 1988.

[5] Astala, K., Area distortion of quasiconformal mappings. *Acta Math.*, 173 (1994), 37–60.

[6] Astala, K. Painleve's theorem and removability properties of planar quasiregular mappings. *Preprint.*

[7] Banuelos, R. & Wang, G., Sharp inequalities for martingales with applications to the Beurling-Ahlfors and Riesz transforms, *Duke Math. J.* 80 (1995) 575–600.

[8] Bers, L., Finite-dimensional Teichmüller spaces and generalizations. *Bull. Amer. Math. Soc.* (N.S.) 5 (1981), no. 2, 131–172

[9] Bojarski, B., Generalized solutions of a system of differential equations of first order and elliptic type with discontinuous coefficients. *Math. Sb.*, 85 (1957), 451–503.

[10] Bowen, R., *Equilibrium States and the Ergodic Theory of Anosov Diffeomorphisms.* Lecture Notes in Math., 470. Springer-Verlag, New York-Heidelberg, 1975.

[11] Burkholder, D., Sharp inequalities for martingales and stochastic integrals. *Asterisque* 157-158 (1988), 75–94.

[12] Donaldson, S.K. & Sullivan D.P., Quasiconformal 4-manifolds. *Acta Math.*, 163 (1989), 181–252.

[13] Douady, A. & Hubbard, J. On the dynamics of polynomial-like mappings. *Ann. Sci. École Norm. Sup.* (4) 18 (1985), no. 2, 287–343.

[14] Earle, C., Kra, I. & Krushkal, S. Holomorphic motions and Teichmüller spaces. *Trans. Amer. Math. Soc.* 343 (1994), no. 2, 927–948.

[15] Eremenko, A. & Hamilton, D. On the area distortion by quasiconformal mappings. *Proc. Amer. Math. Soc.* 123 (1995), no. 9, 2793–2797.

[16] Carleson, L. & Gamelin T., *Complex dynamics.* Universitext: Tracts in Mathematics. Springer-Verlag, New York, 1993.

[17] Gehring, F.W. & Reich E., Area distortion under quasiconformal mappings. *Ann. Acad. Sci. Fenn. Ser. A. I.*, 388 (1966), 1–14.

[18] Gehring, F.W., The L^p-integrability of the partial derivatives of a quasiconformal mapping. *Acta Math.* 130(1973), 265–277.

[19] Gehring, F.W., Spirals and the universal Teichmüller space. *Acta Math.* 141 (1978), 99–113.

[20] Iwaniec, T. & Martin G., Quasiregular mappings in even dimensions. *Acta Math.*, 170 (1992), 29–81.

[21] Lehto O. & Virtanen K., *Quasiconformal mappings in the plane.* Second edition. Springer-Verlag, New York-Heidelberg, 1973.

[22] Lehto O., *Univalent functions and Teichmüller spaces* Springer-Verlag, New York-Heidelberg, 1987.

[23] Mañé R., Sad P. & Sullivan D., On the dynamics of rational maps. *Ann. Sci. École Norm. Sup.*, 16 (1983), 193–217.

[24] Minsky, Y., The classification of punctured-torus groups. *Preprint.*

[25] Morrey, Ch. B., Multiple integrals in the calculus of variations Springer-Verlag, New York-Heidelberg, 1966.

[26] Mori, A., On an absolute constant in the theory of quasiconformal mappings. *J. Math. Soc. Japan*, 8 (1956), 156–166.

[27] Nesi, V. Quasiconformal mappings as a tool to study certain two-dimensional G-closure problems. *Arch. Rational Mech. Anal.* 134 (1996), no. 1, 17–51.

[28] Reimann, H.M., Ordinary differential equations and quasiconformal mappings. *Invent. Math.* 33 (1976), 247-270.

[29] Reimann, H.M., On the parametric representation of quasiconformal mappings. *Symposia Mathematica*, Vol. XVIII (Convegno sulle Transformazioni Quasiconformi e Questioni Connesse, INDAM, Rome, 1974), pp. 421–428. Academic Press, London, 1976.

[30] Reimann, H.M., Invariant extension of quasiconformal deformations. *Ann. Acad. Sci. Fenn. Ser. A. I. Math.*, 10 (1985), 477–492.

[31] Reimann, H.M., Functions of bounded mean oscillation and quasiconformal mappings. *Comment. Math. Helv.* 49 (1976), 260–276.

[32] Shah Dao-Shing, Parametric representation of quasiconformal mappings, *Science Record* 3 (1959) 400–407.

[33] Semenov, V.I., On one-parameter groups of quasiconformal homeomorphisms in Euclidean space *Sibirsk. Math. Zh.* 17 (1976), 177–192; English transl. *Sibirian Math. J.* 17 (1976), 140–152.

[34] Slodkowski, Z., Holomorphic motions and polynomial hulls. *Proc. Amer. Math. Soc.*, 111 (1991), 347–355.

[35] Slodkowski, Z., Extensions of holomorphic motions. *Ann. Scuola Norm. Sup. Pisa Cl. Sci. (4)* 22 (1995), 185–210.

[36] Sullivan, D. Quasiconformal homeomorphisms and dynamics. I. Solution of the Fatou-Julia problem on wandering domains. *Ann. of Math.* (2) 122 (1985), no. 3, 401–418.

[37] Šverák, V., Rank-one convexity does not imply quasiconvexity. *Proc. Roy. Soc. Edinburgh Sect. A* 120 (1992), 185–189.

[38] Thurston, W.P., *The geometry and topology of three-manifolds.* - Lecture notes, Princeton University 1978-79.

[39] Thurston, W.P., Zippers and univalent functions. *The Bieberbach conjecture* (West Lafayette, Ind., 1985), 185–197, Math. Surveys Monographs, 21, Amer. Math. Soc., Providence, R.I., 1986.

[40] Tukia, P. Quasiconformal extension of quasisymmetric mappings compatible with a Möbius group. *Acta Math.* 154 (1985), no. 3-4, 153–193.

University of Jyväskylä
Department of Mathematics
P.O. Box 35
FIN-40351 Jyväskylä
Finland

Astala@math.jyu.fi

Local Minimality Results Related to the Bloch and Landau Constants

Albert Baernstein II and Jade P. Vinson

1 Introduction

Let $H(\mathbb{D})$ denote the set of functions holomorphic in the unit disk $\mathbb{D} \subset \mathbb{C}$. The theorem of Bloch, [Bh] from 1925, asserts existence of an absolute constant B such that for each $f \in H(\mathbb{D})$ there is a disk of radius $B|f'(0)|$ which is the one-one image under f of some subdomain of $\mathbb{D}$. Such a disk is called a "schlicht disk" for f. The best, i.e., largest, value of B is called *Bloch's constant.*

Bloch's theorem implies existence of another absolute constant, L, such that, for each $f \in H(\mathbb{D})$, $f(\mathbb{D})$ contains a disk of radius $L|f'(0)|$. The constant L was introduced by E. Landau [Lu] in 1929. Accordingly, the largest possible value for L is known as *Landau's constant.* From the definitions, it is clear that $L \geq B$.

The exact values of B and L are not known. The best known estimates seem to be

$$\frac{\sqrt{3}}{4} + 2 \cdot 10^{-4} < B \leq \frac{\Gamma(1/3)\Gamma(11/12)}{\Gamma(1/4)(1+\sqrt{3})^{1/2}} = .47... \tag{1.1}$$

$$\frac{1}{2} + 10^{-335} < L \leq \frac{\Gamma(1/3)\Gamma(5/6)}{\Gamma(1/6)} = .54.... \tag{1.2}$$

For Bloch's constant, the lower bound $B \geq \frac{\sqrt{3}}{4} > .43$ was proved by Ahlfors [A1] in 1938, as an application of his method of "ultrahyperbolic metrics". See also [A4]. Heins [Hei] showed that strict inequality holds for B, but did not give an explicit improvement. Such an improvement was supplied by Bonk [Bo] in 1990, who established the lower bound $\frac{\sqrt{3}}{4} + 10^{-14}$. Bonk derives his bound from the solution of an extremal problem involving

[1]The first author was supported in part by NSF grants DMS-9206319 and DMS-9501293. The second author was supported by a Barry Goldwater Scholarship, and by a Compton Scholarship from Washington University.

the real parts of derivatives of certain "Bloch functions". See also [Ha2, p.136]. Chen and Gauthier [CG] refined Bonk's method to increase Bonk's 10^{-14} to $2 \cdot 10^{-4}$.

For L, Ahlfors's ultrahyperbolic method produces the non-strict lower bound $\frac{1}{2}$. [A4, p.20, prob.2]. Pommerenke [P] offered a proof that $L > \frac{1}{2}$. The lower bound in (1.2) is due to Yanagihara [Y2].

The upper bound for B in (1.1) is due to Ahlfors and Grunsky [AG]. For L, the upper bound in (1.2) is due independently to R.M. Robinson (1937, unpublished) and to Rademacher [R]. It was conjectured in [AG], resp.[R], that the upper bounds proved therein for B, resp. L, are their true values.

This paper has two goals. One is to present some modest evidence in favor of the Ahlfors-Grunsky and Rademacher conjectures. The other is to highlight some methods which are used in the proofs of our results, notably second variations of quasiconformal mappings. A number of authors have used this method to attack various problems of complex analysis, but we feel that it potentially has many more uses, especially to concrete extremal problems like those of finding the Bloch and Landau constants. To the newcomer, second variations can look scary. Accordingly, we have supplied semi-detailed derivations of the main formulas needed for our proofs, even though it would have posssible to cite most of them from the published literature.

To explain our results, we first examine the conjectured extremal functions. They are associated with hexagonal lattices. In this paper, the "lattice generated by z_1, z_2, z_3," where $z_1, z_2, z_3 \in \mathbb{C}$, is understood to be the set $\{z_1 + m(z_2 - z_1) + n(z_3 - z_1) : m, n \in \mathbb{Z}\}$. Thus, our lattices do not necessarily contain the origin.

Let $\mathcal{F}$ be the interior of the hyperbolic triangle in $\mathbb{D}$ with vertices at $1, e^{2\pi i/3}, e^{4\pi i/3}$. Thus, $\mathcal{F}$ is the subdomain of $\mathbb{D}$ whose boundary is the union of the hyperbolic geodesics connecting 1 to $e^{2\pi i/3}$, $e^{2\pi i/3}$ to $e^{4\pi i/3}$, and $e^{4\pi i/3}$ to 1. Let F be the conformal map of $\mathcal{F}$ onto the interior of the Euclidean triangle with vertices $1, e^{2\pi i/3}, e^{4\pi i/3}$ which fixes each of $1, e^{2\pi i/3}, e^{4\pi i/3}$. Then $F(0) = 0$ and $F'(0) > 0$. By successive reflections, F extends holomorphically to a map of $\mathbb{D}$ onto $\mathbb{C} \setminus \mathcal{A}$, where $\mathcal{A}$ is the lattice generated by the points $1, e^{2\pi i/3}, e^{4\pi i/3}$. The points of $\mathcal{A}$ are the vertices of a tiling of the plane into equilateral triangles. The function F is a universal covering map of $\mathbb{D}$ onto $\mathbb{C} \setminus \mathcal{A}$. It is the unique universal covering map of $\mathbb{D}$ onto $\mathbb{C} \setminus \mathcal{A}$ with $F(0) = 0$ and $F'(0) > 0$. The largest disk contained in $F(\mathbb{D})$ has radius 1. It turns out that that $F'(0) = \frac{\Gamma(1/6)}{\Gamma(1/3)\Gamma(5/6)}$. Thus, F furnishes the upper bound for L in (1.2), and is a function conjectured by Rademacher to be extremal for the Landau problem. Of course, any function $f(z) = aF(\alpha z) + b$, with a, b complex constants and α a unimodular constant would also be extremal.

For each $a \in \mathcal{A}$, the "Voronoi cells" $\{z \in \mathbb{C} : |z - a| \leq |z - a'|, \ \forall a' \in \mathcal{A}\}$ are regular hexagons. Consequently, $\mathcal{A}$, and lattices obtained from it by similarity mappings, are called hexagonal lattices. For short, we'll call them hex lattices.

A function conjectured by Ahlfors and Grunsky to be extremal for the Bloch problem is obtained by conformally mapping the interior of a hyperbolic triangle in $\mathbb{D}$ with center at the origin and angles $\pi/6, \pi/6, \pi/6$ onto the interior of the Euclidean triangle with vertices $1, e^{2\pi i/3}, e^{4\pi i/3}$, then again extending to $\mathbb{D}$ by reflection. This function, which we'll call F_1 and precisely specify in Section 7, maps $\mathbb{D}$ onto the whole plane. It has a branch point of order two, i.e., its derivative has a simple zero, at each point of ${F_1}^{-1}(\mathcal{A})$. Thus, the largest schlicht disk contained in the image of F_1 has radius 1, while $F_1'(0)$ equals the reciprocal of the number on the right side of (1.1).

For information about conformal maps of hyperbolic triangles, the reader may consult, for example, [H, Sec.17.6]. Formula (17.6.26) of [H] expresses the maps of most interest to us in terms of hypergeometric functions, but contains mistakes. The proof in [H] shows that the formula should read

$$f(z) = z\frac{\Gamma(1-\frac{1}{n})\Gamma(\frac{1}{2}(1+\alpha)+\frac{1}{n})F(\frac{1}{2}(1-\alpha)+\frac{1}{n},\frac{1}{2}(1-\alpha),1+\frac{1}{n};z^n)}{\Gamma(1+\frac{1}{n})\Gamma(\frac{1}{2}(1+\alpha)-\frac{1}{n})F(\frac{1}{2}(1-\alpha)-\frac{1}{n},\frac{1}{2}(1-\alpha),1-\frac{1}{n};z^n)}.$$

Minda's article [Mi] contains additional historical information and some Bloch- Landau type theorems in different settings. On [Mi, p.56] is corrected an error in the book of Goluzin about the value of $F'(0)$, where F reassumes now its meaning as proposed Landau extremal. Goluzin's error apparently propagated to [H, p.388], where is stated $F'(0) = 3/(2\Gamma(1/3)) = .56\cdots$ instead of the correct $F'(0) = \Gamma(1/3)\Gamma(5/6)/\Gamma(1/6) = .54\cdots$.

Another unknown constant, we'll call it B_u, is called the *schlicht Bloch constant*. It is defined to be the largest number B_u such that, for each *univalent* $f \in H(\mathbb{D})$, $f(\mathbb{D})$ contains a disk of radius at least B_u. It is known that $.57 < B_u < .66$. Recent works in which the schlicht Bloch constant plays a role include [BC] and [Be].

Landau's constant can be characterized in numerous ways. Let $R(E)$ denote the radius of the largest open disk contained in a set $E \subset \mathbb{C}$. Let $\mathcal{K}$ denote the set of all universal covering maps of $\mathbb{D}$ onto some plane domain, and let $\mathcal{K}_d$ denote the subset of $\mathcal{K}$ consisting of universal covering maps of $\mathbb{D}$ onto domains of the form $\mathbb{C} \setminus A$, where A is a discrete subset of $\mathbb{C}$. Define constants $L_1, ..., L_4$ as follows .

$$L_1 = \sup\{|f'(0)|/R(f(\mathbb{D})) : f \in H(\mathbb{D})\}.$$

$$L_2 = \sup\{(1-|z|^2)|f'(z)|/R(f(\mathbb{D})) : z \in \mathbb{D},\ f \in H(\mathbb{D})\}.$$

$$L_3 = \sup\{(1-|z|^2)|f'(z)|/R(f(\mathbb{D})) : z \in \mathbb{D},\ f \in \mathcal{K}\}.$$

$$L_4 = \sup\{(1-|z|^2)|f'(z)|/R(f(\mathbb{D})) : z \in \mathbb{D},\ f \in \mathcal{K}_d\}.$$

Then $L_1 = L$. Clearly, $L_1 \le L_2$, and $L_4 \le L_3 \le L_2$. By precomposing f with Möbius transformations, one sees that in fact $L_1 = L_2$. The principle of subordination, see, for example, [HK, p.74], implies that $L_3 = L_2$. Let $f \in \mathcal{K}$. Draw a grid in the plane consisting of closed squares of sidelength $1/n$. Let A_n be the set of centers of squares S in the grid for which S intersects the complement of $f(\mathbb{D})$. For large n, let f_n be the u.c.m. of $\mathbb{D}$ onto $\mathbb{C} \setminus A_n$ with $f_n(0) = f(0)$ and $\arg f_n{}'(0) = \arg f'(0)$. Then, as $n \to \infty$, f_n converges locally uniformly to f in $\mathbb{D}$, and $R(\mathbb{C} \setminus A_n) \to R(f(\Omega))$. Hence, $L_3 = L_4$.

Thus, each of the four constants L_i equals L. In particular, Rademacher's conjecture may be stated as an extremal problem for discrete subsets of $\mathbb{C}$.

Rademacher's Conjecture. *For each discrete subset $A \subset \mathbb{C}$, each universal covering map f of $\mathbb{D}$ onto $\mathbb{C} \setminus A$, and each $z \in \mathbb{D}$, holds*

$$(1 - |z|^2)|f'(z)| \le |F'(0)| R(\mathbb{C} \setminus A). \tag{1.3}$$

Recall that F is a u.c.m. of $\mathbb{D}$ onto the complement of the hex lattice $\mathcal{A}$, and that $F(0) = 0$ is the center of one of the equilateral triangles into which the plane is decomposed by $\mathcal{A}$. A logical first step is to see if (1.3) holds for $f = F$. Theorem 1, whose statement we defer until Section 2, implies that it does.

Corollary to Theorem 1. $\sup_{z \in \mathbb{D}} (1 - |z|^2)|F'(z)| = |F'(0)|$.

Theorem 1 asserts roughly that on each of the triangles into which $\mathcal{A}$ divides $\mathbb{C}$ the hyperbolic density of $\mathbb{C} \setminus \mathcal{A}$ at a point z decreases as z moves closer to the center of the triangle. The corollary turns out, via the Kronecker limit formula, to be equivalent to a result of Osgood, Phillips and Sarnak [OPS] about the determinant of the Laplacian of flat tori, and is implied by a theorem of H.Montgomery [Mo]. Further discussion appears at the end of Section 2.

Having verified (1.3) for $f = F$, it next seems natural to consider the case when A is a lattice conformally different from $\mathcal{A}$. We are unable to settle this case completely. We can, however, show that (1.3) is true for lattices A sufficiently close to hexagonal and points z such that $f(z)$ is sufficiently close to the barycenter (= center of mass) of one of the triangles formed by three neighboring points of A. To make this statement precise, we introduce some notation. Let T be the Möbius transformation which maps $\mathbb{D}$ onto the open upper half plane $\mathbb{H}$ with $T(0) = e^{i\pi/3}$, $T(1) = \infty$. Then

$$T(\sigma) = e^{-i\pi/3} \frac{\sigma - e^{2\pi i/3}}{\sigma - 1}, \quad T^{-1}(\tau) = \frac{\tau - e^{i\pi/3}}{\tau - e^{-i\pi/3}}. \tag{1.4}$$

For $\sigma \in \mathbb{D}$, write $\tau = T(\sigma)$, and let

$$A(\sigma) = \{m + n\tau : m, n \in \mathbb{Z}\}$$

be the lattice generated by $0, 1$, and τ. Let $\Omega(\sigma) = \mathbb{C} \setminus A(\sigma)$, and let $f(\cdot, \sigma)$ be a u.c.m. of $\mathbb{D}$ onto $\Omega(\sigma)$ for which

$$f(0, \sigma) = \frac{1}{3}(1 + \tau).$$

Then $f(0, \sigma)$ is the barycenter of a triangle whose vertices are three neighboring points of $A(\sigma)$. We may take

$$f(z, 0) = e^{-i\pi/6} \frac{1}{\sqrt{3}} (F(z) - e^{4\pi i/3}).$$

Note that $A(0)$ is also a hex lattice, and that $|f'(z, 0)| = \frac{1}{\sqrt{3}} |F'(z)|$. Write $R(\sigma) = R(\Omega(\sigma))$.

Theorem 2. *There exists $\epsilon > 0$ such that*

$$(1 - |z|^2|)|f'(z, \sigma)| \leq |F'(0)| R(\sigma), \quad |z| < \epsilon, \ |\sigma| < \epsilon. \tag{1.5}$$

Equality holds if and only if $z = 0$ and $\sigma = 0$.

The prime in (1.5) denotes differentiation with respect to z. For $z = 0$ this is a theorem of Yanagihara [Y1], while for $\sigma = 0$ the corollary to Theorem 1 of the present paper asserts that (1.5) holds for all $z \in \mathbb{D}$. Our proof of Theorem 2, like Yanagihara's proof of his theorem, is based on second variations of quasiconformal maps. Their theory is relevant because $f(\cdot, \sigma)$ can be constructed by quasiconformal deformation of F. The main new element which enables us to improve on Yanagihara's result is estimation of a certain Poincaré series associated with the full modular group.

The first author is grateful to Scott Wolpert for pointing out to him that the second q.c. variation has also played a role in Teichmüller theory. For example, in [A2] and [W1] the second variation is used to prove results about curvature of the Weil-Petersson metric in the Teichmüller space of compact surfaces of genus $g \geq 2$. It is conceivable that the points of view of those papers, which emphasize "harmonic" Beltrami differentials, could lead to further results about Bloch and Landau constants in the lattice case. One way to begin such an effort would be, in the construction of u.c.m.'s of $\mathbb{C}$ minus lattices in Section 5, to replace the linear homeomorphisms U with homeomorphisms V of $\mathbb{C}$ which satisfy, for complex parameters s of small modulus, the equation $\overline{\partial} V = s\rho^{-2} \partial U$, where ρ is the hyperbolic density of $\Omega(0)$. Additional variational techniques for study of hyperbolic metrics can be found in [W2].

The general Rademacher conjecture (1.3) can be restated as a minimum problem for hyperbolic densities. As in (1.3), let A be a discrete subset of $\mathbb{C}$ and f be a u.c.m. of $\mathbb{D}$ onto $\mathbb{C} \setminus A$. The hyperbolic density $\rho(\cdot, A)$ of $\mathbb{C} \setminus A$ may be defined as follows. For $z \in \mathbb{C} \setminus A$, take any $\zeta \in \mathbb{D}$ with $f(\zeta) = z$. Then

$$\rho(z, A) = \frac{2}{(1 - |\zeta|^2)|f'(\zeta)|}. \tag{1.6}$$

The Riemannian metrics $\rho(z,A)\,|dz|$ have constant Gaussian curvature -1 in $\mathbb{C}\setminus A$. Set $u=\log\rho$. Then, in $\mathbb{C}\setminus A$, $u=u(\cdot,A)$ satisfies the p.d.e. $\Delta u=e^{2u}$. As $z\to a\in A$, $u(z,A)\to\infty$. More precisely, for some complex constant c,

$$u(z,A)=-\log|z-a|-\log\log|1/(z-a)|+c+o(1). \tag{1.7}$$

The expansion (1.7) can be found in [He2, p.104] or [Hei, p.32]. In [He2] the result is stated for hyperbolic densities of the sphere with finitely many punctures, but the analysis is valid near an isolated boundary point of any domain. We remark here that [He2] and [Hal] are good general references for universal covering maps and hyperbolic metrics as used in this paper.

From (1.7), one can show that, in the whole plane, $u(\cdot,A)$ is a distributional solution of

$$\Delta u=e^{2u}-2\pi\sum_{a\in A}\delta_a. \tag{1.8}$$

Conjecture (1.3) is equivalent to the following assertion: In the class of discrete sets A for which $R(\mathbb{C}\setminus A)$ has a fixed value R_0, the minimum value of $u(z,A)$ is achieved when A is a hex lattice with $R(\mathbb{C}\setminus A)=R_0$, and z is a barycenter of one of the equilateral triangles in the tiling of the plane induced by A. The intuition is: Among all sets A with the property that each point of $\mathbb{C}$ is within distance R_0 to A, the hex lattice is most "spread out."

In the theories of packing, covering, and geometry of numbers one finds extremal geometrical problems about discrete subsets of $\mathbb{C}$ for which hex lattices furnish extremals. See, for example, [C], [CS], and [Ro]. There seems however, to have been not much prior work on extremal analytic problems involving discrete sets and solutions of pde's. The theorem of Montgomery mentioned earlier does provide an analytic situation for which the hex lattice is extremal.

Here, on the other hand, is a minimum problem for pde's and discrete sets, closely related to Montgomery's Theorem and parallel to the Bloch-Landau problems, for which the hex lattice is not extremal. For $z\in\mathbb{C}$ and $t>0$, define

$$P(z,t,A)=\frac{1}{4\pi t}\sum_{a\in A}\exp(-|z-a|^2/4t).$$

Then u satisfies the heat equation $\Delta_z P=\partial_t P$, with initial temperature $P(z,0,A)=\sum_{a\in A}\delta_a$. If A_0 is a hex lattice, it's still true that, for each fixed t, $P(z,t,A_0)$ is minimal when z is the center of an A_0-triangle. But if A is any non-rectangular and non-hex lattice, it is shown in [B2] that for all sufficiently small t the minimum over z of $P(z,t,A)$ is smaller than the minimum of $P(z,t,A_0)$, where A_0 is a hex-lattice with $R(\mathbb{C}\setminus A)=R(\mathbb{C}\setminus A_0)$. This answers in the negative a question raised in [B1].

Another unsolved problem involving lattices is a conjecture of Schmutz [S], which asserts that among all once-punctured tori of fixed area the maximal "length spectrum" is achieved when the torus is $\mathbb{C}/A$ with its usual flat metric, with A a hex lattice. Partial results have been proved by Kühlein [K].

Let us turn for a moment back to B. Can the Ahlfors-Grunsky conjecture also be reduced to a problem about mappings associated with discrete subsets of $\mathbb{C}$? We suspect that it can be, but have not formulated such a result. For each lattice $A(\sigma)$ there are holomorphic maps f of $\mathbb{D}$ onto $\mathbb{C}$ which have a branch point of order two over each point of $f^{-1}(A(\sigma))$. In Section 7, we discuss an analogue of Theorem 2 which asserts local extremality of the Ahlfors-Grunsky function within a class of such maps, and also note that a Bloch analogue of Theorem 1 is true. The local extremality result is, like Theorem 2, a two-parameter extension of a one-parameter resut in [Y1].

Section 2 contains the proof of Theorem 1. In Sections 3 and 4 we discuss some topics related to the variational theory of qc maps. As noted above, most of what we derive is already known; these sections are included on the chance that neophyte variationists might find them useful. In Section 5 we use qc maps to construct the u.c.m.'s $f(\cdot, \sigma)$, and reduce the proof of Theorem 2 to the estimation of a certain singular integral. In section 6, elliptic functions are used to express this integral as a Poincaré series, for which the required estimate is carried out, with the help of a computer. The second author thanks Danyel Fisher for help in explaining the C source codes for this computation.

It is a great pleasure for us to dedicate this paper to Fred Gehring. The first author thanks Fred for inspiring him to become a player in the quasiconformal game, and for friendship and good cheer stretching back 25 years, to the first author's first talk at a meeting, where Fred was presiding officer. The second author has not yet made Fred's acquaintance, but hopes to do so on some future occasion.

2 A monotonicity theorem

Recall that $A(0) = \{m + ne^{i\pi/3} : m, n \in \mathbb{Z}\}$ denotes the hex lattice generated by $0, 1$, and $e^{i\pi/3}$. Throughout the paper, we shall denote by z_0 the barycenter of the triangle with vertices $0, 1$, and $e^{i\pi/3}$. Thus

$$z_0 = \frac{1}{3}(1 + e^{i\pi/3}) = \frac{1}{2} + i\frac{\sqrt{3}}{6}.$$

As noted in section 1, $f(z, 0) = e^{-i\pi/6}\frac{1}{\sqrt{3}}(F(z) - e^{4\pi i/3})$ is a universal covering map of $\mathbb{D}$ onto $\Omega(0) = \mathbb{C} \setminus A(0)$, with $f(0, 0) = z_0$. With the notation of (1.6), let

$$u(z) = \log \rho(z, A(0)).$$

Then u is the logarithm of the density of the hyperbolic metric of $\Omega(0)$. Let T denote the triangle (interior plus perimeter) with vertices $0, \frac{1}{2}, z_0$. Using invariance of u under various translations and reflections, it is easy to see that T is a fundamental domain for u. Thus, to determine the behavior of u in the whole plane, it suffices to determine it in T.

Theorem 1. *In $T \setminus \{0\}$ holds*

$$\frac{\partial u}{\partial x} \leq 0 \quad \text{and} \quad \frac{\partial u}{\partial y} \leq 0. \tag{2.1}$$

Strict inequality holds for $\frac{\partial u}{\partial x}$ except on the line $[\frac{1}{2}, z_0]$, and strict inequality holds for $\frac{\partial u}{\partial y}$ except on the line $(0, \frac{1}{2}]$ and at the point z_0.

Corollary. $u(z) \geq u(z_0)$, $\forall z \in \Omega(0)$. *Equality holds if and only if $z = z_0 + a$ or $z = (1 + e^{i\pi/3} - z_0) + a$ for some $a \in A(0)$.*

An equivalent formulation of the Corollary is that the hyperbolic derivative of $f(z,0)$, or of $F(z)$, is maximized over $z \in \mathbb{D}$ when $z = 0$. This is the "Corollary to Theorem 1" stated in Section 1. The point $z = (1+e^{i\pi/3} - z_0)$ is the reflection of z_0 in the line through 1 and $e^{i\pi/3}$.

Another consequence of Theorem 1 is as follows. Let $z \in \Omega(0)$. Consider the tiling of the plane into equilateral triangles whose vertices are the points of $A(0)$. Take one of these triangles which contains z, and let z_1 be its center. Suppose that $z \neq z_1$, and let α denote the unit vector starting at z in the direction z_1. Then

$$\nabla u(z) \cdot \alpha < 0.$$

Similarly, if z is neither a center point nor a point of $A(0)$, then the derivative of u at z in the direction of a point of $A(0)$ nearest to z is strictly positive.

Proof of Theorem 1. Let $v = \frac{\partial u}{\partial y}$. Since $\Delta u = e^{2u}$ in $\Omega(0)$, v satisfies in $\Omega(0)$ the pde

$$\Delta v = 2e^{2u} v. \tag{2.2}$$

Let D be the interior of the triangle with vertices $0, 1, z_0$. Then v is in C^∞ on the closure of D, except at the points $0, 1$. Since u is symmetric in the real axis, it follows that $v = 0$ on the segment $(0, 1)$. Near the boundary point 0 of $\Omega(0)$, u has the form

$$u(z) = -\log|z| - \log\log|1/z| + R(z), \tag{2.3}$$

where R denotes a term symmetric with respect to the real axis. From inspection of [He2, section 5], one can show that $\frac{\partial R}{\partial y}(z) = o(y|z|^{-2})$ as $z \to 0$, so that

$$\limsup_{z \to 0} v(z) = 0, \tag{2.4}$$

as $z \to 0$ from within D. Since u is symmetric in the line $\operatorname{Re} z = \frac{1}{2}$, (2.4) also holds as $z \to 1$ from within D. Also, u is invariant under reflection in each of the six lines connecting z_0 to a median or a vertex of the triangle with vertices $0, 1$ and $e^{i\pi/3}$. Hence, u must have a critical point at z_0, so that $v(z_0) = 0$.

Since u is symmetric in the line through 0 and z_0, the gradient of u on the segment $(0, z_0)$ is parallel to $(0, z_0)$.

Claim. $|\nabla u| > 0$ *on* $(0, z_0)$, *and* $u(z)$ *decreases as* z *moves from* 0 *to* z_0 *along* $(0, z_0)$.

A consequence of the claim is that $v < 0$ on $(0, z_0)$. By symmetry, we'll also have $v < 0$ on $(1, z_0)$. Since $v = 0$ on $(0, 1)$, the strong maximum principle, applied to the equation (2.2), will imply that $v < 0$ everywhere in D. The assertions in Theorem 1 about $\frac{\partial u}{\partial y}$ will then follow from the strict negativity of v in D and the strict decrease of u on $(0, z_0)$ asserted by the claim.

Proof of claim. Let S denote the half-strip $\{x+iy : -\infty < x < 0, 0 < y < \pi\}$, and Φ be the conformal map of S onto the interior of T which sends the boundary points $\infty, 0, \pi i$ to $0, \frac{1}{2}, z_0$, in that order. By reflection, Φ extends to a conformal map of the full strip $0 < y < \pi$ onto D. Let $u_1 = u \circ \Phi$ and $v_1 = \frac{\partial u_1}{\partial x}$. The claim asserts that $v_1(x + i\pi) < 0$ for $-\infty < x < 0$.

By means of the Schwarz relection principle, one easily shows that u_1 has a C^∞ extension from the strip $0 < y < \pi$ to the whole plane with the set $\{(2n+1)\pi i : n \in \mathbb{Z}\}$ deleted. This extension is symmetric with respect to all the lines $y = (2n+1)\pi$. Let L denote the left half plane $\operatorname{Re} z < 0$. We'll show that in fact

$$v_1(z) < 0, \quad z \in L. \tag{2.5}$$

To prove (2.5), we first examine the behavior of v_1 near its singularities. The map Φ is the composition of e^z with a Schwarz-Cristoffel map. From this, we see that, for some positive real constant c,

$$\Phi'(z) = ce^{z/6}(1+e^z)^{-1}, \quad 0 < y < \pi. \tag{2.6}$$

The behavior of u near $z = 0$ is given by (2.3). Near z_0, the reflection symmetries of u noted above show that $u(z) = u(z_0) + \text{const}|z - z_0|^2 + O(z - z_0)^3$. Using (2.6), one computes that

$$v_1(x+iy) = -\frac{1}{6} + O(1/x), \quad x \to -\infty,\ y \in \mathbb{R},$$

and that $v_1(z) = o(1)$ as $z \to \pi i$. By symmetry and smoothness, it is also true that $v_1(iy) = 0$ for $y \in \mathbb{R} \setminus \{(2n+1)\pi i : n \in \mathbb{Z}\}$.

In $\mathbb{C} \setminus \{(2n+1)\pi i : n \in \mathbb{Z}\}$, v_1 satisfies the pde

$$\Delta v_1 = 2e^{2u_1}|\Phi'|^2 v_1 + e^{2u_1}\frac{\partial}{\partial x}(|\Phi'|^2).$$

By differentiation of (2.6), one verifies that the second term on the right is positive in L. Thus, v_1 is a subsolution in L for an operator of the form $\Delta - V$, with $V \geq 0$. Moreover, v_1 is $2\pi i-$ periodic, and, as we just saw, $v_1(x+iy) \to 0$ as $x \to 0-$, $v_1(x+iy) \to -1/6$ as $x \to -\infty$. The strong maximum principle applies to v_1 in L. We conclude that $v_1 < 0$ in L. The claim, and with it Theorem 1, are proved.□

The proof of Theorem 1 has some features in common with an argument of Hempel [He1], which shows that the density of the hyperbolic metric of the twice punctured plane $\mathbb{C} \setminus \{-1, 0\}$ is a symmetric decreasing function on each circle $|z| = r$. See also [Ha1, p.702] and [Lai]. The symmetric decrease result was extended to hyperbolic metrics on all circularly symmetric plane domains by Weitsman [W1]. In [W2], Weitsman proves another, very powerful, theorem in which hyperbolic metrics on arbitrary plane domains are shown to be dominated, in a certain sense, by metrics on associated circularly symmetric domains.

Another feature of Theorem 1 is that it is only partly a theorem about hyperbolic densities. For any increasing continuous function ϕ, the same proof gives the same monotonicity properties for the solution of any pde $\Delta u = \phi(u)$ in $\Omega(0)$ which has appropriate behavior at the punctures $a \in A(0)$.

We explain here how Theorem 1 is related to results in [OPS] and [Mo]. For a compact two-dimensional Riemannian manifold M, let λ_j, $j \geq 1$, denote the nonzero eigenvalues of the (positive) Laplace-Beltrami operator Δ_M on M, listed according to multiplicity. The trace K of the heat kernel and the zeta function Z of Δ_M are defined by

$$K(t, M) = 1 + \sum_{j=1}^{\infty} \exp(-t\lambda_j), \quad t > 0,$$
$$Z(s, M) = \sum_{j=1}^{\infty} \lambda_j{}^{-s}, \quad \operatorname{Re} s > 1. \tag{2.7}$$

The zeta function can be analytically continued to a function holomorphic in $\mathbb{C} \setminus \{1\}$, with a simple pole at $s = 1$. The determinant of the Laplacian of M, det'(M), is defined to be

$$\det{}' M = e^{-Z'(0,M)}.$$

Recall that $A(\sigma)$, $\sigma \in \mathbb{D}$, denotes the lattice generated by $0, 1$, and $\tau = T\sigma$, where T is given by (1.4). Corollary 1.3 of [OPS], see also [Sa], asserts that among all M of genus 1 and area 1, the maximum value of det'M is achieved by the equlilateral torus $M = \mathbb{C}/A(0)$, endowed with the flat metric scaled so that M has area 1. General results in [OPS] enable the authors to reduce to the case when M is some $A(\sigma)$ with appropriately scaled flat metric. They then observe that the Kronecker limit formula, see, e.g., [La] or [Q], implies that

$$\det{}'A(\sigma) = \operatorname{Im}\tau|\eta(\tau)|^4 := L(\tau),$$

where $\eta(\tau) = e^{\pi i\tau/12}\prod_{n=1}^{\infty}(1-e^{2\pi in\tau})$, $\tau \in \mathbb{H}$, is the Dedekind eta function. To complete the proof of their Corollary 1.3, they show that L has a local maximum at $\tau = e^{i\pi/3} = T(0)$, then say that a numerical check using the product shows that the local maximum is in fact a global max over $\mathbb{H}$.

Now let $F_1 = F \circ T^{-1}$, with F as in Section 1. Then F_1 is a u.c.m. of $\mathbb{H}$ onto $\mathbb{C} \setminus \mathcal{A}$, where $\mathcal{A}$ is the lattice generated by $1, e^{2\pi i/3}$, and $e^{4\pi i/3}$. By comparing mapping properties in the hyperbolic triangle $0 < \operatorname{Re} z < 1$, $|z - \frac{1}{2}| > \frac{1}{2}$, one finds that $F_1 = G \circ \lambda$, where λ is the elliptic modular function and G is a conformal map of $\mathbb{H}$ onto the Euclidean triangle with vertices $1, e^{2\pi i/3}, e^{4\pi i/3}$. When the constants for G are determined, it turns out that

$$(F_1{}')^3 = (2\pi)^3\Gamma^{-9}(1/3)i\frac{(\lambda')^3}{\lambda^2(1-\lambda)^2}. \tag{2.8}$$

From [Ran, pp.196, 231], it follows, after correcting a typo, that

$$\eta^{12} = 2^{-4}\pi^{-3}i\frac{(\lambda')^3}{\lambda^2(1-\lambda)^2}. \tag{2.9}$$

From (2.8), (2.9), and comparison of arguments as τ approaches ∞ along the positive imaginary axis, one obtains

$$F_1{}' = c\eta^4, \quad c = e^{4\pi i/3}2^{1/3}(2\pi)^2\Gamma^{-3}(1/3). \tag{2.10}$$

The Corollary stated after Theorem 1 is equivalent to the statement that $\operatorname{Im} z|F_1{}'(z)|$ is maximal over $\mathbb{H}$ at $z = e^{i\pi/3}$. Thus, our Corollary, established by analytic means, is equivalent to the maximization result for η established numerically in [OPS].

Consider next, for $\sigma \in \mathbb{D}$ and $t > 0$,

$$W(t,\sigma) := \sum_{a\in A(\sigma)} \exp(-\pi t|a|^2/\operatorname{Im}\tau) \quad \tau = T\sigma.$$

In [Mo], Montgomery proves that, for each fixed t, W is minimized over $\mathbb{D}$ at $\sigma = 0$.

When $\mathbb{C}/A(\sigma)$ has the area 1 flat metric, its eigenvalues are $4\pi^2|a|^2/\operatorname{Im}\tau$, $a \in A(\sigma)$. Writing $Z(s,\sigma)$, $K(t,\sigma)$ for the associated zeta functions and heat kernels, (2.7) gives

$$Z(s,\sigma) = (2\pi)^{-2s}(\operatorname{Im}\tau)^s \sum_{a\in A(\sigma)}{}' |a|^{-2s}, \quad \operatorname{Re} s > 1,$$

$$K(t,\sigma) = W(4\pi t, \sigma), \quad t > 0.$$

Formula (3) of [Mo] can be written as

$$(4\pi)^s\Gamma(s)Z(s,\sigma) = \frac{1}{s-1} - \frac{1}{s} + \int_1^\infty (W(t,\sigma)-1)(t^{s-1}+t^{-s})\,dt, \quad \operatorname{Re} s > 1. \tag{2.11}$$

The integral on the right hand side extends to an entire function of s. Thus, (2.11) furnishes the analytic continuation of Z. (2.11) also implies the functional equation

$$(4\pi)^s\Gamma(s)Z(s,\sigma) = (4\pi)^{1-s}\Gamma(1-s)Z(1-s,\sigma), \quad s \in \mathbb{C}. \tag{2.12}$$

Montgomery's inequality $W(t,0) \le W(t,\sigma)$, together with (2.11), yields

$$Z(s,0) \le Z(s,\sigma), \quad 1 < s < \infty. \tag{2.13}$$

Letting $s \to 1$, and using (2.12), one obtains the OPS inequality $\det' A(\sigma) \le \det' A(0)$. Thus, Montgomery's Theorem implies both the OPS inequality and the Corollary to our Theorem 1.

Montgomery's proof that $W(t,0) \le W(t,\sigma)$ is very computational. It would be interesting to find another proof using pde's and maximum principles, as in our proof of Theorem 1.

3 Variations of QC maps

The object of this section and the next is to explain how to differentiate families of qc maps $g(z,\sigma)$, where σ denotes a complex parameter, with respect to the two complex variables. To do this requires familiarity with a certain Calderon-Zygmund singular integral operator often called nowadays the "Ahlfors-Beurling operator." The operator T defined below is a variant of the AB operator. We assume the reader knows the rudiments of singular integral theory, and will attempt to explain how the theory is applied to q.c. maps. Even with the semi-details provided, to follow the presentation will likely take effort. Sources of information about singular integrals or the AB transform include [A3], [L], and [St].

We use ∂ and $\overline{\partial}$ to denote formal complex derivatives with respect to z and $\overline{z}$, respectively. Our study is based on operators P and T defined as follows. Let

$$K(z,\zeta) = \frac{z(z-1)}{\zeta(\zeta-1)(\zeta-z)} = \frac{z-1}{\zeta} - \frac{z}{\zeta-1} + \frac{1}{\zeta-z}.$$

Write $m = -\frac{1}{\pi}\cdot$ Lebesgue measure on $\mathbb{C}$. For suitable functions f on $\mathbb{C}$, define, for $z \in \mathbb{C}$,

$$Pf(z) = \int_{\mathbb{C}} f(\zeta)K(z,\zeta)\,dm(\zeta),$$

$$Tf(z) = \partial Pf(z) = \int_{\mathbb{C}} f(\zeta)[\frac{1}{(\zeta-z)^2} - \frac{1}{\zeta(\zeta-1)}]\,dm(\zeta).$$

The second integral is to be understood in the sense of Cauchy principal values. Note that $\overline{\partial}Pf = f$, and $Pf(0) = Pf(1) = 0$.

The following formulas show the behavior of P and T under inversions. Set $F(z) = z^2\overline{z}^{-2}f(1/z)$. Then $f(z) = z^2\overline{z}^{-2}F(1/z)$, and

$$Pf(z) = -z^2PF(1/z), \qquad Pf(1/z) = -z^{-2}PF(z), \tag{3.1}$$

$$\begin{aligned} Tf(z) &= -2zPF(1/z) + TF(1/z), \\ Tf(1/z) &= -2z^{-1}PF(z) + TF(z). \end{aligned} \tag{3.2}$$

Standard properties of T include the facts that if f is infinitely differentiable on an open set then so is Tf, and if f is locally in L^p for some $1 < p < \infty$ then so is Tf. To study QC variations it is helpful to have a more precise result of this kind.

Proposition 1. *Suppose that $f \in L^p(\mathbb{D})$, for some $1 < p < \infty$, that $k \geq 0$, and that*

$$f(z) = \sum_{j=0}^{k} z^{-j} f_j(z), \quad z \in \mathbb{D}, \tag{3.3}$$

where each $f_j \in C^\infty(\mathbb{D})$, and f_j has a zero of order at least j at 0. Set $f = 0$ in $\mathbb{C}\setminus\mathbb{D}$. Then

(a) *In $\mathbb{D}$, $\frac{1}{z}Pf(z)$ and $Tf(z)$ also have the form (3.3), with k replaced by $k+1$. The coefficient functions F in the sums representing $\frac{1}{z}Pf(z)$ and $Tf(z)$ can be chosen so that for each pair of compact subsets S_1, S_2 with $S_1 \subset S_2 \subset \mathbb{D}$, the supremum on S_1 of each derivative of F of order m can be bounded in terms of the suprema of the derivatives of the f_j of order $m+1$ on S_2.*

(b) *In $\mathbb{D}$, $z(Pf)(1/z)$ and $(Tf)(1/z)$ are in C^∞. On each compact $S \subset \mathbb{D}$, the supremum of each of their partial derivative of order m can be bounded in terms of the L^p norm of f on $\mathbb{D}$.*

(c) *On each compact $S \subset \mathbb{C}$, the L^p norms of $\frac{1}{z}Pf(z)$ and of Tf are bounded in terms of the L^p norm of f on $\mathbb{D}$ and the suprema of the first order derivatives of f on $|z| \leq \frac{1}{2}$.*

Proof. To obtain (a), fix a positive integer m and write each f_j as a finite linear combination of monomials in ζ and $\bar{\zeta}$, plus a function h_j with a zero of of order at least $m+j+2$ at the origin. Then $H_j(\zeta) \equiv \zeta^{-k}h(\zeta)$ has a C^{m+1} extension to $\mathbb{D}$, and an integration by parts argument shows that $P(H_j) \in C^{m+1}(\mathbb{D})$. Moreover, $P(H_j)(z) = O(z)$ as $z \to 0$. Hence $\frac{1}{z}P(H_j)(z)$ and TH_j are in $C^m(\mathbb{D})$. For monomials $\xi_{p,q}(\zeta) = \zeta^p\bar{\zeta}^q\chi_{\mathbb{D}}$ with p, q integers, $p+q \geq -1$, $q \neq -1$, we have, if $p \geq q+2$,

$$\begin{aligned} P(\xi_{p,q})(z) &= \frac{1}{q+1}(z^p\bar{z}^{q+1} - z^{p-q-1}), \quad z \in \mathbb{D} \\ &= 0, \quad z \in \mathbb{D}^c, \end{aligned}$$

while if $p \leq q+1$,

$$\begin{aligned} P(\xi_{p,q})(z) &= \frac{1}{q+1}(z^p\bar{z}^{q+1} - z), \quad z \in \mathbb{D} \\ &= \frac{1}{q+1}(z^{p-q-1} - z), \quad z \in \mathbb{D}^c. \end{aligned}$$

When the expressions above are differentiated with respect to z or are divided by z, the resulting expressions have the form asserted by (a). This argument shows that the coefficient functions are in $C^m(\mathbb{D})$. The boundedness statements are proved by keeping track of boundedness of the pieces.

The proof of (b) is accomplished straightforwardly from the definitions of T and P. As noted above, that Tf satisfies (c) is a standard fact about singular integrals. To verify that $\frac{1}{z}Pf(z)$ satisfies (c), one can use (a) to bound this function in $|z| \leq \frac{1}{4}$, and the definition of P to bound it in $|z| \geq \frac{1}{4}$. □

Suppose now that $\mu \in C^\infty(D)$, with $||\mu||_{L^\infty(\mathbb{D})} < 1$. Set $\mu(z) = 0$ for $|z| \geq 1$, and define μ_1 on $\mathbb{C}$ by $\mu_1(z) = z^2\bar{z}^{-2}\bar{\mu}(1/\bar{z})$. Let g be the qc homeomorphisn of $\mathbb{C}$ which fixes $0, 1, \infty$ and satisfies

$$\bar{\partial}g = \mu\partial g, \text{ in } \mathbb{D}, \quad \bar{\partial}g = \mu_1\partial g, \text{ in } \mathbb{D}^c. \tag{3.4}$$

Then $g \in C^\infty(\mathbb{D} \cup \mathbb{D}^c)$, and $g(z) = 1/\bar{g}(1/\bar{z})$. In $\mathbb{D}^c$, g has the form $g(z) = z^2G(1/z)$, where $G \in C^\infty(\mathbb{D})$ with $G(0) = 0$. Hence $\partial g(z) = 2zG(1/z) - \partial G(1/z)$, so that $\partial g(z) = O(1)$ as $z \to \infty$. Moreover, if we set $f(z) = \mu_1(z)\partial g(z)$, then $F(z) \equiv z^2\bar{z}^{-2}f(1/z) = \bar{\mu}(\bar{z})[\frac{2}{z}G(z) - \partial G(z)]$. Thus, F has the form (3.3), with $k = 1$. From (3.4) it follows that $\bar{\partial}g$ is also bounded at ∞. Moreover, ∂g and $\bar{\partial}g$ are in $L^p{}_{loc}(\mathbb{C})$ for p sufficiently close to 2. From these facts, we obtain the representations $g(z) = z + P(\bar{\partial}g)(z)$ and

$$\partial g = 1 + T(\bar{\partial}g) = 1 + T(\mu\partial g) + T(\mu_1\partial g). \tag{3.5}$$

In (3.5), replace the $\partial g's$ on the right hand side by the complete right hand side, then repeat. This yields

$$\begin{aligned} \partial g &= 1+T\mu+T\mu_1+T\mu T\mu\partial g+T\mu_1 T\mu_1\partial g \\ &\quad +(T\mu T\mu_1\partial g+T\mu_1 T\mu\partial g) \qquad (3.6) \\ &= 1+T\mu+T\mu_1+T\mu T\mu+T\mu_1 T\mu_1+(T\mu T\mu_1+T\mu_1 T\mu)+\sum_{i=1}^{8} R_i. \end{aligned}$$

Here and later we omit composition signs and parentheses if it is clear (we hope) from context that they should be there. In (3.6), each remainder term R_i has the form $Tf_1Tf_2Tf_3\partial g$, where each f_j is either μ or μ_1. From Proposition 1, it follows that the suprema of the R_i on a compact subset of $\mathbb{D}$ can be bounded in terms of $||\partial g||_{L^p(\mathbb{D})}$ and of suprema of partial derivatives of g on a slightly larger subset of $\mathbb{D}$.

Let $\nu \in C^\infty(\mathbb{D})$ with $||\nu||_{L^\infty(\mathbb{D})} \leq 1$. Set $\nu(z) = 0$ for $|z| \geq 1$, and let $\nu_1(z) = z^2\overline{z}^{-2}\overline{\nu}(1/\overline{z})$. For $\sigma \in \mathbb{D}$, let $g(\cdot,\sigma)$ be the qc homeomorphism of $\mathbb{C}$ which fixes $z = 0, 1, \infty$, and has complex dilatation $\sigma\nu + \overline{\sigma}\nu_1$. Then each $g(\cdot,\sigma)$ enjoys the properties of the g discussed previously in this section. Furthermore, $g(\cdot,\cdot)$ is a C^∞ function on $\mathbb{D}\times\mathbb{D}$ and on $\mathbb{D}^c\times\mathbb{D}$. From (3.6), we obtain

$$\begin{aligned} \partial g(z,\sigma) &= 1+\sigma T\nu+\overline{\sigma}T\nu_1+\sigma^2 T\nu T\nu \\ &\quad +\overline{\sigma}^2 T\nu_1 T\nu_1+|\sigma|^2(T\nu T\nu_1+T\nu_1 T\nu)+\sum_{i=1}^{8} R_i. \end{aligned} \qquad (3.7)$$

In (3.7), the functions involving T are evaluated at z. Each R_i is a monomial of degree three in σ, $\overline{\sigma}$ times a function of z and σ whose supremum on each compact subset of $\mathbb{D}$ is controlled by $||\partial g(\cdot,\sigma)||_{L^p(\mathbb{D})}$ and of suprema of partial derivatives of g in the z variable on a slightly larger subset of $\mathbb{D}$. Since these suprema are continuous functions of σ, we have proved

Proposition 2. *In (3.7), for each pair of compact subsets S_1, S_2 of $\mathbb{D}$, holds*

$$\sup_{z\in S_1,\,\sigma\in S_2} |\sum_{i=1}^{8} R_i(z,\sigma)| \leq |\sigma|^3 C(S_1,S_2),$$

where $C(S_1,S_2)$ depends only on ν, S_1, and S_2.

4 Taylor polynomials

We continue to assume, as at the end of section 3, that $g(z,\sigma)$, $z \in \mathbb{C}$, $\sigma \in \mathbb{D}$, is the family of qc automorphisms of $\mathbb{C}$ fixing $\mathbb{D}$ and $z = 0, 1, \infty$ which is generated by $\nu \in C^\infty(\mathbb{D})$ with $||\nu||_{L^\infty(\mathbb{D})} \leq 1$.

We noted in section 3 that $g \in C^\infty(\mathbb{D}^2)$. From Proposition 1, or from (3.7) and Proposition 2, it follows that each of the functions of z in (3.7) multiplying the monomials in $\sigma, \overline{\sigma}$ is in $C^\infty(\mathbb{D})$. Thus, the second order Taylor decomposition of $\partial g(z,\sigma)$ around $(0,0)$ is given by

$$\begin{aligned} \partial g(z,\sigma) \;=\; & 1 + T\nu(0)\sigma + T\nu_1(0)\overline{\sigma} \\ & +(\partial T\nu)(0)z\sigma + (\overline{\partial}T\nu)(0)\overline{z}\sigma + (\partial T\nu_1)(0)z\overline{\sigma} \\ & +(\overline{\partial}T\nu_1)(0)\overline{z\sigma} \\ & +(T\nu T\nu)(0)\sigma^2 + (T\nu_1 T\nu_1)(0)\overline{\sigma}^2 \\ & +(T\nu T\nu_1 + T\nu_1 T\nu)(0)|\sigma|^2 + R(z,\sigma), \end{aligned} \tag{4.1}$$

where $R(z,\sigma)(|z|^3+|\sigma|^3)^{-1}$ is bounded on compact subsets of $\mathbb{D}^2$. Lest the reader fret that the pure z terms got lost, we note that $g(z,0)=z$, so that all the terms of order at least 1 in the Taylor expansion of $\partial g(z,0)$ are zero.

To prove Theorem 2, we require formulas for the coefficients of σ, $\overline{\sigma}$, σz, $\overline{\sigma z}$ and $|\sigma|^2$. Here they are.

$$T\nu(0) = \int_{\mathbb{D}} \Big(\frac{1}{\zeta^2} - \frac{1}{\zeta(\zeta-1)}\Big)\nu(\zeta)\, dm(\zeta). \tag{4.2}$$

$$T\nu_1(0) = \int_{\mathbb{D}} \frac{1}{\zeta(\zeta-1)}\overline{\nu}(\overline{\zeta})\, dm(\zeta). \tag{4.3}$$

$$(\partial T\nu)(0) = 2\int_{\mathbb{D}} \nu(\zeta)\zeta^{-3}\, dm(\zeta). \tag{4.4}$$

$$(\overline{\partial}T\nu_1)(0) = 0. \tag{4.5}$$

$$\begin{aligned} &(T\nu T\nu_1 + T\nu_1 T\nu)(0) = \\ &\int_{\mathbb{D}} dm(w) \int_{\mathbb{D}} \frac{\nu(w)\overline{\nu}(\overline{s})}{w^2 s^2}\Big[(1-sw)^{-2}\Big(\frac{1}{1-w} - \frac{s^3w^2}{1-s}\Big) \\ &\qquad - \frac{1+ws}{(1-s)(1-w)}\Big]\, dm(s). \end{aligned} \tag{4.6}$$

The integral in (4.3) is absolutely convergent. Those in (4.2) and (4.4) are Cauchy principal values; for $\nu \in L^1(\mathbb{D})$, the CPV in (4.2) exists when ν is differentiable at the origin, the CPV in (4.4) exists when ν is twice differentiable near the origin. In (4.6), the integration is first performed with respect to s. For each fixed w the integral is absolutely convergent. Then the integration with respect to w is a CPV. Existence of this CPV can be shown by examining the behavior of the $dm(s)$ integral as $w \to 0$.

Formula (4.2) follows from the definition of T given at the beginning of section 3, while (4.3) follows from the definition of T together with a change of variable. (4.5) is a consequence of the fact that $T\nu_1$ is holomorphic in $\mathbb{D}$. When we formally differentiate in the definition of $T\nu$, the result is (4.4), but since $T\nu$ is itself a singular integral, we should proceed with caution. Here then is a rigorous proof of (4.4), valid when $\nu \in C^2(\mathbb{D}) \cap L^1(\mathbb{D})$.

Let $\phi \in C^\infty(\mathbb{D})$ with $\phi = 1$ on $|z| \leq \frac{1}{2}$, $\phi = 0$ on $|z| \geq 3/4$, $0 \leq \phi \leq 1$ in $\mathbb{D}$. Define $f_1 = \nu\phi$, $f_2 = \nu(1-\phi)$, and set $f_1 = 0$ in $\mathbb{C} \setminus \mathbb{D}$. From the definition of P at the beginning of section 3 we have, for constants a and b,

$$P\nu(z) = az + b + \int_{\mathbb{C}} f_1(z+w)\frac{dm(w)}{w} + \int_{\mathbb{D}} f_2(\zeta)\frac{dm(\zeta)}{\zeta - z}.$$

Two differentiations with respect to z give

$$\begin{aligned}(\partial T\nu)(0) &= (\partial^2 P\nu)(0) \\ &= \int_{\mathbb{D}} (\partial^2 f_1)(w)\frac{dm(w)}{w} + 2\int_{\mathbb{D}} f_2(\zeta)\frac{dm(\zeta)}{\zeta^3}.\end{aligned} \tag{4.7}$$

From Green's theorem one obtains the relation

$$\frac{1}{2\pi i}\int_{\partial\Omega} f(\zeta)\,d\bar{\zeta} = \int_{\Omega} \partial f(\zeta)\,dm(\zeta),$$

where $\partial\Omega$ denotes the positively oriented boundary of the bounded domain Ω. Thus, writing $\mathbb{D}_\epsilon = \{\epsilon < |\zeta| < 1\}$, two integrations by parts give

$$\begin{aligned}\int_{\mathbb{D}_\epsilon} (\partial^2 f_1)(w)\frac{dm(w)}{w} &= \int_{\mathbb{D}_\epsilon} 2f_1(w)\frac{dm(w)}{w^3} \\ &\quad -\frac{1}{2\pi i}\int_{|w|=\epsilon} \left(\frac{f_1(w)}{w^2} + \frac{\partial f_1(w)}{w}\right) d\bar{w}.\end{aligned} \tag{4.8}$$

Assume that $0 < \epsilon < \frac{1}{2}$. Then in the line integral f_1 can be replaced by ν. Since $f_1 + f_2 = \nu$, from (4.8) it follows that

$$\begin{aligned}&\int_{\mathbb{D}_\epsilon} (\partial^2 f_1)(w)\frac{dm(w)}{w} + 2\int_{\mathbb{D}_\epsilon} f_2(\zeta)\frac{dm(\zeta)}{\zeta^3} \\ &= \int_{\mathbb{D}_\epsilon} 2\nu(w)\frac{dm(w)}{w^3} - \frac{1}{2\pi i}\int_{|w|=\epsilon} \left(\frac{\nu(w)}{w^2} + \frac{\partial\nu(w)}{w}\right) d\bar{w}.\end{aligned} \tag{4.9}$$

Writing $\nu(w)$ as a linear function plus a remainder which is $O(w^2)$, one checks that as $\epsilon \to 0$ the line integral in (4.9) is $O(\epsilon)$. Thus, (4.4) follows from (4.7) and the passage $\epsilon \to 0$ in (4.9).

Proof of (4.6). In the iterated integral defining $(T\nu_1 T\nu)(0)$, the product of the integrands is in $L^1(\mathbb{C})$. After reversal of order of integration and changes of variable, we obtain the absolute convergent integral representation

$$\begin{aligned}&(T\nu_1 T\nu)(0) = \\ &\int_{\mathbb{D}^2} \nu(w)\bar{\nu}(\bar{s})\left[\frac{s}{(s-1)(1-sw)^2} - \frac{1}{s(s-1)w(w-1)}\right] dm(s)\,dm(w).\end{aligned} \tag{4.10}$$

On the other hand, for $(T\nu T\nu_1)(0)$, the integral defining $T\nu_1(w)$ is absolutely convergent for $w \in \mathbb{D}$, but the second integration, over $\mathbb{D}$, is a CPV. We have

$$\begin{aligned}
&(T\nu T\nu_1)(0) \qquad\qquad (4.11)\\
&= \lim_{\epsilon\to 0}\int_{\mathbb{D}_\epsilon}\nu(w)\Big(\frac{1}{w^2}-\frac{1}{w(w-1)}\Big)\,dm(w)\int_{\mathbb{D}^c}\nu_1(\zeta)\Big(\frac{1}{(w-\zeta^2)}-\frac{1}{\zeta(\zeta-1)}\Big)\,dm(\zeta)\\
&= \lim_{\epsilon\to 0}\int_{\mathbb{D}_\epsilon}dm(w)\int_{\mathbb{D}}\nu(w)\overline{\nu}(\overline{s})\frac{1}{s^2w^2(1-w)}\Big(\frac{1}{(1-sw)^2}-\frac{1}{1-s}\Big)\,dm(s).
\end{aligned}$$

The sum of the functions multiplying $\nu(w)\overline{\nu}(\overline{s})$ in (4.10) and (4.11) is equal to the function multiplying $\nu(w)\overline{\nu}(\overline{s})$ in (4.6). Since (4.10) can also be written as an integral over $s \in \mathbb{D}$ followed by integration over $w \in \mathbb{D}_\epsilon$ followed by $\lim_{\epsilon\to 0}$, (4.6) is proved.

We've identified the coefficients in the expansion (4.1) of $\partial g(z,\sigma)$ which are of chief interest to us. Now we consider the second order Taylor polynomials of $\log \partial g(z,\sigma)$ and of $\log|\partial g(z,\sigma)|$. Since $g(\cdot,\sigma)$ is smooth on $\mathbb{D}$, its Jacobian, and hence $\partial g(z,\sigma)$, does not vanish. Thus, the branch of $\log \partial g(z,\sigma)$ which vanishes for $(z,\sigma)=(0,0)$ is well defined and C^∞ in $\mathbb{D}^2$. Let b_i, $i = 1,\dots,9$ be the constants such that

$$\begin{aligned}
\log \partial g(z,\sigma) \;=\; & b_1\sigma + b_2\overline{\sigma} + b_3 z\sigma + b_4\overline{z}\sigma + b_5 z\overline{\sigma} + b_6\overline{z\sigma} \qquad (4.12)\\
& + b_7\sigma^2 + b_8\overline{\sigma}^2 + b_9|\sigma|^2 + O(|z|^3+|\sigma|^3).
\end{aligned}$$

The coefficients b_i with $1 \le i \le 6$ are the same as the corresponding ones for ∂g. Also,

$$b_9 = (T\nu T\nu_1 + T\nu_1 T\nu)(0) - T\nu(0)T\nu_1(0).$$

From (4.2), (4.3),

$$T\nu(0)T\nu_1(0) = \lim_{\epsilon\to 0}\int_{\mathbb{D}_\epsilon}dm(w)\int_{\mathbb{D}}\frac{\nu(w)\overline{\nu}(\overline{s})}{w^2s^2}\frac{s}{(1-w)(1-s)}\,dm(s). \quad (4.13)$$

Adding the multipliers of $\nu(w)\overline{\nu}(\overline{s})$ in (4.6) and (4.13), we obtain

$$b_9 = \int_{\mathbb{D}^2}\frac{\nu(w)\overline{\nu}(\overline{s})}{ws}\frac{1-w-w(1-s)}{(1-s)(1-w)(1-sw)^2}\,dm(w)\,dm(s). \quad (4.14)$$

The integral in (4.14) is absolutely convergent. From (4.12), we obtain

$$\begin{aligned}
\log|\partial g(z,\sigma)| = \;& \mathrm{Re}\,[(b_1+\overline{b}_2)\sigma + (b_3+\overline{b}_6)z\sigma + (b_4+\overline{b}_5)\overline{z}\sigma + (b_7+\overline{b}_8)\sigma^2]\\
& + \mathrm{Re}\,b_9\,|\sigma|^2 + O(|z|^3+|\sigma|^3). \qquad (4.15)
\end{aligned}$$

From (4.2), (4.3) follows

$$b_1 + \overline{b}_2 = T\nu(0) + \overline{T\nu_1}(0) = \int_{\mathbb{D}}\frac{\nu(\zeta)}{\zeta^2}\,dm(\zeta). \quad (4.16)$$

From (4.4), (4.5) follows

$$b_3 + \overline{b}_6 = 2\int_{\mathbb{D}} \frac{\nu(\zeta)}{\zeta^3}\, dm(\zeta). \tag{4.17}$$

Take conjugates in (4.14), change w and s to their conjugates, then interchange them. Add the result to (4.14). From the identity

$$1 - s - s(1-w) + 1 - w - w(1-s) = 2(1-s)(1-w)$$

follows the absolutely convergent integral formula

$$\operatorname{Re} b_9 = \int_{\mathbb{D}^2} \frac{\nu(w)\overline{\nu}(\overline{s})}{ws} \frac{1}{(1-sw)^2}\, dm(w)\, dm(s). \tag{4.18}$$

We state the results of the last few calculations as a proposition.

Proposition 3. *In the expansion (4.15), the coefficients of σ, $z\sigma$, and $|\sigma|^2$ are given by (4.16) - (4.18).*

5 Proof of Theorem 2. Part 1

Let T reassume its meaning as a Möbius transformation from $\mathbb{D}$ onto $\mathbb{H}$:

$$T(\sigma) = e^{-i\pi/3}\frac{\sigma - e^{2\pi i/3}}{\sigma - 1}, \quad T^{-1}(\tau) = \frac{\tau - e^{i\pi/3}}{\tau - e^{-i\pi/3}}. \tag{5.1}$$

We construct now the u.c.m. $f(\cdot,\sigma)$ of $\mathbb{D}$ onto the plane minus the lattice $A(\sigma)$. Define, for $\sigma \in \mathbb{D}$, $z \in \mathbb{C}$, and $\tau = T(\sigma)$,

$$U(z,\sigma) = \frac{1}{i\sqrt{3}}[(\tau - e^{-i\pi/3})z + (e^{i\pi/3} - \tau)\overline{z}].$$

For fixed σ, $U(\cdot,\sigma)$ is the real-linear map of the plane with $U(1,\sigma) = 1$, $U(e^{i\pi/3},\sigma) = \tau$. Hence, $U(\cdot,\sigma)$ carries the hex lattice $A(0)$ onto the lattice $A(\sigma)$. We have

$$\overline{\partial} U(z,\sigma) = -\sigma \partial U(z,\sigma), \quad z \in \mathbb{C}, \quad \sigma \in \mathbb{D}. \tag{5.2}$$

Recall that F is the u.c.m. of $\mathbb{D}$ onto the plane minus the hex lattice $\mathcal{A}$, with $F(0) = 0$, $F'(0) > 0$, and that $f(\cdot,0) = e^{-i\pi/6}\frac{1}{\sqrt{3}}(F - e^{4\pi i/3})$. Define

$$\begin{aligned} \nu(z) &= -\overline{F'}(z)/F'(z), \quad z \in \mathbb{D}, \quad \nu(z) = 0, \quad z \in \mathbb{C}\setminus\mathbb{D}, \\ \nu_1(z) &= z^2\overline{z}^{-2}\nu(1/\overline{z}), \quad z \in \mathbb{C}. \end{aligned} \tag{5.3}$$

Let $g(\cdot,\sigma)$ be the solution in $\mathbb{C}$ of the Beltrami equation

$$\overline{\partial} g(\cdot,\sigma) = \sigma\nu\partial g(\cdot,\sigma), \quad \text{in} \quad \mathbb{D}, \quad \overline{\partial} g(\cdot,\sigma) = \sigma\nu_1\partial g(\cdot,\sigma), \quad \text{in} \quad \mathbb{D}^c$$

which fixes $z = 0, 1, \infty$. As discussed in Section 4, $g \in C^\infty(\mathbb{D}^2)$.

Define

$$f(\cdot,\sigma) = U(\cdot,\sigma) \circ f(\cdot,0) \circ g^{-1}(\cdot,\sigma). \tag{5.4}$$

By the chain rule for complex dilatations, $f(\cdot,\sigma)$ is defined and holomorphic in $\mathbb{D}$. It is a u.c.m. of $\mathbb{D}$ onto $A(\sigma)$. Moreover, $f(0,\sigma) = \frac{1}{3}(1+\tau)$. To compute $|f'(\cdot,\sigma)|$, we use the fact that the Jacobian determinant of a holomorphic function is the square of the modulus of its derivative, the chain rule for Jacobians, the formula $Jac(h) = |\partial h|^2 - |\bar\partial h|^2$, and the relation $\tau - e^{-i\pi/3} = -\frac{i\sqrt{3}}{\sigma-1}$ to compute

$$|f'(g(z,\sigma),\sigma)| = \frac{|F'(z)|}{\sqrt{3}|(\sigma-1)\partial g(z,\sigma)|}. \tag{5.5}$$

Next, we examine $R(\sigma)$, the radius of the largest disk contained in $\Omega(\sigma) = \mathbb{C} \setminus A(\sigma)$. For $\tau \in \mathbb{H}$, the circle $C(\tau)$ passing through the three points $\{0,1,\tau\}$ has radius

$$\frac{|\tau(1-\tau)|}{2\,\mathrm{Im}\,\tau}. \tag{5.6}$$

Let $\mathcal{F}_1 = T(\mathcal{F})$. Then $\mathcal{F}_1$ is the interior of the hyperbolic triangle in $\mathbb{H}$ bounded by the lines $\mathrm{Re}\, z = 0$, $\mathrm{Re}\, z = 1$ and the circle $|z - \frac{1}{2}| = \frac{1}{2}$. If $\tau \in \mathbb{H}$ is in the closure of $\mathcal{F}_1$, one can check that $C(\tau)$ bounds a disk of maximal radius contained in $\Omega(\sigma)$. From (5.1) and (5.6), we obtain

$$R(\sigma) = \frac{|1-\sigma^3|}{\sqrt{3}(1-|\sigma|^2)|1-\sigma|}, \qquad \sigma \in \mathbb{D} \cap \mathrm{closure}(\mathcal{F}). \tag{5.7}$$

If τ is not in the closure of $\mathcal{F}_1$ then the interior of $C(\tau)$ contains points of $A(\sigma)$, and $R(\sigma)$ is smaller than the radius of $C(\tau)$. Choose a new basis for $A(\sigma)$ by letting ω_1 be a point of smallest modulus in $A(\sigma) \setminus \{0\}$ and ω_2 be a point of $A(\sigma) \setminus \omega_1\mathbb{Z}$ for which ω_2/ω_1 has minimal positive imaginary part. Let $\tau' = \omega_2/\omega_1$. Then $\tau' \in \mathrm{closure}(\mathcal{F}_1)$, and ${\omega_1}^{-1}A(\sigma)$ is the lattice generated by 0, 1,and τ'. Let $\sigma' = T^{-1}(\tau')$. Then $\sigma' \in \mathrm{closure}(\mathcal{F})$, and

$$|f'(z,\sigma)| = |\omega_1 f'(z,\sigma')|, \quad R(\sigma) = |\omega_1|\, R(\sigma'). \tag{5.8}$$

To prove the putative inequality (1.5) for all $(z,\sigma) \in \mathbb{D}^2$ it suffices by (5.8) to prove it when $\sigma \in \mathcal{F}$. In the proof of Theorem 2 we shall assume that $\mathcal{F}$ contains the disk $|\sigma| < \epsilon$. Then $R(\sigma)$ is given by (5.7). To prove Theorem 2, we must show the quantity in (5.5) multiplied by $1 - |g(z,\sigma)|^2$ and divided by the quantity in (5.7) has a local maximum at $(0,0)$. It will be more convenient to work with the logarithm of the reciprocal.

Define Q on $\mathbb{D}^2$ by

$$Q(z,\sigma) = \log \frac{|(1-\sigma^3)\partial g(z,\sigma)|}{(1-|\sigma|^2)|F'(z)|(1-|g(z,\sigma)|^2} \tag{5.9}$$

$$= \log \frac{|1-\sigma^3|}{1-|\sigma|^2} + \log \frac{|F'(z)|^{-1}}{1-|z|^2} + \log \frac{1-|z|^2}{1-|g(z,\sigma)|^2}| + \log|\partial g(z,\sigma)|.$$

Then $Q \in C^\infty(\mathbb{D}^2)$, and Theorem 2 is the assertion that Q has a strict local minimum at $(0,0)$. To show this, we first record some symmetry relations.

$$F(e^{2\pi i/3}z) = e^{2\pi i/3}F(z), \quad g(e^{2\pi i/3}z, e^{-2\pi i/3}\sigma) = \kappa(\sigma)g(z,\sigma), \tag{5.10}$$

where $\kappa(\sigma) = g(e^{2\pi i/3}, e^{-2\pi i/3}\sigma)$.

The first relation follows from the uniqueness theory of conformal maps. The second follows from the fact that the two sides have the same complex dilatation quotient almost everywhere in $\mathbb{C}$ and take the same values at $z = 0, 1$, and ∞. Note that $|\kappa(\sigma)| = 1$. From (5.9), (5.10), we deduce

$$Q(e^{2\pi i/3}z, e^{-2\pi i/3}\sigma) = Q(z,\sigma), \quad (z,\sigma) \in \mathbb{D}^2. \tag{5.11}$$

From (5.11) it follows that $\nabla Q(0,0) = 0$. It follows also that if we write the quadratic part of the Taylor expansion of Q in the form

$$a_1|z|^2 + a_2|\sigma|^2 + \operatorname{Re}\left[a_3\sigma^2 + a_4z^2 + a_5z\sigma + a_6z\overline{\sigma}\right],$$

then $a_3 = a_4 = a_6 = 0$.

From $g(z,0) = z$, $g(0,\sigma) = 0$, it follows that $g(z,\sigma) = z + O(|z|^2 + |\sigma|^2)$, and hence, as $(z,\sigma) \to (0,0)$,

$$\log \frac{1-|z|^2}{1-|g(z,\sigma)|^2}| = O(|z|^3 + |\sigma|^3).$$

From (5.9) and subsequent facts, we obtain

$$\begin{aligned} Q(z,\sigma) &= \log\frac{1}{1-|\sigma|^2} + \log\frac{1}{1-|z|^2} + \log|F'(0)|^{-1} \\ &\quad + \log|\partial g(z,\sigma)| + O(|z|^3 + |\sigma|^3) \\ &= \log|F'(0)|^{-1} + a_1|z|^2 + a_2|\sigma|^2 + \operatorname{Re}(a_5z\sigma) + O(|z|^3 + |\sigma|^3). \end{aligned}$$

Finally, using (4.15), (4.17), (4.18), we obtain the second order Taylor expansion of Q at $(0,0)$:

$$\begin{aligned} Q(z,\sigma) &= \log|F'(0)|^{-1} + |z|^2 + (1+\alpha)|\sigma|^2 \\ &\quad + 2\operatorname{Re}(\beta z\sigma) + O(|z|^3 + |\sigma|^3), \end{aligned} \tag{5.12}$$

where

$$\alpha = \int_{\mathbb{D}^2} \frac{\nu(w)\overline{\nu}(\overline{\sigma})}{ws}\frac{1}{(1-sw)^2}\,dm(w)\,dm(s), \tag{5.13}$$

$$\beta = \int_{\mathbb{D}} \frac{\nu(\zeta)}{\zeta^3}\,dm(\zeta). \tag{5.14}$$

Recall from Section 4 that dm is Lebesgue measure multiplied by $-1/\pi$. Set $c_n = \int_{\mathbb{D}} \nu(w)w^n\,dm(w)$. Expanding the kernel in (5.13) in power series, we obtain

$$\alpha = \sum_{n=-1}^{\infty} (n+2)|c_n|^2.$$

In particular, $\alpha \geq 0$. To complete the proof of Theorem 2, it will be sufficient, in view of (5.12), to prove that $|\beta| < 1$. In the next section, we shall see that in fact $.96 < |\beta| < .98$.

6 Estimation of β.

First, we record some automorphy properties of the u.c.m. $F : \mathbb{D} \to \mathcal{A}$. Let G_1 denote the full modular group of Möbius transformations $z \to \frac{az+b}{cz+d}$, with $a, b, c, d \in \mathbb{Z}$, $ad - bc = 1$, acting on $\mathbb{H}$, and let $G = T^{-1}G_1T$ be its conjugate by T, which acts on $\mathbb{D}$. The rotation $P(z) := e^{2\pi i/3}z$ belongs to G, has order three, and has a fixed point at the origin. Let $z_1 = -1/(2+\sqrt{3})$. Then z_1 is the intersection of the real axis with the geodesic S of $\mathbb{D}$ connecting $e^{2\pi i/3}$ and $e^{4\pi i/3}$. Let V denote the elliptic Möbius transformation of order two whose fixed points are z_1 and $1/z_1$. Then V is the conjugate by T of the element of G_1 of order two with fixed point at $\frac{1}{2}(1+i)$. From [Ran, Theorem 1.2.6], it follows that G is freely generated by P and V. From symmetry properties of F, we have, for $z \in \mathbb{D}$, and writing $Pz = P(z)$, $Vz = V(z)$,

$$F(Pz) = PF(z), \quad F(Vz) = -F(z) - 1. \tag{6.1}$$

Hence

$$F'(Pz) = F'(z), \quad F'(Vz)(V'z) = -F'(z). \tag{6.2}$$

To establish the second property in (6.1), one may note that V is the product of a reflection in S with a reflection in the real axis.

Each $\gamma \in G$ may be expressed uniquely in the form

$$\gamma = P^{p_0}VP^{p_1}\cdots VP^{p_n}, \tag{6.3}$$

with $0 \leq p_i \leq 2$ for $0 \leq i \leq n$, $p_i > 0$, for $0 < i < n$, for some $n \geq 0$. Write $s = \sum_{i=0}^{n} p_i$. Define

$$\chi(\gamma) = e^{4\pi i s/3}. \tag{6.4}$$

Then χ is a homomorphism of G onto the subgroup $\{e^{2\pi i/3}, e^{4\pi i/3}, 1\}$ of $\partial\mathbb{D}$. From (6.2), it follows that

$$(F' \circ \gamma)^2 (\gamma')^2 = \chi(\gamma)(F')^2, \quad \gamma \in G. \tag{6.5}$$

Recall that $\nu = -\overline{F'}/F.'$ Thus, $\nu \circ \gamma = (\gamma')^2|\gamma'|^{-2}\overline{\chi}(\gamma)\nu$. Let $\mathcal{F}_0$ be the intersection of $\mathcal{F}$ with the open sector $|\arg z| < \pi/3$. The union of $\mathcal{F}_0$ with an appropriate subset of its boundary is a fundamental domain for G acting on $\mathbb{D}$. Hence, we have formally

$$\beta \;=\; \sum_{\gamma\in G}\int_{\mathcal{F}_0} \nu(\gamma z)(\gamma z)^{-3}|\gamma'(z)|^2\, dm(z)$$

$$= \int_{\mathcal{F}_0} \nu(z) \sum_{\gamma \in G} (\gamma'(z))^2 \gamma(z)^{-3} \overline{\chi}(\gamma)\, dm(z).$$

The integral on the right does not change when $\mathcal{F}_0$ is replaced by $e^{2\pi i/3}\mathcal{F}_0$ or $e^{4\pi i/3}\mathcal{F}_0$. Hence, defining

$$K(z) = \frac{1}{3} \sum_{\gamma \in G} (\gamma'(z))^2 (\gamma(z))^{-3} \overline{\chi}(\gamma), \tag{6.6}$$

we have formally

$$\beta = \int_{\mathcal{F}} \nu(z) K(z)\, dm(z). \tag{6.7}$$

Now Möbius transformations γ from $\mathbb{D}$ onto $\mathbb{D}$ satisfy $|\gamma'(z)| \le |\gamma'(0)|(1-|z|)^{-2}$. Moreover [T, p.518], the number $n(r, G)$ of points of the orbit $G(0)$ in $|z| \le r$ satisfies $n(r, G) = O(1-r)^{-1}$ as $r \to 1$. From these facts, it follows that the series defining K converges absolutely and uniformly on compact subsets of $\mathbb{D} \setminus G(0)$. For $\epsilon > 0$,

$$|K(z)| \le C(\epsilon)(1-|z|)^{-4}, \quad z \in \mathcal{F} \cap (|z| > \epsilon). \tag{6.8}$$

Also, by an argument with areas, $\int_{\mathcal{F}\cap(|z|>\epsilon)} |K(z)|\, dx\, dy < \infty$. From these estimates, one can justify the derivation of (6.7), and one sees that K is holomorphic in $\mathbb{D} \setminus G(0)$, with triple poles at each point of $G(0)$. Moreover, K satisfies the same automorphy conditions (6.5) as does $(F')^2$.

Next, recall that $\mathcal{A}$ is the lattice generated by $1, e^{2\pi i/3}$, and $e^{4\pi i/3}$. The points of $\mathcal{A}$ are the vertices of a tiling of the plane into equilateral triangles. Let $\mathcal{B}$ denote the set of centers of these triangles, and write $\mathcal{B} = \mathcal{B}_e \cup \mathcal{B}_o$, where $b \in \mathcal{B}_e$, respectively $b \in \mathcal{B}_o$, if b is the image of the origin under an even, respectively odd, number of reflections in sides of the $\mathcal{A}$–tiling. Then $\mathcal{B} = F(G(0))$, and $F(\gamma 0) \in \mathcal{B}_e$ if and only if the number n in (6.3) is even.

Define

$$H(w) = \sum_{b \in \mathcal{B}_e} \frac{1}{(w-b)^3} - \sum_{b \in \mathcal{B}_o} \frac{1}{(w-b)^3}. \tag{6.9}$$

It is easy to verify that $H(e^{2\pi i/3}w) = H(w)$ and $H(-w-1) = H(w)$. From (6.1), and the fact that P and V generate G, it follows that

$$H \circ F(\gamma z) = H \circ F(z), \quad z \in \mathbb{D}, \quad \gamma \in G. \tag{6.10}$$

Thus, $H \circ F$ is a meromorphic G-automorphic function in $\mathbb{D}$. Its poles are at the points of $G(0)$. At the pole $z = 0$, $H \circ F(z) = z^{-3}F'(0)^{-3} + O(1)$. The function $K/(F')^2$ is also G-automorphic with poles at the points of $G(0)$, and at $z = 0$ has the same form as $H \circ F$, except that $F'(0)$ appears with power -2. Let

$$f := \frac{K}{(F')^2} - F'(0) H \circ F.$$

The singularity of f at 0 is removable. Since f is G-automorphic, its singularities at all the points of $G(0)$ are removable, so that f is holomorphic and G-automorphic in $\mathbb{D}$. Let $f_1 = f \circ T^{-1}$. Then f_1 is holomorphic in $\mathbb{H}$, and is automorphic with respect to the full modular group G_1. Let us examine $f_1(z)$ as $z \to \infty$ with $0 \leq \operatorname{Re} z \leq 1$. It is clear that $H \circ F \circ T^{-1} = O(1)$, while (6.8) implies $K \circ T^{-1}(z) = O(\operatorname{Im} z)^4$. From the fact that $F \circ T^{-1}$ maps $T\mathcal{F}$ onto an equilateral triangle, it follows that $|F' \circ T^{-1}|^2$ has the form $ce^{-2\pi y/3}(\operatorname{Im} z)^4(1 + O(e^{-2\pi y/3}))$. Hence, $f(z) = O(e^{2\pi y/3})$. From the theory of automorphic functions, see, for example [Ap], it follows that f is constant.

Let us call this constant C_0. Then

$$K = C_0(F')^2 + F'(0)\,(F')^2\, H \circ F. \tag{6.11}$$

Insert (6.11) in (6.7). Recall that $\nu = -\overline{F'}/F'$, and that F maps $\mathcal{F}$ 1-1 onto $F(\mathcal{F})$, which is the Euclidean triangle with vertices $1, e^{2\pi i/3}, e^{4\pi i/3}$, and that $m = -1/\pi\cdot$ Lebesgue measure. Let $B_\epsilon = F(|z| < \epsilon)$. Then

$$e^{-i\pi/3}\beta = \frac{3\sqrt{3}}{4\pi}C_0 - F'(0) \lim_{\epsilon \to 0} \int_{F(\mathcal{F})\setminus B_\epsilon} H(w)\, dm(w). \tag{6.12}$$

Let $w = M(z) := e^{4\pi i/3} + \sqrt{3}e^{i\pi/6}z$. Then $z = z_0 + \bar{z}_0 w$, where $z_0 = \frac{1}{2} + i\frac{\sqrt{3}}{6}$, and

$$\begin{aligned} H(M(z)) &= \frac{1}{(\sqrt{3})^3 i} \sum_{b \in A(0)} \left(\frac{1}{(z - z_0 - b)^3} - \frac{1}{(z - 2z_0 - b)^3}\right) \\ &= \frac{1}{2(\sqrt{3})^3 i}(\mathcal{P}'(z - 2z_0) - \mathcal{P}'(z - z_0)), \end{aligned} \tag{6.13}$$

where $\mathcal{P}$ denotes the Weierstrass $\mathcal{P}$-function with period set $A(0)$:

$$\mathcal{P}(z) = \frac{1}{z^2} + \sum_{b \in A(0), b \neq 0} \left(\frac{1}{(z - b)^2} - \frac{1}{b^2}\right).$$

Let S be the parallellogram with vertices $0, 1, 1 + e^{i\pi/3}, e^{i\pi/3}$, and E_ϵ be the union of $M^{-1}(B_\epsilon)$ and its reflection in the line connecting $e^{i\pi/3}$ and 1. Using (6.13), properties of $\mathcal{P}$ and Green's Theorem, one obtains

$$\begin{aligned} \int_{F(\mathcal{F})\setminus B_\epsilon} H(w)\, dm(w) &= \frac{i}{2\sqrt{3}} \int_{S\setminus E_\epsilon} \mathcal{P}'(z - z_0)\, dm(z) \\ &= \frac{1}{4\pi\sqrt{3}} \int_{\partial E_\epsilon} \mathcal{P}(z - z_0)\, d\bar{z}. \end{aligned} \tag{6.14}$$

In the line integral, ∂E_ϵ denotes the union of two near-circles, one surrounding z_0, the other $2z_0$, each oriented clockwise. As $\epsilon \to 0$, the integral over the curve around $2z_0$ is $o(1)$. Undoing the map $w = M(z)$, we find, with ∂B_ϵ also oriented clockwise,

$$\begin{aligned}\int_{\partial E_\epsilon} \mathcal{P}(z - z_0)\, d\overline{z} &= i\sqrt{3} \int_{\partial B_\epsilon} \frac{1}{\overline{w}^2}\, d\overline{w} + o(1) \\ &= -i\sqrt{3} \int_{|z|=\epsilon} \frac{\overline{F'(z)}}{\overline{F^2(z)}}\, d\overline{z} + o(1).\end{aligned}$$

For any nonconstant function F holomorphic in a neighborhood of 0, development in powers of z and $\overline{z}$ shows that the integral with respect to $d\overline{z}$ is $o(1)$. From (6.12) and (6.14) we deduce

$$e^{-i\pi/3}\beta = \frac{3\sqrt{3}}{4\pi} C_0. \tag{6.15}$$

From its definition as the constant value of the function f, we have

$$C_0 = \lim_{z\to 0}\left[\frac{K(z)}{(F'(z))^2} - F'(0)\, H(F(z))\right]. \tag{6.16}$$

Define

$$K_0 = \frac{1}{3}\sum_{\gamma\in G}{}' (\gamma'(0))^2(\gamma(0))^{-3}\overline{\chi}(\gamma), \tag{6.17}$$

where $\sum'$ signifies that the identity map and the rotations $z \to e^{2\pi i/3}z$, $z \to e^{4\pi i/3}z$ are to be excluded from the sum. Also, let c_1, c_2 be the coefficients in the expansion

$$F(z) = c_1 z + c_2 z^4 + O(z^7), \quad z \to 0.$$

Then, up to terms which are $o(1)$ as $z \to 0$,

$$\frac{K(z)}{F'(z)^2} = {c_1}^{-2} z^{-3} + {c_1}^{-2}\left(K_0 - 8\frac{c_2}{c_1}\right) + \cdots,$$

$$F'(0)\, H(F(z)) = {c_1}^{-2} z^{-3} - 3{c_1}^{-2}\frac{c_2}{c_1} + 2^{-1}3^{-3/2} c_1 \mathcal{P}'(z_0)\, i + \cdots.$$

In the second expansion, we made use of the facts that $\mathcal{P}'(z) = -\frac{2}{z^3} + 0(z)$ as $z \to 0$, and that $\mathcal{P}'(-z_0) = -\mathcal{P}'(z_0)$.

Hence,

$$C_0 = {c_1}^{-2} K_0 - 5{c_1}^{-2}\frac{c_2}{c_1} - 2^{-1}3^{-3/2} c_1 \mathcal{P}'(z_0)\, i. \tag{6.18}$$

From the expression of F in terms of hypergeometric functions [H, p.382], as corrected in Section 1 of the present paper, one computes

$$c_1 = \frac{\Gamma(1/6)}{\Gamma(1/3)\Gamma(5/6)} = 3\cdot 2^{-5/3}\pi^{-2}\Gamma^3(1/3) = 1.8407\cdots, \tag{6.19}$$

$$\begin{aligned}\frac{c_2}{c_1} &= -2^{-4}3^{-4}\left(\frac{\Gamma(1/6)\Gamma(1/3)}{\Gamma(5/6)\Gamma(2/3)}\right)^3 \\ &= -2^{-12}3^{1/2}\pi^{-9}\Gamma^{18}(1/3) = -.716\cdots.\end{aligned} \tag{6.20}$$

Thus,

$$c_1{}^{-2}\frac{c_2}{c_1} = -2^{-9}2^{1/3}3^{-3/2}\pi^{-5}\Gamma^{12}(1/3). \tag{6.21}$$

We claim that

$$\mathcal{P}'(z_0) = 2^{-3}\pi^{-3}\Gamma^9(1/3)i, \tag{6.22}$$

and that

$$-12.87 < K_0 < -12.81. \tag{6.23}$$

If so, then

$$2^{-1}3^{-3/2}c_1\mathcal{P}'(z_0)\,i = -2^{-6}2^{1/3}3^{-1/2}\pi^{-5}\Gamma^{12}(1/3),$$

$$C_0 = c_1{}^{-2}K_0 + 2^{-9}2^{1/3}3^{-3/2}\pi^{-5}\Gamma^{12}(1/3)(5+3\cdot 2^3) = .2951K_0 + 6.130.$$

Thus, C_0 is positive real, with

$$2.3321 < C_0 < 2.3498. \tag{6.24}$$

From (6.15), it follows that $\arg\beta = \frac{\pi}{3}$, with

$$.9643 < |\beta| < .9716. \tag{6.25}$$

Theorem 2 is proved, once we prove (6.22) and (6.23).

Proof of (6.22). Write $e_1 = \mathcal{P}(1/2)$. Using general properties of elliptic functions and symmetries of our particular function $\mathcal{P}$, one sees that $e_1 > 0$. Write $S = \{x \in \mathbb{R} : e_1 \le x < \infty\}$. Then $\mathcal{P}$ is a 1-1 map of the interior of the triangle with vertices $0, 1, e^{i\pi/3}$ onto the plane with the three slits S, $e^{2\pi i/3}S$, $e^{4\pi i/3}S$ removed. Let Φ denote the conformal map of $\mathbb{D}$ onto the interior of the Euclidean triangle with vertices $1, e^{2\pi i/3}, e^{4\pi i/3}$ which fixes each of $1, e^{2\pi i/3}, e^{4\pi i/3}$. Then

$$\Phi(z) = 2\sqrt{3}\pi\Gamma^{-3}(1/3)\int_0^z \frac{d\zeta}{(1-\zeta^3)^{2/3}}, \quad z \in \mathbb{D}.$$

We have the identity

$$\mathcal{P}(z_0 + \bar{z}_0\Phi(z)) = ce^{\pi i/3}\frac{z}{(1-z^3)^{2/3}}, \quad c = 2^{2/3}e_1, \tag{6.26}$$

since each side conformally maps $\mathbb{D}$ onto the threefold slit plane, and the two maps agree at the points $-1, e^{i\pi/3}, e^{-i\pi/3}$. Differentiate (6.26) and set $z = 0$. The result is $\mathcal{P}'(z_0) = i\sqrt{3}c/\Phi'(0)$. In (6.26), take $z = x \in (0,1)$, multiply both sides by $(1-x^3)^{2/3}$, and let $x \to 1$. From $\mathcal{P}(z) = (z-1)^{-2} + O(1)$, we obtain the equation $c = 3/(\Phi'(0))^2$. Hence,

$$\mathcal{P}'(z_0) = i\,3^{3/2}(\Phi'(0))^{-3}.$$

The right hand side of the last equation equals the right hand side in (6.22).

Proof of 6.23. For $k = 0, 1, 2$, define $V_k = P^{k-1}VP^{1-k}$, where V and P are the Möbius transformations defined at the beginning of this section. The V_k are elliptic Möbius transformations of order two which map $\mathbb{D}$ onto $\mathbb{D}$. Each V_k fixes one of the points of $\partial\mathcal{F}$ closest to the origin. In the factorization (6.3) of $\gamma \in G$, use the relations $P^kV = V_{k-1}P^k$ to move all the P terms succesively to the right. After reductions mod 3, γ has the form

$$\gamma = V_{k_1} \cdots V_{k_n} P^{k_{n+1}}, \tag{6.27}$$

where the k_i are $0, 1$, or 2, $k_i \neq k_{i+1}$ for $1 \le i \le n-1$.

From uniqueness in (6.3), it follows that the representation in (6.27) is unique. Let G_2 be the subgroup of G generated by V_k, $k = 0, 1, 2$. By (6.27), each $\gamma \in G$ has a unique representation $\gamma = \gamma_1 P^k$, where $\gamma_1 \in G_2$ and $k = 0, 1$ or 2. It is also easy to see that

$$\gamma'(0)^2\gamma(0)^{-3}\overline{\chi}(\gamma) = \gamma_1'(0)^2\gamma_1(0)^{-3}.$$

Hence,

$$K_0 = \sum_{\gamma \in G_2 \setminus \{0\}} \gamma'(0)^2\gamma(0)^{-3}. \tag{6.28}$$

For $0 < R < 1$, let $K_0(R)$ be the sum of the terms on the right side of (6.28) for which $\gamma(\mathcal{F})$ meets the disk $|z| < R$. Let $R_0 = 2 - \sqrt{3}$. Then $|z| < R_0$ is the largest open disk centered at the origin which is contained in $\mathcal{F}$. The images of $\mathcal{F}$ under distinct elements of G_2 are disjoint. This fact, together with the subharmonicity estimate $|\gamma'(0)|^2 < (\pi {R_0}^2)^{-1}\int_{|z|\le R_0} |\gamma'(z)|^2\,dx\,dy$, leads to the estimate

$$|K_0 - K_0(R)| < {R_0}^{-2}R^{-3}(1-R^2) < 13.929R^{-3}(1-R^2). \tag{6.29}$$

In Appendix A, we describe a program for computing $K_0(R)$. For $R = .999$, the program yields

$$-12.8393 < K_0(.999) < -12.8391. \tag{6.30}$$

When $R = .999$, the right side of (6.29) is less than .028. With (6.30) we obtain

$$-12.8671 < K_0 < -12.8113.$$

This confirms (6.23).

7 Results related to B.

In this section we discuss analogues of Theorems 1 and 2 which provide some evidence for the Ahlfors-Grunsky conjecture. In this section, let $\mathcal{F}_1$ denote the interior of the hyperbolic triangle in $\mathbb{D}$ with vertices at R_1, $e^{2\pi i/3}R_1$, $e^{4\pi i/3}R_1$, where

$$R_1 = (1+\sqrt{3})^{-\frac{1}{2}}. \tag{7.1}$$

Then $\partial\mathcal{F}_1$ is the union of three equally long hyperbolic geodesics in $\mathbb{D}$ for which each pair meets at angle $\pi/6$. The rotation $P(z) = e^{2\pi i/3}z$ maps the sides one upon another. As in the Introduction, denote by F_1 the conformal map of $\mathcal{F}_1$ onto the interior of the Euclidean triangle with vertices $1, e^{2\pi i/3}, e^{4\pi i/3}$ which sends $R_1, R_1e^{2\pi i/3}, R_1e^{4\pi i/3}$ to $1, e^{2\pi i/3}, e^{4\pi i/3}$, respectively. Define $\nu = -\overline{F_1'}/F_1'$. Let $g(z,\sigma)$ and $f(z,\sigma)$ be defined as they were near the beginning of Section 5, with the same linear function U but with the new function ν. For each $\sigma \in \mathcal{F}$, $f(\cdot,\sigma)$ is a holomorphic mapping of $\mathbb{D}$ onto $\mathbb{C}$ which has a branch point of order two over each preimage of each point of the lattice $A(\sigma)$. The radius of the largest schlicht disk of $f(\cdot,\sigma)$ is given by (5.7). Define $Q(z,\sigma)$ as in (5.9), with F_1 in place of F, and with the new version of g. Define also, for $z \in \mathbb{C}$,

$$u(z) = \log \frac{2}{(1-|\zeta|^2)|f'(\zeta,0)|},$$

where ζ is a point of $\mathbb{D}$ with $f(\zeta,0) = z$. Then u is well defined, and $u = \infty$ at points of the hex lattice $A(0)$.

The conclusion of Theorem 1 is true for the function u just defined. The only modification required in the proof is that the expression (2.3) for the behavior of u near its singular point $z = 0$ has to be changed to

$$u(z) = -\frac{1}{2}\log|z| + h(z),$$

where h is continuous at $z = 0$, and $\frac{\partial h}{\partial y}(z) = o(y/|z|^2)$ as $z \to 0$. It follows that the hyperbolic derivatives of the Ahlfors-Grunsky functions $f(z,0)$ and F_1 are maximized over $\mathbb{D}$ at $z = 0$.

The "Bloch" Q defined in this section has, like its Landau counterpart, a local minimum at $(z,\sigma) = (0,0)$. To see this, observe that the second order Taylor development of the Bloch Q again has the form (5.12), with F replaced by F_1, the coefficients of $|z|^2$ and $|\sigma|^2$ still, respectively, equal to and greater than 1, and the coefficient of $z\sigma$ replaced by a new number β_1, which is given by (5.14) with the new function ν. Thus, local minimality of Q at $(0,0)$ again will follow if $|\beta_1| < 1$.

For $k = 0,1,2$, let z_k be the midpoints of the sides of $\partial\mathcal{F}_1$, with $\arg z_k = \frac{\pi}{3}(2k+1)$, V_k be the elliptic Möbius transformation of order two with fixed points $z_k, 1/z_k$, G_1 be the group of Möbius transformations of $\mathbb{D}$ generated by the V_k, and

$$K_1(z) = \sum_{\gamma \in G_1} (\gamma')^2(z)\gamma(z)^{-3}.$$

The images of $\mathcal{F}_1$ under distinct elements of G_1 are disjoint, and fill up $\mathbb{D}$ except for a set of area zero. We have $(F_1{}')^2(\gamma z)(\gamma')^2(z) = (F_1{}')^2(z)$ for each $\gamma \in G_1$. Formula (6.7) holds, with β, K, and ν replaced by β_1, K_1, and the new ν. Also, an analogue of (6.11) holds:

$$K_1(z) = C_1(F_1{}')^2(z) + F_1{}'(0)\, H \circ F_1(z)(F_1{}')^2(z), \quad z \in \mathbb{D}. \tag{7.2}$$

In (7.2), H is again defined by (6.9) and C_1 is some constant. The derivation of (7.2) is like that of (6.11) but simpler. It relies on the fact that a holomorphic G_1-automorphic function has a fundamental domain with compact closure in $\mathbb{D}$, and therefore such a function is constant. The arguments which led to (6.15) and (6.18) are again valid, and give

$$e^{-i\pi/3}\beta_1 = \frac{3\sqrt{3}}{4\pi}C_1, \tag{7.3}$$

$$C_1 = d_1{}^{-2}K_{1,0} - 5d_1{}^{-2}\frac{d_2}{d_1} - 2^{-1}3^{-3/2}d_1\mathcal{P}'(z_0)i, \tag{7.4}$$

where

$$K_{1,0} = \sum_{\gamma \in G_1 \setminus \{0\}} \gamma'(0)^2\gamma(0)^{-3}, \quad \text{and} \quad F_1(z) = d_1 z + d_2 z^4 + \cdots. \tag{7.5}$$

The expression of F_1 in terms of hypergeometric functions [H, p.382], as corrected in Section 1, and the formula (6.22) for $\mathcal{P}'(z_0)$, lead to

$$d_1 = (1+\sqrt{3})^{1/2}\frac{\Gamma(1/4)}{\Gamma(1/3)\Gamma(11/12)} = 2.1277, \tag{7.6}$$

$$d_2/d_1 = -2^{-6}3^{-3}(1+\sqrt{3})^{3/4}\left(\frac{\Gamma(1/4)\Gamma(1/3)}{\Gamma(2/3)\Gamma(11/12)}\right)^3 = -.82008, \tag{7.7}$$

$$\mathcal{P}'(z_0) = 28.65i. \tag{7.8}$$

To estimate $K_{1,0}$, let a denote the modulus of the midpoints of the sides of $\partial\mathcal{F}_1$. Then

$$a = \frac{1}{R_1}[1 + R_1{}^2 + (1 + R_1{}^2 + R_1{}^4)^{\frac{1}{2}}] = .2337. \tag{7.9}$$

To prove (7.9), observe that the Möbius transformation $w = (z+a)/(1+az)$ must map the hyperbolic geodesic through $R_1e^{2\pi i/3}, -a$, and $R_1e^{4\pi i/3}$ into the imaginary axis. Solution of $\mathrm{Re}[(R_1e^{2\pi i/3}+a)/(1+aR_1e^{2\pi i/3})] = 0$ for a leads to (7.9).

In Appendix B, we describe, for $n \geq 1$, a program for computing certain partial sums $K_{1,0}(n)$ of $K_{1,0}$. When $n = 8$, the result is

$$-21.0781 < K_{1,0}(8) < -21.0780. \tag{7.10}$$

Moreover, all $\gamma \in G_1$ whose contribution to $K_{1,0}$ is not included in $K_{1,0}(8)$ map $\mathcal{F}_1$ into $R_2 < |z| < 1$, with $R_2 > .99999$. The subharmonicity argument used to establish (6.29) yields

$$|K_{1,0} - K_{1,0}(8)| < a^{-2}R_2{}^{-3}(1 - R_2{}^2) < .0005.$$

Hence,

$$-21.0786 < K_{1,0} < -21.0775. \tag{7.11}$$

From (7.3), (7.4), (7.6)-(7.8) and (7.11), it follows that C_1 is positive real, and that $\arg \beta_1 = \frac{\pi}{3}$, with

$$2.098 < C_1 < 2.132, \quad .86 < |\beta_1| < .89. \tag{7.12}$$

Thus $|\beta_1| < 1$, as required for the local minimality of Q at $(z, \sigma) = (0, 0)$.

Appendix A. Estimation of $K_0(R)$.

If $0 < R \leq R_0 = 2 - \sqrt{3}$, the partial sum $K_0(R)$ of the series (6.28) for K_0 equals zero. Assume that $R_0 < R < 1$. To compute $K_0(R)$ by computer, we shall use the fact that the tiling $\{\gamma(\mathcal{F}) : \gamma \in \mathcal{G}_{\in}\}$ of $\mathbb{D}$ has the graph theoretic structure of a very simple tree. Consider the graph whose nodes are triangles of the tiling. Connect two nodes if the triangles share a side. Think of the single node $\mathcal{F}$ as constituting the zero'th generation. There is a 1-1 correspondence between triangles in the tiling and injective paths in the graph, obtained by starting in $\mathcal{F}$ and jumping across adjacent sides into new triangles until the quarry triangle is reached. The literal minded may confirm this correspondence by observing that the discussion leading up to (6.28) shows that each $\gamma \in G_2 \setminus \{0\}$ has a unique factorization of the form $\gamma = V_{k_1}...V_{k_n}$ for some $n \geq 1$, with $k_i \in \{0, 1, 2\}$, $k_i \neq k_{i+1}$ for $1 \leq i \leq n-1$. To visualize how the factorization is related to the geometry, the reader may show that if $\gamma = V_{k_1}, ..., V_{k_n} := \gamma_0 V_{k_n}$, then $\gamma(\mathcal{F})$ is the reflection of $\gamma_0(\mathcal{F})$ across its side $\gamma_0(S_{k_n})$, where S_k denotes the side of $\mathcal{F}$ mapped onto itself by V_k.

For $\gamma \in G_2$, write $h(\gamma) = \gamma'(0)^2\gamma(0)^{-3}$. To find the sum of the $h(\gamma)$ over the γ for which $\gamma(\mathcal{F})$ intersects $|z| < R$, the computer starts in $\mathcal{F}$ with initial sum zero. It then follows an injective path in the graph, at each node testing whether or not the next triangle, call it $\mathcal{F}'$, intersects $|z| < R$. If $\mathcal{F}'$ does intersect, the computer calculates $h(\gamma)$, where $\gamma(\mathcal{F}) = \mathcal{F}'$, and adds it to the tally. If $\mathcal{F}'$ does not intersect, the computer retreats to look for paths to nodes whose contributions have not yet been tallied but should be.

Eventually the contributions of all appropriate triangles are tallied, exactly once, the final sum equals $K_0(R)$, and the program stops.

Here is a summary of the program. The source code may be obtained by email from ftp://wuarchive.wustl.edu/ tr-math/software/landauin.c

LANDAUIN.C

MAIN

input: R, between R_0 and 1.
output: *sum*, the sum of $\gamma'(0)^2/\gamma(0)^3$ over all triangles which intersect $|z| \leq R$.
steps: initialize *sum* = *0*.
  define constants $V_0 = P^{-1} \circ V \circ P$, $V_1 = V$, $V_2 = P \circ V \circ P^{-1}$.
  set $\gamma = I$.
  for *side* = 0 to 2
    do *FIND_ALL_TRIANGLES* (γ, *side*, *sum*) $\rightarrow$ *sum*.
  print "The final sum is" *sum*.

FIND_ALL_TRIANGLES

input: $\gamma \in G_2$.
  side, a side of $\gamma(\mathcal{F})$.
  sum, the sum of all triangles counted so far.
output: *sum*, the updated sum of all the triangles counted.
steps: compose $\gamma = \gamma \circ V_{side}$.
  increment $sum = sum + \gamma'(0)^2/\gamma(0)^3$.
  for $i = 1$ to 2
    do *CLOSEST_APPROACH* (γ, *side* + i) $\rightarrow$ *distance*
    if *distance* $< R$
      do *FIND_ALL_TRIANGLES* (γ, *side* + i, *sum*) $\rightarrow$ *sum*.

CLOSEST_APPROACH

input: $\gamma \in G_2$.
  side, a side of $\gamma(\mathcal{F})$.
output: *distance*, the closest approach of this side to the origin.
steps: set $p_0 = \gamma(e^{2\pi i(side)/3})$ and $p_1 = \gamma(e^{2\pi i(side+1)/3})$.
  set $a_0 = \arg(p_0)$, $a_1 = \arg(p_1)$.
  return $distance = \sec \frac{1}{2}(a_1 - a_0) - \tan \frac{1}{2}(a_1 - a_0)$.

Appendix B. Estimation of $K_{1,0}(n)$.

For the tiling $\{\gamma(\mathcal{F}_1) : \gamma \in G_1\}$ of $\mathbb{D}$ the associated edge graph is less simple than in the Landau case of Appendix A. For example, starting from a neighbor triangle of $\mathcal{F}_1$, by appropriate counterclockwise jumps one can traverse a simple closed path in the graph, of length 30, which contains $\mathcal{F}_1$ in its interior. Thus, for the program below, it is more difficult to explain in old-fashioned language what the computer is doing. We'll simply give a summary of the program, preceded by a brief glossary of terms to be used, and leave further decipherment to the interested reader. The source code may be obtained by email from ftp://wuarchive.wustl.edu/ tr-math/software/blochin.c

The *stage* of $\mathcal{F}_1$ is defined to be zero. Each of the 30 triangles which share a vertex with $\mathcal{F}_1$ is said to have stage 1. For $n \geq 2$, a triangle in the tiling is defined to be of stage n if it shares a vertex with a triangle of stage $n-1$ but with no triangle of stage $< n-1$. The union of all triangles of stage $\leq n$ is denoted Ω_n. The *inradius* of Ω_n is the radius of the largest disk centered at the origin and contained in Ω_n. $K_{1,0}(n)$ is the partial sum of $K_{1,0}$ obtained by summing over all triangles of level $\leq n$. It turns out that the inradius of $\Omega(8)$ is $> .99999$. The *affinity* of a vertex is, roughly, a number which helps us keep track of triangles within the same level.

BLOCHIN.C

MAIN

input: n, an integer ≥ 1.
output: *sum*, the sum of $\gamma'(0)^2/\gamma(0)^3$ over all triangles of stage at most n.
  inradius , the inradius of Ω_n.
steps: initialize *sum* = 0.
  initialize *inradius* = 1.
  set γ = *identity*.
  for *side* = 0 to 2
    do *FIND_ALL_TRIANGLES* (γ, *stage* = 0, *side*, *affinity* = 1, *sum*) $\to$ *sum*
  print "$K_{1,0}(n)$ =" *sum*.
  print "The inradius of Ω_n is " *inradius*.

FIND_ALL_TRIANGLES

input: $\gamma \in G_1$.
  stage, the stage of γ.
  side, an element of $\mathbb{Z}_3$ representing a side of γ.
  affinity, the affinity of the most counterclockwise vertex of *side*.
  sum, the sum over all triangles counted so far.
output: *sum*, the updated sum of all the triangles counted.
steps: set *stage* = *stage* + 1.
  compose $\gamma = \gamma \circ V_{side}$.
  set *affinity* = *affinity* + 1
  increment $sum = sum + \gamma'(0)^2/\gamma(0)^3$.
  while (*affinity* < 11) **do**
    set *side* = *side* + 2.
    compose $\gamma = \gamma \circ V_{side}$.
    set *affinity* = *affinity* + 1.
    increment $sum = sum + \gamma'(0)^2/\gamma(0)^3$.
    if (*stage* < *n*) **do**
      if (*affinity* = 11)
        then *FIND_ALL_TRIANGLES* (γ, *stage*, *side* + 1, 3, *sum*) $\to$ *sum*.
        else *FIND_ALL_TRIANGLES* (γ, *stage*, *side* + 1, 2, *sum*) $\to$ *sum*.
    if (*stage* = *n*)
      do *CLOSEST_APPROACH* (γ, *side* + 1) $\to$ *distance*
      set *inradius* = min(*inradius*, *distance*).

CLOSEST_APPROACH

input: $\gamma \in G_1$.
side, a side of γ.
output: *distance*, the closest approach of this side to the origin.
steps: **set** p_1, p_2 = the vertices of *side*.
set q_1, q_2 = the intersections of the extension of *side* with $\partial\mathbb{D}$.
for $i = 1$ to 2.
set $a_i = \arg(p_i)$ and $b_i = \arg(q_i)$, where $a_1 < a_2$, $b_1 < b_2$.
set *distance* $= \sec \frac{1}{2}(b_2 - b_1) - \tan \frac{1}{2}(b_2 - b_1)$.
if not $(a_1 < \frac{1}{2}(b_1 + b_2) < a_2)$
then *distance* $= \min(|p_1|, |p_2|)$.
return *distance*.

REFERENCES

[A1] L.V. Ahlfors, *An extension of Schwarz's lemma*, Trans. Amer. Math. Soc. **43** (1938), 359–364.

[A2] L.V. Ahlfors, *Curvature properties of Teichmüller space of Riemann surfaces*, J. Anal. Math. **9** (1961), 161–176.

[A3] L.V. Ahlfors, *Lectures on Quasiconformal Mapping*, Van Nostrand, 1966.

[A4] L.V. Ahlfors, *Conformal Invariants*, McGraw-Hill, 1973.

[AG] L.V. Ahlfors and H. Grunsky, *Über die Blochsche Konstante*, Math Z. **42** (1937), 671–673.

[Ap] T.M. Apostol, *Modular functions and Dirichlet Series in Number Theory*, Second Edition, Springer, 1990.

[B1] A. Baernstein II, *Landau's constant, and extremal problems involving discrete subsets of* $\mathbb{C}$ in Linear and Complex Analysis Problem Book 3, Part II, Lecture Notes in Mathematics 1543, 404–407, edited by V.P. Havin and N.K. Nikolski, Springer, 1994.

[B2] A. Baernstein II, *A minimum problem for heat kernels of flat tori*, in Extremal Riemann Surfaces, Contemp. Math. 201, 227–243, edited by J. Quine and P. Sarnak, A.M.S., 1997.

[BC] R. Bañuelos and T. Carroll, *Brownian motion and the fundamental frequency of a drum*, Duke Math. J. **75** (1994), 575-602.

[Be] D. Betsakos, *Harmonic Measure, Domains of Fixed Inradius, and Polarization*, Ph.D. thesis, Washington University, 1996.

[Bh] A. Bloch, *Les théorèmes M.Valiron sur les fonctions entières et la théorie de l'uniformisation*, Ann. Fac. Sci. Univ. Toulouse(3) **17** (1925), 1–22.

[Bo] M. Bonk, *On Bloch's constant*, Proc. Amer. Math. Soc. **110** (1990), 889–894.

[C] J.W.S. Cassels, *An Introduction to the Geometry of Numbers*, Springer, 1959.

[CG] H. Chen and P.M. Gauthier, *On Bloch's constant*, preprint, 1996.

[CS] J.H. Conway and N.J.A. Sloane, *Sphere Packing, Lattices and Groups*, Second Edition, Springer, 1993.

[Ha1] W.K. Hayman, *Subharmonic Functions*, Volume 2, Academic Press, 1989.

[Ha2] W.K. Hayman, *Multivalent Functions*, Second Edition, Cambridge U.P., 1994.

[HK] W.K. Hayman and P.B. Kennedy, *Subharmonic Functions*, Volume 1, Academic Press, 1976.

[Hei] M. Heins, *On a class of conformal metrics*, Nagoya Math. J. **21** (1962), 1–60.

[He1] J. Hempel, *The Poincaré metric on the twice punctured plane and the theorems of Landau and Schottky*, J. London Math. Soc. (2) **20** (1979), 435–445.

[He2] J. Hempel, *On the uniformization of the n-punctured sphere*, Bull. London Math. Soc. **20** (1988), 97–115.

[H] E. Hille, *Analytic Function Theory*, Vol.II, Ginn, 1962.

[IT] Y. Imayoshi and M. Taniguchi, *An Introduction to Teichmüller Spaces*, Springer-Verlag, 1992.

[K] S. Kühlein, *Partial solution of a conjecture of Schmutz*, Arch. Math. **67** (1996), 164–172.

[Lai] W.T. Lai, *The exact value of Hayman's constant in Landau's theorem*, Scientia Sinica **22** (1979), 129–134.

[La] S. Lang, *Elliptic Functions*, Second Edition, Springer-Verlag, 1987.

[Lu] E. Landau, *Über die Blochsche Konstante und zwei verwandte Weltkonstanten*, Math. Z. **30** (1929), 608–634.

[L] O. Lehto, *Univalent Functions and Teichmüller Spaces*, Springer-Verlag, 1987.

[Mi] C.D. Minda, *Bloch constants*, J. d'Analyse Math. **41** (1982), 54–84.

[Mo] H.L. Montgomery, *Minimal theta functions*, Glasgow Math. J. **30** (1988), 75–85.

[OPS] B. Osgood, R. Phillips, and P. Sarnak, *Extremals of determinants of Laplacians*, J. Funct. Anal. **80** (1988), 148–210.

[P] Ch. Pommerenke, *Normal functions*, in Proc. NRL Conf. on Classical Function Theory, pp. 77–93, edited by F.Gross, Math. Res. Center, Naval Research Lab., 1970.

[Q] J. Quine, S.H. Heydari, and R.Y. Song, *Zeta regularized products*, Trans. Amer. Math. Soc. **338** (1993), 213–231.

[R] H. Rademacher, *On the Bloch-Landau constant*, Amer. J. Math. **65** (1943), 387–390.

[Ran] R.A. Rankin, *Modular Forms and Functions*, Cambridge U.P., 1977.

[Ro] C.A. Rogers, *Packing and Covering*, Cambridge U.P., 1964.

[Sa] P. Sarnak, *Extremal geometries*, in Extremal Riemann Surfaces, Contemp. Math. **201**, 1–7, edited by J. Quine and P. Sarnak, A.M.S., 1997.

[S] P. Schmutz, *Arithmetic Groups and the length spectrum of Riemann surfaces*, Duke Math. J. **84** (1996), 199–215.

[St] E.M. Stein, *Singular Integrals and Differentiability Properties of Functions*, Princeton, 1970.

[T] M. Tsuji, *Potential Theory in Modern Function Theory*, Second Edition, Chelsea, 1975.

[W1] S. Wolpert, *Chern forms and the Riemann tensor for the moduli space of curves*, Invent. Math. **85** (1986), 119–145.

[W2] S. Wolpert, *The hyperbolic metric and the universal curve*, J. Differential Geometry **31** (1990), 417–472.

[Y1] H. Yanagihara, *Quasi-conformal variations and local minimality of the Ahlfors-Grunsky function*, J. Analyse Math. **51** (1988), 30–61.

[Y2] H. Yanagihara, *On the locally univalent Bloch constant*, J. d'Analyse Math. **65** (1995), 1–17.

Washington University, St. Louis, MO 63130
al@math.wustl.edu, jpvinson@artsci.wustl.edu

The Apollonian Metric of a Domain in $\mathbb{R}^n$

A.F. Beardon

1 Introduction

This paper introduces a new Möbius invariant metric α_D, which we call the *Apollonian metric*, of a domain D with the following desirable properties:

(1) α_D is defined for any proper subdomain D of $\overline{\mathbb{R}}^n$ $(= \mathbb{R}^n \cup \{\infty\})$;

(2) α_D is Möbius invariant, and

(3) α_D is monotonic with respect to the domain D.

Strictly speaking, α_D is a pseudo-metric on D but we shall refer to it as a metric because in most cases it is indeed a metric. More precisely, we will prove the following result.

Theorem 1.1 *For any subdomain D of $\overline{\mathbb{R}}^n$, α_D is a pseudo-metric on D. If ∂D does not lie in any $(n-1)$-dimensional sphere in $\overline{\mathbb{R}}^n$, then α_D is a metric on D.*

We also show that, in addition to the properties given above, α_D provides a lower bound for the hyperbolic distance between any two points of a simply connected plane domain D (in contrast to the more familiar bounds on the *density* of the hyperbolic metric of D). Explicitly, we will prove the following result.

Theorem 1.2 *Let D be a simply connected plane hyperbolic domain with hyperbolic metric ρ_D and Apollonian metric α_D. Then, for all z and w in D, $\alpha_D(z,w) \leqslant 2\rho_D(z,w)$, and the constant 2 is best possible.*

If a metric is of interest, then so are its isometries; thus we consider the subgroup of those conformal automorphisms f of a plane domain D that satisfy

$$\alpha_D\big(f(z), f(w)\big) = \alpha_D(z,w).$$

I know of no example of an α_D-isometry that is not a Möbius map and, as an example, we have the following result.

Theorem 1.3 *Suppose that E is a compact subset of the extended negative real axis that contains at least three points, and let D be its complement. Then any conformal α_D-isometry is a Möbius map.*

We prove Theorem 1.1 in §2. In §3, §4 and §5, we compare the Apollonian metric with the quasi-hyperbolic, the hyperbolic and the Klein-Hilbert metrics, respectively. The proof of Theorem 1.2 is in §4. We consider convex plane domains in §6, and in §7 we prove Theorem 1.3. Finally, in §8 we use the Apollonian metric to give a covering theorem for meromorphic functions and, using an argument suggested by T.K. Carne, we show in §9 that Theorem 1.2 is equivalent to Koebe's 1/4-Theorem. I am grateful to the referee for his or her careful reading of the manuscript, and for contributing Theorem 6.2 and the second proof of Theorem 1.2

2 The Apollonian metric of a domain

For our purposes a ring domain D is the complement in $\overline{\mathbb{R}}^n$ of two disjoint non-empty compact balls, and the modulus $\mathrm{mod}(D)$ of D is $\log(R_2/R_1)$, where D is mapped by a Möbius map onto the set $\{x : R_1 < |x| < R_2\}$. The modulus of a ring domain is known to be invariant under Möbius maps.

There are two ways to define the Apollonian metric and we begin with the more geometric way. Let x and y be distinct points of a subdomain D of $\overline{\mathbb{R}}^n$; then, for sufficiently small positive values of k and k', the two balls

$$\left\{w \in \overline{\mathbb{R}}^n : \frac{|w-x|}{|w-y|} < k\right\}, \quad \left\{w \in \overline{\mathbb{R}}^n : \frac{|w-y|}{|w-x|} < k'\right\} \qquad (2.1)$$

lie within D. Here, $|u - v|$ denotes the Euclidean distance between u and v in $\mathbb{R}^n$, and we are employing the usual conventions regarding ∞. Now denote by $\mathcal{B}_x$ the closure of the maximal ball of the first type in (2.1) that lies within D, and similarly for $\mathcal{B}_y$ (using the second ball in (2.1)), and note that

(a) if $\infty \in D$, then one of $\mathcal{B}_x$ and $\mathcal{B}_y$ may contain ∞ and so be the complement of an open Euclidean ball in $\mathbb{R}^n$;

(b) the interiors of $\mathcal{B}_x$ and $\mathcal{B}_y$ are disjoint (for if not, their union is $\overline{\mathbb{R}}^n$);

(c) the complement in $\overline{\mathbb{R}}^n$ of $\mathcal{B}_x \cup \mathcal{B}_y$ is an open ring domain $\mathcal{A}_{x,y}$ which is empty only when ∂D lies in an $(n-1)$-dimensional sphere in $\overline{\mathbb{R}}^n$ with respect to which x and y are inverse points. We call $\mathcal{B}_x$ and $\mathcal{B}_y$ the *Apollonian balls* for x and y, respectively, and $\mathcal{A}_{x,y}$ the *Apollonian ring* for x and y, each with respect to the domain D.

For distinct points x and y in D, we now define

$$\alpha_D(x,y) = \begin{cases} \mathrm{mod}\,(\mathcal{A}_{x,y}) & \text{if } \mathcal{A}_{x,y} \neq \emptyset\,; \\ 0 & \text{if } \mathcal{A}_{x,y} = \emptyset\,, \end{cases}$$

and, of course, $\alpha_D(x,x) = 0$. Regardless of whether or not α_D is a metric, it is immediate that *α_D is Möbius invariant* in the sense that if g is a Möbius map acting in $\overline{\mathbb{R}}^n$, then

$$\alpha_{g(D)}\big(g(x), g(y)\big) = \alpha_D(x,y)$$

and we shall use this in our proof of Theorem 1.1. It is also immediate that α_D is monotonic in the sense that $U \subset V$ implies that $\alpha_V \leqslant \alpha_U$ on U.

Before starting our proof of Theorem 1.1, we discuss a general construction of metrics and pseudo-metrics (see [10] and [11]). Let Ω be any non-empty set, let $\mathcal{F}$ be a non-empty class of positive real-valued functions on Ω, and define

$$\lambda_\Omega(x,y) = \log \sup \left\{ \frac{f(x)}{f(y)} : f \in \mathcal{F} \right\} = \sup \left\{ \log \frac{f(x)}{f(y)} : f \in \mathcal{F} \right\} \leqslant +\infty .$$

In order that λ_Ω be finite, we assume that *$\mathcal{F}$ satisfies the Harnack condition*; that is, for each x and y in Ω,

$$\sup \left\{ \frac{f(x)}{f(y)} : f \in \mathcal{F} \right\} < +\infty .$$

It is immediate that $\lambda_\Omega(x,x) = 0$ and that λ_Ω satisfies the triangle inequality, but λ_Ω need not be non-negative, nor need it be symmetric. To circumvent this, we introduce the function

$$\mu_\Omega(x,y) = \max\{\lambda_\Omega(x,y), \lambda_\Omega(y,x)\} = \sup \left\{ \left| \log \frac{f(x)}{f(y)} \right| : f \in \mathcal{F} \right\} ,$$

and it is clear that if $\mathcal{F}$ satisfies the Harnack condition, then μ_Ω is a pseudo-metric on Ω. If, in addition to this, *$\mathcal{F}$ separates Ω* (that is, if x and y are distinct points of Ω, then there exists some f in $\mathcal{F}$ such that $f(x) \neq f(y)$) then then μ_Ω is a metric on Ω. These facts are recorded in the next lemma.

Lemma 2.1 *If $\mathcal{F}$ satisfies the Harnack condition on Ω then μ_Ω is a pseudo-metric on Ω. If, in addition, $\mathcal{F}$ separates Ω then μ_Ω is a metric on Ω.*

We shall use Lemma 2.1 to construct, for any subdomain D of $\overline{\mathbb{R}}^n$, a pseudo-metric β_D, and we shall then show that $\alpha_D = \beta_D$. It is this equivalence of two superficially different definitions that makes the Apollonian metric interesting.

The proof of Theorem 1.1. We recall that the Euclidean distance in $\mathbb{R}^n$ has been denoted by $|x-y|$, and we now denote the chordal distance in $\overline{\mathbb{R}}^n$ by $\chi(x,y)$; thus, for x and y in $\mathbb{R}^n$,

$$\chi(x,y) = \frac{2|x-y|}{(1+|x|^2)^{1/2}(1+|y|^2)^{1/2}} , \quad \chi(x,\infty) = \frac{2}{(1+|x|^2)^{1/2}} .$$

Observe that
$$\chi(0,a) = |a|\chi(a,\infty)\,. \tag{2.2}$$

Given a domain D, let $\mathcal{F}$ be the class of functions
$$f_{a,b}(x) = \frac{\chi(x,a)}{\chi(x,b)}\,,$$
indexed by the pairs (a,b) in $\partial D \times \partial D$. For each x and y in D, the function
$$\frac{f_{a,b}(x)}{f_{a,b}(y)} = \frac{\chi(x,a)\chi(y,b)}{\chi(x,b)\chi(y,a)}$$
a continuous positive function of (a,b) over the compact set $\partial D \times \partial D$ and so is bounded above there; thus $\mathcal{F}$ satisfies the Harnack condition. Lemma 2.1 now implies that
$$\beta_D(x,y) = \sup\left\{\left|\log\frac{\chi(x,a)\chi(y,b)}{\chi(x,b)\chi(y,a)}\right| : a,b \in \partial D\right\} \tag{2.3}$$
is a pseudo-metric on D.

It is easy to see that $\beta_D = \alpha_D$. As cross-ratios (formed by using the chordal metric) are invariant under Möbius transformations (see [2], p.32), the pseudo-metric β_D is Möbius invariant. Given x and y, there is a Möbius g acting in $\overline{\mathbb{R}}^n$ with $g(x) = 0$ and $g(y) = \infty$, and we may suppose that $\{x : |x| < r\}$ and $\{x : |x| > R\}$ are the largest open balls centred at 0 and ∞, respectively, which lie in $g(D)$. Then, with a and b in $\partial g(D)$, we have, using (2.2),
$$\begin{aligned}
\beta_D(x,y) &= \beta_{g(D)}(0,\infty)\\
&= \sup_{a,b}\left|\log\frac{\chi(0,a)\chi(\infty,b)}{\chi(0,b)\chi(\infty,a)}\right|\\
&= \sup_{a,b}\left|\log\frac{|a|}{|b|}\right|\\
&= \log\frac{R}{r}\\
&= \alpha_{g(D)}(0,\infty)\\
&= \alpha_D(x,y)
\end{aligned}$$
as required. This proves that α_D is a pseudometric on D. If ∂D does not lie in any $(n-1)$-dimensional sphere then, for $x \neq y$, $\mathcal{A}_{x,y} \neq \emptyset$ so that $\alpha_D(x,y) > 0$. This completes the proof of Theorem 1.1.

We end this section with a useful and illustrative result.

Lemma 2.2 (i) *If* $D = \overline{\mathbb{R}}^n - \{\zeta,\infty\}$ *then*
$$\alpha_D(x,y) = \left|\log\frac{|x-\zeta|}{|y-\zeta|}\right|\,. \tag{2.4}$$

(ii) *If* $D \subset \overline{\mathbb{R}}^n - \{\zeta, \infty\}$ *then*

$$\alpha_D(x,y) \geqslant \left| \log \frac{|x-\zeta|}{|y-\zeta|} \right| .$$

Proof. To prove (i) recall that $\alpha_D = \beta_D$, where β_D is given by (2.3). Now the only possibilities for (a,b) in (2.3) are (ζ,∞) and (∞,ζ), and these yield (2.4). Clearly, (ii) follows from (i) and the monotonicity of α_D with respect to D.

3 Comparison between the Apollonian and quasihyperbolic metrics

In this section we recall two other metrics used in complex analysis and compare these with the metric α_D. First, if $n \geqslant 3$, the only domains in $\overline{\mathbb{R}}^n$ that support a hyperbolic metric are the open balls and half-spaces and, as an alternative to the hyperbolic metric, Gehring and Palka [5] introduced the *quasihyperbolic metric* of a domain in $\mathbb{R}^n$. The quasihyperbolic metric q_D of a domain D is the Riemannian metric derived from the line element $ds = |dx|/\delta_D(x)$, where

$$\delta_D(x) = \inf\{|x-y| : y \in \partial D\} = \operatorname{dist}(x, \partial D) .$$

In the case of the upper half-space $H = \{(x_1, \dots, x_n) : x_n > 0\}$, the hyperbolic and quasihyperbolic metrics are the same (for they have the same density, namely $1/x_n$), and we shall now show that these also coincide with the Apollonian metric of H.

Lemma 3.1 *Let* H *be the upper half-space in* $\mathbb{R}^n$; *then* $q_H = \rho_H = \alpha_H$.

Proof. As both ρ_H and α_H are Möbius invariant, as any two points in H can be mapped by a Möbius self-map of H to the points $u = (0, \dots, 0, \ell)$ and $v = (0, \dots, 0, 1)$, where $\ell > 1$, and as $\rho_H(u,v) = \log \ell$, it suffices to prove that $\alpha_H(u,v) = \log \ell$. Now

$$\alpha_H(u,v) = \sup\left\{ \left| \log \frac{\chi(u,a)\chi(v,b)}{\chi(u,b)\chi(v,a)} \right| : a,b \in \partial H \right\}$$

and, taking a and b to be finite (as we may), we obtain

$$\alpha_H(u,v) = \sup\left\{ \left| \frac{1}{2} \log \frac{(|a|^2+\ell^2)(|b|^2+1)}{(|a|^2+1)(|b|^2+\ell^2)} \right| : a,b \in \partial H \right\} .$$

A simple calculation shows that this supremum is attained when $a = 0$ and $b = \infty$; thus $\alpha_H(u,v) = \log \ell$. This completes the proof of Lemma 3.1.

Next, we recall another metric. For any domain D in $\mathbb{R}^n$, the function

$$j_D(x,y) = \log\left(1 + \frac{|x-y|}{\delta_D(x,y)}\right) ,$$

where $\delta_D(x,y) = \min\{\delta_D(x), \delta_D(y)\}$, is a metric in D (see [12], p.28). As $j_D \leqslant q_D$ (see [5], Lemma 2.1, and [12], p.34), j_D provides an explicit lower bound of the quasihyperbolic distance between two points; we shall now compare α_D with j_D.

Theorem 3.2 *For any domain D in $\mathbb{R}^n$, $\alpha_D \leqslant 2j_D \leqslant 2q_D$. Further, each of the three inequalities $\alpha_D \leqslant 2j_D$, $j_D \leqslant q_D$ and $\alpha_D \leqslant 2q_D$ is best possible.*

Proof. Take any x and y in D, construct the open ball D_x with centre x, radius $\delta_D(x)$ and similarly for y, and let k_1 and k_2 be the largest numbers satisfying

$$\mathcal{B}'_x = \left\{w : \frac{|w-x|}{|w-y|} < k_1\right\} \subset D_x\ , \quad \mathcal{B}'_y = \left\{w : \frac{|w-y|}{|w-x|} < k_2\right\} \subset D_y\ .$$

The (maximal) Apollonian balls $\mathcal{B}_x$ and $\mathcal{B}_y$ for x and y, respectively, contain the two balls $\mathcal{B}'_x$ and $\mathcal{B}'_y$ so it follows that if we denote the complement of $\mathcal{B}'_x \cup \mathcal{B}'_y$ by A, then

$$\alpha_D(x,y) \leqslant \mathrm{mod}(A) = \log\frac{1}{k_1k_2} = \log\frac{1}{k_1} + \log\frac{1}{k_2}\ .$$

The balls $\mathcal{B}'_x$ and D_x are tangent at the point w on ∂D_x furthest from y, and as

$$k_1 = \frac{|w-x|}{|w-y|} = \frac{\delta_D(x)}{\delta_D(x) + |x-y|}\ ,$$

we see that

$$\log\frac{1}{k_1} = \log\left(1 + \frac{|x-y|}{\delta_D(x)}\right) \leqslant j_D(x,y)\ .$$

As the same holds for k_2, we see that $\alpha_D \leqslant 2j_D$. Because $j_D \leqslant q_D$ ([12], p.34), the inequalities in Theorem 3.2 have been verified.

We end this section with three examples (which show that the inequalities in Theorem 3.2 are best possible).

Example 3.1: The upper half-space H. From Lemma 3.1 and the inequality $j_H \leqslant \rho_H \leqslant 2j_H$ (see [12], Lemma 2.41, p.29) we have

$$j_H \leqslant \rho_H = \alpha_H = q_H \leqslant 2j_H\ . \tag{3.1}$$

Working in dimension 2, if $x > 0$ we have $j_H(i, x+i) = \log(1+x)$, $\rho_H(i, x+i) \sim x$ as $x \to 0$, and $\rho_H(i, x+i) \sim 2\log x$ as $x \to +\infty$ (see [2], p.130).

These imply that that the two inequalities $j_H \leqslant \rho_H$ and $\rho_H \leqslant 2j_H$ are best possible and, from (3.1), so too are the inequalities $\alpha_D \leqslant 2j_D$ and $j_D \leqslant q_D$ in Theorem 3.2.

Example 3.2: The unit ball B**.** In this case ρ_B has density $2(1-|x|^2)^{-1}$ and the chain of inequalities is

$$j_B \leqslant q_B \leqslant \rho_B = \alpha_B \leqslant 2j_B \ . \tag{3.2}$$

The two outer inequalities come from Theorem 3.2, the remaining inequality is trivial, and $\rho_B = \alpha_B$ because $\rho_H = \alpha_H$ and both ρ and α are Möbius invariant. As

$$\rho_B(0,x) = \log\frac{1+|x|}{1-|x|}\ , \quad q_B(0,x) = \log\frac{1}{1-|x|} = j_B(0,x)\ , \tag{3.3}$$

we see (by letting $x \to 0$) that the inequality $\alpha_D \leqslant 2q_D$ in Theorem 3.2 is best possible. Observe also that (3.3) implies that the two outer inequalities in (3.2) are also best possible (as is the central inequality).

In view of the inequalities given in Theorem 3.2, the following example is of interest for it shows that there are no inequalities of the same type but in the opposite direction.

Example 3.3. Let $E = \{iy : |y| \leqslant 1\}$, let $\mathbb{R}^+ = \{x : x \geqslant 0\}$, and let D be the complement of $E \cup \mathbb{R}^+$ in the complex plane. We claim that if $z_n = n+i$ and $w_n = n-i$ then, as $n \to \infty$,

$$\alpha_D(z_n,w_n) \to 0\ , \quad j_D(z_n,w_n) = \log 3\ , \quad q_D(z_n,w_n) \to +\infty\ . \tag{3.4}$$

The second assertion in (3.4) is immediate. To verify the first, let

$$g_n(z) = \frac{z-z_n}{z-w_n}\ .$$

Then $g_n(\mathbb{R}^+)$ is an arc of the unit circle and, for z on E and $n > 3$, $|g_n(z) - 1| \leqslant 3/n$. It follows that

$$\alpha_D(z_n,w_n) = \alpha_{g_n(D)}(0,\infty) \leqslant \log\left(\frac{1+3/n}{1-3/n}\right) \to 0\ .$$

To verify the third assertion in (3.4), we join z_n to w_n by a geodesic arc γ for the quasihyperbolic metric (these are known to exist; see [5]), and we let γ_0 be the initial segment of this geodesic arc lying between z_n and the point where γ first hits the ray R given by $\{x+ix : x > 0\}$. Now $q_D(z_n,w_n)$ is greater than the quasihyperbolic length of γ_0 and, as the quasihyperbolic density at a point z on γ_0 is $1/\mathrm{Im}[z]$, it follows that $q_D(z_n,w_n)$ exceeds the hyperbolic length $\ell(\gamma_0)$ of γ_0 measured in the upper half-plane model H. We deduce that $q_D(z_n,w_n) \to +\infty$ as $n \to \infty$ and this completes our discussion of Example 3.3.

4 Comparison between the Apollonian and hyperbolic metrics

This section is devoted to proving Theorem 1.2. The hyperbolic metric $\rho_{\mathbb{D}}$ of the unit disc $\mathbb{D}$ in the complex plane has density $2|dz|/(1-|z|^2)$, and the hyperbolic metric ρ_D of a simply connected plane domain D (other than $\mathbb{C}$) is obtained by transferring $\rho_{\mathbb{D}}$ to ρ_D by any conformal map of $\mathbb{D}$ onto D. Before proving Theorem 1.2, we derive a preliminary result which, incidentally, shows that the constant 2 in Theorem 1.2 is best possible.

Lemma 4.1 *Let W be the simply connected domain on the sphere whose complement is the closed segment $[r, R]$, where $0 < r < R < +\infty$. Then*

$$\alpha_W(0,\infty) = 2\rho_W(0,\infty) = \log\frac{R}{r}\ .$$

The proof of Lemma 4.1. The definition of α_W in terms of the modulus of a ring domain shows immediately that $\alpha_W(0,\infty) = \log(R/r)$. As the hyperbolic distance is invariant under the univalent map

$$z \mapsto \sqrt{\frac{z-r}{R-z}}$$

of W onto the upper half-plane H, we see that

$$2\rho_W(0,\infty) = 2\rho_H(i, i\sqrt{r/R}) = \log(R/r) = \alpha_W(0,\infty)$$

as required.

The proof of Theorem 1.2. We take any simply connected domain D, and any two points in D. As both the hyperbolic and Apollonian metrics are Möbius invariant, we may assume that these two points are 0 and ∞, and we let $\{z : |z| < r\}$ be the largest open disc centred at the origin that lies in D, and $\{z : |z| > R\}$ the largest disc centred at ∞ and lying in D. As $\alpha_D(0,\infty) = \log(R/r)$, we have to prove that

$$2\rho_D(0,\infty) \geqslant \log(R/r)\ .$$

The complement K of the domain D is a compact connected set which lies in the closed annulus $\{z : r \leqslant |z| \leqslant R\}$, and which meets both of its bounding circles. We now apply circular symmetrization to D and K; specifically, for each t in the interval $[r, R]$ we replace the set $K \cap \{|z| = t\}$ by a closed arc of $\{|z| = t\}$ which has the same linear measure but which is situated symmetrically with respect to the negative real axis. This symmetrization produces a new compact set K_0 that is symmetric about the negative real axis, and which contains the segment $[-R, -r]$, and a new simply connected domain D_0, namely the complement of K_0. We now apply Theorem 1.3 in [13] which relates the hyperbolic metrics of D and D_0. Denoting the densities of these hyperbolic metrics by λ_D and λ_{D_0}, respectively, Theorem 1.3, [13], states that

$$\lambda_D(z) \geqslant \lambda_{D_0}(|z|) .$$

Consequently, the integral of λ_D along any path that joins 0 to ∞ in D is not less than the integral of λ_{D_0} along the positive real axis and this, in turn, is not less than the hyperbolic distance between 0 and ∞, measured in the domain D_0. We deduce that

$$\rho_D(0,\infty) \geqslant \rho_{D_0}(0,\infty) .$$

Now the hyperbolic metric is monotonic with respect to the domain in the sense that if $D_1 \subset D_0$, then $\lambda_0 \leqslant \lambda_1$, where λ_j is the density of the hyperbolic metric in D_j. This inequality extends from densities to distances measured in D_1 and D_0 so, if we now let D_1 be the complement of the closed interval $[-R,-r]$, we have

$$\rho_{D_0}(0,\infty) \geqslant \rho_{D_1}(0,\infty) = \tfrac{1}{2}\log(R/r) ,$$

by virtue of Lemma 4.1. This completes the proof of Theorem 1.2.

The proof of Theorem 1.2 just given shows clearly, and geometrically, what the extremal domains are. We now give an elementary proof (free of symmetrization) along the lines suggested (independently) by T.K. Carne and the referee.

A second proof of Theorem 1.2. As ρ_D and α_D are invariant under Möbius maps, the inequality $2\rho_D \geqslant \alpha_D$ will follow once we have proved that $2\rho_{g(D)} \geqslant \alpha_{g(D)}$ for some Möbius map g. Now take z_1 and z_2 in D. There are points a and b in ∂D such that

$$\alpha_D(z_1,z_2) = \left|\log\frac{\chi(z_1,a)\chi(z_2,b)}{\chi(z_1,b)\chi(z_2,a)}\right| ,$$

and, by choosing g suitably, we may assume that $a=\infty$ and $b=0$. It follows then, that we may assume that D is simply connected, that 0 and ∞ are on the boundary of D, and that

$$\alpha_D(z_1,z_2) = \left|\log\frac{|z_2|}{|z_1|}\right| .$$

As D is simply connected the density λ_D of ρ_D satisfies $\lambda_D(z) \geqslant [2\delta_D(z)]^{-1}$ and, as $0 \notin D$, we also have $\delta_D(z) \leqslant |z|$.

Now let γ be the hyperbolic geodesic joining z_1 and z_2 in D. Then

$$\begin{aligned} 2\rho_D(z_1,z_2) &\geqslant \int_\gamma \frac{|dz|}{\delta_D(z)} \\ &\geqslant \int_\gamma \frac{|dz|}{|z|} \\ &\geqslant \left|\int_\gamma \frac{dz}{z}\right| \\ &\geqslant \left|\log\frac{|z_2|}{|z_1|}\right| \\ &= \alpha_D(z_1,z_2)\,, \end{aligned}$$

and the proof is complete.

5 Comparison between the Apollonian and the Klein-Hilbert metrics

In this section we compare the Apollonian metric α_D with the Klein-Hilbert metric κ_D of a bounded convex subdomain D of $\mathbb{R}^n$. Given x and y in D, let the Euclidean line through x and y meet ∂D at the points a and b, where these are labelled so that a, x, y, b occur in this order on L; then the Klein-Hilbert metric is given by

$$\kappa_D(x,y) = \log\frac{|x-b|.|y-a|}{|x-a|.|y-b|}$$

if $x \neq y$ and, of course, $\kappa_D(x,x) = 0$ (see [4] and [7] for a discussion of the metric κ). As

$$\frac{|x-b|.|y-a|}{|x-a|.|y-b|} = \frac{\chi(x,b)\chi(y,a)}{\chi(x,a)\chi(y,b)}\,,$$

so we see immediately that, for such domains,

$$\kappa_D \leqslant \alpha_D\,. \tag{5.1}$$

One interesting application of (5.1) (with Theorem 1.2) is that if D is a bounded convex plane domain, and if C is a chord of D of length $2d$ which (for convenience) is centred at the origin, then, for z on C,

$$\rho_D(0,z) \geqslant \frac{1}{2}\log\frac{d+|z|}{d-|z|}$$

or, equivalently,

$$|z| \leqslant d\tanh\left[\rho_D(0,z)\right]\,.$$

We end this section with the remark that the Klein-Hilbert metric has the property that, for distinct points x and y in D, the set of points w satisfying

$$\kappa_D(x,y) = \kappa_D(x,w) + \kappa_D(w,y)$$

may contain an *open* subset of D. The same is true of the Apollonian metric; we omit the details.

6 Convex plane domains

In this section we prove the following two results.

Theorem 6.1 *Let D be a bounded convex plane domain. Then*

$$\tfrac{1}{2}\alpha_D(z,w) \leqslant \rho_D(z,w) \leqslant 4\sinh\left[\tfrac{1}{2}\alpha_D(z,w)\right] .$$

Theorem 6.2 *Let D be a bounded plane domain. If $\alpha_D \leqslant \rho_D$ then D is convex.*

Theorem 6.1 shows that, when D is convex, ρ_D and α_D are equivalent metrics. Theorem 6.2 and its proof were suggested by the referee.

The proof of Theorem 6.1. Of course, the first inequality comes from Theorem 1.2 so we need only prove the second inequality. Choose z and w in D, let $\mathcal{B}_z$ and $\mathcal{B}_w$ be the maximal Apollonian discs for $\{z,w\}$ with respect to the domain D (these are Euclidean discs), and let Ω be the convex hull of $\mathcal{B}_z \cup \mathcal{B}_w$. As $\mathcal{B}_z$ and $\mathcal{B}_w$ are also the maximal Apollonian discs for $\{z,w\}$ with respect to the domain Ω,

$$\alpha_D(z,w) = \alpha_\Omega(z,w) ,$$

and, as $\Omega \subset D$,

$$\rho_D(z,w) \leqslant \rho_\Omega(z,w) .$$

It suffices, therefore, to prove the result for Ω. There are two possibilities to consider, namely

(1) the Euclidean radii of $\mathcal{B}_z$ and $\mathcal{B}_w$ are equal (in which case Ω is bounded by two parallel line segments and two arcs of the circles bounding $\mathcal{B}_z$ and $\mathcal{B}_w$), and

(2) the Euclidean radii of $\mathcal{B}_z$ and $\mathcal{B}_w$ are unequal.

We shall first show how Case (2) reduces to Case (1). In Case (2), there are two Euclidean straight lines, say L and L', which are both tangent to $\mathcal{B}_z$ and $\mathcal{B}_w$, and which meet at the two distinct points ∞ and, say,

ζ. We construct a Möbius map g that maps ζ and ∞ to points -1 and 1, respectively, and which also maps the discs $\mathcal{B}_z$ and $\mathcal{B}_w$ to two discs of equal size lying inside the unit disc $\mathbb{D}$. Observe that $g(\Omega)$ contains the Euclidean convex hull Σ of $g(\mathcal{B}_z)$ and $g(\mathcal{B}_w)$, so that if we replace $g(\Omega)$ by Σ, we increase the hyperbolic metric. Also, as $g(\mathcal{B}_z)$ and $g(\mathcal{B}_w)$ are the maximal Apollonian discs for the domain Σ, the Apollonian metric is unaltered; thus we see that, in Case (2),

$$\rho_\Omega = \rho_{g(\Omega)} \leqslant \rho_\Sigma \,, \quad \alpha_\Omega = \alpha_{g(\Omega)} = \alpha_\Sigma \,.$$

It therefore suffices to prove the result for Σ instead of Ω; equivalently, we may assume that Case (1) holds.

Given that Case (1) holds, we may apply a suitable map of the form $z \mapsto cz + d$, and so assume that $z = a$ and $w = -a$, where $a > 0$, that $\mathcal{B}_z$ and $\mathcal{B}_w$ each have Euclidean radius 1, and that Ω is the convex hull of $\mathcal{B}_z$ and $\mathcal{B}_w$ (recall that z and w are inverse points with respect to each of $\mathcal{B}_z$ and $\mathcal{B}_w$; the reader may find a diagram helpful here). Recalling the well-known upper bound of $2/\delta_\Omega(z)$ on the density of ρ_Ω we find, by integrating along the segment $[-a, a]$, that

$$\rho_\Omega(z, w) = \rho_\Omega(a, -a) \leqslant 4a \,.$$

The two circles bounding $\mathcal{B}_z$ and $\mathcal{B}_w$ have unit radius and a and $-a$ as inverse points, and their defining equations (for a complex parameter t) are

$$\frac{|t-a|}{|t+a|} = \frac{1}{\sqrt{1+a^2}+a} \,, \quad \frac{|t+a|}{|t-a|} = \frac{1}{\sqrt{1+a^2}+a} \,,$$

respectively. It follows that

$$\alpha_\Omega(z, w) = \alpha_\Omega(-a, a) = 2\log(\sqrt{1+a^2}+a) = 2\sinh^{-1} a \,.$$

We deduce that

$$\rho_\Omega(z, w) \leqslant 4a = 4\sinh\left[\tfrac{1}{2}\alpha_\Omega(z, w)\right]$$

as required. This completes the proof of Theorem 6.1.

Example 6.1. Let D be the infinite strip $\{x + iy : |y| < 1\}$. A simple calculation shows that for each positive a,

$$\rho_D(-a, a) = \pi a = \pi \sinh\left[\tfrac{1}{2}\alpha_D(-a, a)\right] \,,$$

so that the upper bound of $\rho_D(z, w)$ in Theorem 6.1 is of the correct general order of magnitude. The constant appearing in this bound, namely 4, is not best possible; indeed, A. Rhodes has shown that the best possible constant lies between π and 3·7.

The proof of Theorem 6.2. The idea is to show that if $\alpha_D \leqslant \rho_D$ then the hyperbolic density λ_D of ρ_D satisfies $\lambda_D(z) \geqslant 1/\delta_D(z)$, and that this implies that D is convex.

Given z in D choose a in ∂D such that $|a-z| = \delta_D(z)$. We may assume that $a=0$ and that $z>0$. Now take any real number h satisfying $0<h<z$; then $z+h \in D$ and

$$\begin{aligned} \rho_D(z,z+h) &\geqslant \alpha_D(z,z+h) \\ &\geqslant \alpha_{\mathbb{C}-\{0\}}(z,z+h) \\ &= \log\left(\frac{z+h}{z}\right) \end{aligned}$$

(see Lemma 2.2). We deduce that

$$\lambda_D(z) = \lim_{h\to 0} \frac{\rho_D(z,z+h)}{h} \geqslant \frac{1}{z},$$

and hence (after removing the normalisation) that for all z in D,

$$\lambda_D(z) \geqslant \frac{1}{\delta_D(z)}. \tag{6.1}$$

It remains to show that (6.1) implies that D is convex and I know of no published proof of this. It was proved by J.R. Hilditch (1984, unpublished) but rather than appeal to this, we shall use the following result of Keogh ([8]).

Lemma 6.3 *Suppose that a plane domain D has the property that for every analytic map f of $\mathbb{D}$ into D, f^* is also an analytic map of $\mathbb{D}$ into D, where $f^*(z) = f(0) + \frac{1}{2}f'(0)z$. Then D is convex.*

Suppose now that (6.1) holds, and let f be any analytic map of $\mathbb{D}$ into D. Then (see [1], Theorem 1.7)

$$\lambda_D(f(0))|f'(0)| \leqslant 2$$

so that

$$\tfrac{1}{2}|f'(0)| \leqslant \frac{1}{\lambda_D(f(0))} \leqslant \delta_D(f(0)).$$

This means that

$$f^*(\mathbb{D}) = \{z : |z-f(0)| < \tfrac{1}{2}|f'(0)|\} \subset D$$

so that, by Lemma 6.3, D is convex. This completes the proof of Theorem 6.2.

7 The proof of Theorem 1.3

Let D be the complement of the set E given in Theorem 1.3. The invariance of α_D under all Möbius maps shows that in proving Theorem 1.3 it suffices to assume that 0 and ∞ are in E, and that $(0,+\infty) \subset D$. Observe now that, because E contains three points on the negative real axis, for any z and w, $\alpha_D(z,w) = 0$ if and only if $w = z$ or $w = \bar{z}$; thus if x is real and $\alpha_D(x,w) = 0$, then $w = x$. Now take any x in $D \cap \mathbb{R}$ and let $y = f(x)$, where f is an α_D-isometry. Then

$$\alpha\big(x, f^{-1}(\bar{y})\big) = \alpha(f(x), \bar{y}) = \alpha(y, \bar{y}) = 0$$

so that $f^{-1}(\bar{y}) = x$. We deduce that $\bar{y} = f(x) = y$, and hence that $f(x) \in D \cap \mathbb{R}$. As f^{-1} is also an isometry, we deduce that

$$f(D \cap \mathbb{R}) = D \cap \mathbb{R} = f^{-1}(D \cap \mathbb{R}) ,$$

so that f permutes the components of $D \cap \mathbb{R}$. One of the components of $D \cap \mathbb{R}$ is $(0,+\infty)$, and we write

$$I = (0,+\infty) , \qquad J = f(I) .$$

We now construct a Möbius map h that leaves the extended real axis $\mathbb{R} \cup \{\infty\}$ invariant, and that maps J onto I, and let $F = hf$. It follows that F is an Apollonian isometry of (D, α_D) onto $(F(D), \alpha_{F(D)})$, and that the complements of D and $F(D)$ are compact subsets, say E and E', respectively, of the negative real axis, each containing 0 and ∞. Note also that F is a bijection of I onto itself. Replacing F by $z \mapsto F(z)/F(1)$, and then replacing F by $z \mapsto 1/F(z)$ if necessary, we may now assume that F is a strictly increasing homeomorphism of $(0,+\infty)$ onto itself with $F(1) = 1$ and $F(I) = I$.

If $x > 1$, then the Apollonian discs for 1 and x, respectively, have boundaries given by

$$\frac{|z-1|}{|z-x|} = \frac{|0-1|}{|0-x|} = \frac{1}{x} , \qquad \frac{|z-1|}{|z-x|} = 1 .$$

The map $z \mapsto (z-1)/(z-x)$ maps these to the circles $|z| = 1/x$ and $|z| = 1$, respectively, and we deduce that $\alpha_D(1,x) = \log x$. As exactly the same argument holds for the domain $F(D)$, and as $F(x) > 1$ when $x > 1$, we have, for $x > 1$,

$$\log x = \alpha_D(1,x) = \alpha_{F(D)}\big(1, F(x)\big) = \log F(x) ,$$

so that $F(x) = x$ whenever $x > 1$. This implies that F is the identity map; thus the original isometry f is Möbius and the the proof of Theorem 1.3 is complete.

8 A covering theorem for meromorphic functions

We begin with a simply connected domain D on the Riemann sphere containing 0 and ∞. Then there are positive numbers r and R, with $r \leqslant R$, such that $\{z : |z| < r\}$ and $\{z : |z| > R\} \cup \{\infty\}$ are the largest discs centred at 0 and ∞, respectively, that lie in D. By definition,

$$\alpha_D(0,\infty) = \log R/r \ .$$

Now let $g(z) = z/\sqrt{Rr}$, and write $g(D) = W$. Then W is a simply connected domain on the Riemann sphere such that that $\{z : |z| < \sqrt{r/R}\}$ and $\{z : |z| > \sqrt{R/r}\} \cup \{\infty\}$ are the largest discs centred at 0 and ∞, respectively, that lie in W. The stereographic projection of $\mathbb{C} \cup \{\infty\}$ onto the unit sphere in $\mathbb{R}^3$ is given by

$$\Phi(z) = \left(\frac{2x}{|z|^2+1}, \frac{2y}{|z|^2+1}, \frac{|z|^2-1}{|z|^2+1}\right) , \qquad z = x + iy$$

and this shows that the discs $\{z : |z| < \sqrt{r/R}\}$ and $\{z : |z| > \sqrt{R/r}\} \cup \{\infty\}$ both have spherical radius θ, where

$$\cos\theta = \frac{R-r}{R+r} ,$$

or, equivalently,

$$\tan\theta \sinh\left[\tfrac{1}{2}\alpha_D(0,\infty)\right] = 1 \ .$$

As $r \geqslant \sqrt{r/R}$ or $R \leqslant \sqrt{R/r}$, we conclude that D itself contains a disc centred either at 0 or at ∞ of spherical radius at least θ, and we have proved the following theorem.

Theorem 8.1 *Suppose that D is a simply connected domain on the Riemann sphere containing 0 and ∞. Then D contains a disc centred at either 0 or ∞ of spherical radius θ, where*

$$\tan\theta \sinh\left[\tfrac{1}{2}\alpha_D(0,\infty)\right] = 1 \ . \tag{8.1}$$

If we apply this to univalent meromorphic functions we immediately obtain the following covering theorem.

Theorem 8.2 *Suppose that f is univalent in the unit disc $\mathbb{D}$, and that $f(z_1) = 0$ and $f(z_2) = \infty$. Then $f(\mathbb{D})$ contains a disc on the Riemann sphere, centred either at 0 or at ∞, of spherical radius θ, where*

$$\tan\theta \sinh\rho_{\mathbb{D}}(z_1, z_2) \geqslant 1 \ .$$

Proof. We take D in Theorem 8.1 to be $f(\mathbb{D})$, so that $f(\mathbb{D})$ contains a disc at 0 or ∞ of spherical radius θ, where

$$\tan\theta \sinh\left[\tfrac{1}{2}\alpha_D(0,\infty)\right] = 1 \ .$$

As Theorem 1.2 implies that

$$2\rho_{\mathbb{D}}(z_1, z_2) = 2\rho_D(fz_1, fz_2) = 2\rho_D(0, \infty) \geqslant \alpha_D(0, \infty) ,$$

the result follows.

In fact, the lower bound for θ in Theorem 8.2 is best possible and is attained by the conformal map

$$f(z) = \phi(w)\left(\frac{\phi(z)^2 - 1}{\phi(z)^2 - \phi(w)^2}\right) , \quad \phi(t) = \frac{1+t}{1-t} , \qquad 0 < w < 1 ,$$

of $\mathbb{D}$ onto the sphere slit along the real segment from $1/\sqrt{\zeta}$ to $\sqrt{\zeta}$, where

$$\zeta = \left(\frac{1+w}{1-w}\right)^2 .$$

We leave the calculations to the reader.

Theorem 8.2 gives quantitative information about any conformal map of $\mathbb{D}$ onto a simply connected domain D. It is, in fact, closely related to Koebe's 1/4-Theorem that if $f : \mathbb{D} \to \mathbb{C}$ is univalent, and if $f(0) = 0$ and $f'(0) = 1$, then $f(\mathbb{D})$ contains the disc $\{z : |z| < 1/4\}$. Intuitively, we may regard the normalisation $f(0) = 0$ and $f'(0) = 1$ in Koebe's Theorem as $f(0) = 0$ and $f(w) = w$ for some infinitesimal w, and Theorem 8.2 answers the question (with a different normalisation) of what can be said if we no longer require w to be infinitesimal. T.K. Carne has shown that Theorem 1.2 (on which Theorem 8.2 depends) is equivalent to Koebe's 1/4-Theorem and his argument is given in the next section. Further, W.K. Hayman has pointed out that the conclusion of Theorem 1.2 may be derived from Theorem 7.1, [6], page 145, and it also follows from the covering result of Kirwan and Schober, [9], that if f is meromorphic and univalent on $\mathbb{D}$, and if $f(0) = 0$, $f'(0) = 1$ and $f(w) = \infty$, then $f(\mathbb{D})$ contains the two discs

$$\left\{z : |z| < \frac{|w|}{(1+|w|)^2}\right\} , \quad \left\{z : |z| > \frac{|w|}{(1-|w|)^2}\right\} .$$

9 Koebe's 1/4-Theorem

The second proof of Theorem 1.2 shows that the inequality $2\rho_D \geqslant \alpha_D$ follows from the inequality $\lambda_D \geqslant [2\delta_D]^{-1}$ which itself follows from Koebe's 1/4-Theorem. In this section we give an argument of T.K. Carne which shows that, conversely, Theorem 1.2 contains Koebe's Theorem.

Take any univalent analytic map $f : \mathbb{D} \to \mathbb{C}$ with $f(0) = 0$ and $f'(0) = 1$. Now choose any ζ not in $f(\mathbb{D})$. Then $f(\mathbb{D}) \subset \mathbb{C} - \{\zeta\}$ so that (from

Lemma 2.2 and Theorem 1.2), for any w in $\mathbb{D}$,

$$2\rho_{\mathbb{D}}(0,w) = 2\rho_{f(\mathbb{D})}(f(0),f(w)) \geqslant \alpha_{f(\mathbb{D})}(f(0),f(w)) \geqslant \left|\log \frac{|\zeta|}{|\zeta - f(w)|}\right| .$$

Now choose w so that $f(w)$ lies in the segment $[0,\zeta]$; then

$$2\rho_{\mathbb{D}}(0,w) \geqslant \log \frac{|\zeta|}{|\zeta| - |f(w)|}$$

and hence

$$1 - \frac{|f(w)|}{|\zeta|} \geqslant \left(\frac{1-|w|}{1+|w|}\right)^2 .$$

This yields

$$\frac{1}{|\zeta|} = \frac{1}{|\zeta|}\lim_{w\to 0}\frac{|f(w)|}{|w|} = \lim_{w\to 0}\frac{|f(w)|}{|w|.|\zeta|} \leqslant 4$$

as required.

An alternative proof of this part (suggested by the referee) is that (exactly as in the proof of Theorem 6.2) if $2\rho_D \geqslant \alpha_D$, then $\lambda_D \geqslant [2\delta_D]^{-1}$. Thus Theorem 1.2 implies that $\lambda_D \geqslant [2\delta_D]^{-1}$, and this, in turn, implies Koebe's Theorem.

REFERENCES

[1] Ahlfors, L.V., *Conformal invariants*, McGraw-Hill, 1973.

[2] Beardon, A.F., *The geometry of discrete groups*, GTM Vol.91, Springer Verlag, 1983.

[3] Beardon, A.F. and Pommerenke, Ch., On the Poincaré metric of plane domains, *J. London Math. Soc.* (2), **18** (1978), 475–483.

[4] de la Harpe, P., Hilbert's metric for simplices, *Geometric Group Theory*, (G.A.Niblo and M.A.Roller, eds.), London Math. Soc. Lecture Notes, Vol. 182, Cambridge Univ. Press, 1993, 97–119.

[5] Gehring, F.W. & Palka, B.P., Quasiconformally homogeneous domains, *Journal d'Analyse Math.*, **30** (1976), 172–199.

[6] Hayman, W.K., *Multivalent functions*, 2nd edition, Cambridge Univ. Press, 1994.

[7] Hilbert, D., Ueber die gerade Linie als kürzeste Verbindung zweier Punkte, *Math. Ann.* **46** (1895), 91–98.

[8] Keogh, F.R., A characterisation of convex domains in the plane, *Bull. London Math. Soc.*, **8** (1976), 183–185; see also *Math. Review*, **57** (1979), # 6392.

[9] Kirwan, W.E. and Schober, G., Extremal problems for meromorphic univalent functions, *J. Analyse Math.*, **30** (1976), 330–348.

[10] Köhn, J., Die Harnacksche Metrik in der Theorie der harmonischen Funktionen, *Math. Zeit.*, **91** (1966), 50–64.

[11] Leutwiler, H., On a distance invariant under Möbius transformations in $\mathbb{R}^n$, *Ann. Acad. Sci. Fenn.*, **12** (1987), 3–17.

[12] Vuorinen, M., *Conformal geometry and quasiregular mapping*, Spring-er Lecture Notes 1319, 1988.
[13] Weitsman, A., Symmetrization and the Poincaré metric, *Annals of Math.*, **124** (1986), 159–169.

Department of Pure Mathematics and Mathematical Statistics
University of Cambridge
16 Mill Lane, Cambridge CB2 1SB
England
A.F.Beardon@pmms.cam.ac.uk

Conjugacy Invariants of Möbius Groups

C. Cao and P. L. Waterman

1 Introduction

We first consider the basic geometric relationships between orthogonal and affine transformations and the Clifford algebra. Using these we show how to find the fixed points of elliptic and parabolic Möbius transformations. Our results also give the fixed points of a general Möbius transformation of $\overline{\mathbb{R}}^3$ using quaternions. Conjugacy invariants such as the cross ratio are then introduced and utilized to obtain invariant forms of higher dimensional Jørgensen-type inequalities. As a consequence of these we obtain tubes around short geodesics and give a lower bound for the radius of inscribed balls in hyperbolic n-space.

2 The Clifford algebra

In [A1] Ahlfors shows how a 2×2 matrix with entries in a Clifford algebra may be used to describe a Möbius transformation of $\overline{\mathbb{R}}^n$. We first briefly review some material on Clifford numbers that is treated in detail in [A1], [A2] and [W2].

The Clifford algebra C_n is the associative algebra over the reals generated by elements $i_1, i_2, \ldots, i_n$ subject to the relations $i_k^2 = -1$ and $i_h i_k = -i_k i_h$, $h \neq k$, and no others. An element of C_n is called a Clifford number. Thus $C_0 = \mathbb{R}$, $C_1 = \mathbb{C}$ and $C_2 = \mathbb{H}$, the quaternions with $i_1 = i$ and $i_2 = j$. Every Clifford number a can be expressed uniquely in the form $a = \sum a_I I$ where $a_I \in \mathbb{R}$ and the sum is over all products $I^m = i_{\nu_1} i_{\nu_2} \ldots i_{\nu_m}$ with $1 \leq \nu_1 < \ldots < \nu_m \leq n$. The null product is included and identified with the real number $i_0 = 1$. The coefficient of the null product is denoted by $a_{\mathbb{R}}$ or $\mathrm{Re}(a)$ and called the real part of a. The sum of all other terms of a is referred to as the imaginary part and we write $a = a_{\mathbb{R}} + a_C = \mathrm{Re}(a) + \mathrm{Im}(a)$.

[1]Research supported in part by grants from the National Science Foundation.
[2]1991 *Mathematics Subject Classification.* 30F40

The Euclidean norm of a Clifford number is given by $|a|^2 = \sum a_I{}^2 = \mathrm{Re}(a)^2 + |\mathrm{Im}(a)|^2$ and the dot product by $a \cdot b = \mathrm{Re}(a\bar{b}) = \mathrm{Re}(\bar{a}b)$.

There are three involutions of $\mathcal{C}_n$:

(1) The major involution $a \to a'$ replaces each i_k by $-i_k$. It determines an automorphism of $\mathcal{C}_n$: $(ab)' = a'b'$, $(a+b)' = a' + b'$.

(2) Reversion: $a \to a^*$ replaces each $I = i_{\nu_1} \dots i_{\nu_m}$ with $I = i_{\nu_m} \dots i_{\nu_1}$. It defines an anti-automorphism: $(ab)^* = b^*a^*$, $(a+b)^* = a^* + b^*$.

(3) Conjugation: $a \to \bar{a}$ is a composition: $\bar{a} = a'^* = a^{*\prime}$, which is again an anti-automorphism.

We identify $\mathbb{R}^{n+1}$ with the subspace $\mathcal{V}$ of $\mathcal{C}_n$ spanned by $1, i_1, i_2, \dots, i_n$. Clifford numbers of the form $x = x_0 + x_1 i_1 + \cdots + x_n i_n$ are called vectors. If $a = \sum a_I I$, we let $\mathcal{V}(a) = a_0 + a_1 i_1 + \cdots + a_n i_n$ and call it the vector part of a. Every non-zero vector x is invertible with $x^{-1} = \bar{x}|x|^{-2}$ so products of non-zero vectors form a multiplicative group known as the Clifford group Γ_n.

A Clifford matrix of dimension n is a matrix $A = \begin{pmatrix} a & b \\ c & d \end{pmatrix}$ which satisfies the conditions:

(1) $a, b, c, d \in \Gamma_n \cup \{0\}$,

(2) $ad^* - bc^* = 1$,

(3) $ab^*, cd^*, c^*a, d^*b \in \mathbb{R}^{n+1}$.

The set of all Clifford matrices is denoted by $\mathrm{SL}(2, \mathcal{C}_n)$. Any A in $\mathrm{SL}(2, \mathcal{C}_n)$ induces a mapping f in $\mathrm{M}(n+1)$, the group of orientation preserving Möbius transformations of $\overline{\mathbb{R}}^{n+1}$, by the formula $A \to f_A = f$ where

$$f_A = (ax+b)(cx+d)^{-1}, \qquad x \in \overline{\mathbb{R}}^{n+1}.$$

Vahlen's Theorem states that $\mathrm{SL}(2, \mathcal{C}_n)$ is a group whose quotient modulo $\pm\begin{pmatrix} 1 & 0 \\ 0 & 1 \end{pmatrix}$ is isomorphic to $\mathrm{M}(n+1)$. See [V], [A1] and [W2]. A Clifford matrix in dimension n is also a Clifford matrix in dimension $n+1$. It automatically extends the corresponding transformation in $\mathrm{M}(n+1)$ to one in $\mathrm{M}(n+2)$, given by the same matrix.

A subgroup G of $\mathrm{GM}(n)$, the group of general (conformal and anti-conformal) Möbius transformation of $\overline{\mathbb{R}}^n$, is said to be elementary if there exists a finite G-orbit in $\overline{\mathbb{R}}^{n+1}$.

Elements of $\mathrm{GM}(n)$ may be classified as follows:

$f \in \mathrm{GM}(n)$ is elliptic if it has a fixed point in H^{n+1}. Such maps are $\mathrm{GM}(n+1)$ conjugate to $x \to Tx$ with $T \in \mathrm{O}(n+1)$.

$f \in \mathrm{GM}(n)$ is parabolic if it has exactly one fixed point, necessarily in $\overline{\mathbb{R}}^n$. Such maps are $\mathrm{GM}(n)$ conjugate to $x \to Tx + a$ with $T \in \mathrm{O}(n), a \in \mathbb{R}^n \backslash \{0\}$. A parabolic which is conjugate to $x \to x + a$ is called strictly parabolic.

$f \in \mathrm{GM}(n)$ is loxodromic if it has exactly two fixed points, both in $\overline{\mathbb{R}}^n$. Such maps are $\mathrm{GM}(n)$ conjugate to $x \to rTx$ with $r > 0, r \neq 1$ and $T \in \mathrm{O}(n)$. If $T = I$, then f is called hyperbolic.

Remark 2.1. The above classification is not completely standard; some authors use a dimension dependent definition of elliptic.

3 Orthogonal transformations

We describe the basic geometry of Clifford multiplication and its relationship to orthogonal transformations, elaborating on results in [A1] and [W2].

Definition 3.1. If $a \in \Gamma_n$ is a product of $2m$ vectors and may not be written as a product of $2m-2$ vectors we write $\mathcal{D}^+(a) = 2m$. If a is real we write $\mathcal{D}^+(a) = 0$.

Lemma 3.2. *If $a \in \Gamma_n$ then $\mathcal{D}^+(\lambda a \lambda^{-1}) = \mathcal{D}^+(a)$ for any $\lambda \in \Gamma_n$.*

Proof.

$$\lambda u_1 u_2 \dots u_{2m} \lambda^{-1} = \lambda u_1 \lambda^* \lambda^{*-1} u_2 \lambda^{-1} \dots \lambda u_{2m-1} \lambda^* \lambda^{*-1} u_{2m} \lambda^{-1}. \qquad \square$$

Remark 3.3. It is important to note that if v is a vector then $\lambda v \lambda^{-1}$ will usually be a product of two vectors and not a vector.

Corollary 3.4. *If $\lambda \in \Gamma_n, \alpha_k \in \mathbb{R}$ and $I_k = I_k^{2m}$ or I_k^{2m-1} then*

$$\lambda \Sigma \alpha_k I_k \lambda^{-1} = \Sigma \beta_k I_k$$

for some $\beta_k \in \mathbb{R}$.

Proof. Follows from Lemma 3.2 on noting that I_k is a product of vectors and $|\lambda a| = |a\lambda| = |\lambda||a|$ for any Clifford number a. $\square$

Definition 3.5. If $a \in \Gamma_n$ is a product of $2m+1$ vectors and may not be written as a product of $2m-1$ vectors we write $\mathcal{D}^-(a) = 2m+1$.

Lemma 3.6. *If $a \in \Gamma_n$ then $\mathcal{D}^-(\lambda a \lambda^*) = \mathcal{D}^-(a)$ for any $\lambda \in \Gamma_n$.*

Proof.

$$\lambda u_1 u_2 \dots u_{2n} u_{2n+1} \lambda^* = \lambda u_1 \lambda^* . \lambda^{*-1} u_2 \lambda^{-1} \dots \lambda u_{2n+1} \lambda^*. \qquad \square$$

Corollary 3.7. *If $\lambda \in \Gamma_n, \alpha_k \in \mathbb{R}$ and $I_k = I_k^{2m}$ or $I_k = I_k^{2m+1}$ then*

$$\lambda \Sigma \alpha_k I_k \lambda^* = \Sigma \beta_k I_k$$

for some $\beta_k \in \mathbb{R}$.

Proof. Similar to Corollary 3.4. □

Lemma 3.8. *If $\{\xi_1, \eta_1, \ldots, \xi_m, \eta_m\}$ are pairwise orthogonal unit vectors, then*

$$\xi_1 \bar{\eta}_1 \cdots \xi_m \bar{\eta}_m = \Sigma \alpha_k I_k \quad \textit{with } I_k = I_k^{2m} \textit{ or } I_k = I_k^{2m-1}.$$

Proof. There exists $\lambda \in \Gamma_n$ such that $\lambda \xi_r \lambda^* = \pm i_{2r-1}$ and $\lambda \eta_r \lambda^* = \pm i_{2r}$ (change of basis via an orthogonal transformation).

Thus $\lambda \xi_1 \lambda^* \overline{\lambda \eta_1 \lambda^*} \cdots \overline{\lambda \eta_m \lambda^*} = \pm i_1 \cdots i_{2m}$ and the result follows from Corollary 3.4. □

A variant on the above gives a simple way of computing Euclidean volume:

Lemma 3.9. *If $v_1, \ldots, v_m \in \mathbb{R}^n$ and*

$$\begin{aligned} p &= v_1 \bar{v}_2 v_3 \bar{v}_4 \ldots \\ &= p_0 + p_1 + \ldots p_m \quad \textit{where } p_k = \Sigma \alpha_j I_j^k \end{aligned}$$

then the volume spanned by $v_1, \ldots, v_m$ is $|p_{m-1} + p_m|$.

Proof. The result follows by observing that $|p_{m-1} + p_m|$ is invariant under orthogonal transformations and that $p_{m-1} + p_m$ is invariant when $\beta_1 v_1 + \ldots \beta_{k-1} v_{k-1} + \beta_{k+1} v_{k+1} + \ldots \beta_m v_m$ is added to v_k. □

Lemma 3.10. *If $\xi_1, \eta_1, \ldots, \xi_m, \eta_m$ are pairwise orthogonal unit vectors then $\mathcal{V}(\xi_1 \bar{\eta}_1 \cdots \xi_m \bar{\eta}_m I)$ is orthogonal to $\xi_1, \eta_1, \ldots, \xi_m, \eta_m$ where $I = I^{2m}$ or $I = I^{2m+1}$.*

Proof.

$$\begin{aligned} \xi_1 \cdot v &= \mathrm{Re}[\bar{\xi}_1 v] \\ &= \mathrm{Re}[\bar{\xi}_1 (\xi_1 \bar{\eta}_1 \cdots \xi_m \bar{\eta}_m I)] \\ &= \mathrm{Re}[\bar{\eta}_1 \cdots \xi_m \bar{\eta}_m I] \\ &= 0 \end{aligned}$$

The result follows by the 'commutativity' of the ξ_r, η_r. □

Lemma 3.11. *Any $\lambda \in \Gamma_n$ with $|\lambda| = 1$ has a decomposition $\lambda = \lambda_1 \cdots \lambda_m$ with $\lambda_r = \cos(\theta_r) + \sin(\theta_r) \eta_r \bar{\xi}_r$ such that the λ_r commute and the ξ_r, η_r are* *pairwise orthogonal unit vectors. The map $w \mapsto \lambda w \lambda^*$ fixes the orthogonal complement of the span of $\{\xi_1, \ldots, \xi_r, \eta_1, \ldots, \eta_r\}$ operating as a rotation through 2θ on the plane spanned by ξ_r, η_r.*

Proof. Follows by considering the Jordan normal form. □

Definition 3.12. $i_k \notin a$ if $a = \sum a_I I$ and i_k does not appear in any basis element I.

Lemma 3.13. *Suppose that $\lambda \in \Gamma_{2m+2k}$, $|\lambda| = 1$ and $\mathcal{D}^+(\lambda) = 2m$ so that $\lambda = \ldots + \alpha I + \ldots$ with $\alpha \neq 0$ for some $I = I^{2m}$ or I^{2m-1}, then a basis for the fixed point set of $w \mapsto \lambda w \lambda^*$ (from $\mathbb{R}^{2m+2k+1} \to \mathbb{R}^{2m+2k+1}$) is given by the $2k+1$ elements:*

(1) $\{v_m = \mathcal{V}(\lambda I i_{n_m}) : i_{n_m} \notin I\}$ *if* $I = I^{2m-1}$.
(2) $\{v_0 = \mathcal{V}(\lambda I)\} \cup \{v_m = \mathcal{V}(\lambda I i_{n_m}) : i_{n_m} \notin I\}$ *if* $I = I^{2m}$.

Proof. Follows from Lemma 3.10 and Lemma 3.11. □

In the full rank case the above simplifies to:

Lemma 3.13*. *If $\lambda \in \Gamma_{2n}$, $|\lambda| = 1$ and $\mathcal{D}^+(\lambda) = 2n$, then $w \mapsto \lambda w \lambda^*$ (from $\mathbb{R}^{2n+1} \to \mathbb{R}^{2n+1}$) fixes η with:*

$$\lambda = \cos(\theta_1)\cos(\theta_2)\ldots\cos(\theta_n) + \ldots + \sin(\theta_1)\sin(\theta_2)\ldots\sin(\theta_n)\eta\, i_1 i_2 \ldots i_{2n}.$$

In the case of rotations of $\mathbb{R}^3$ the above further simplifies to:

Lemma 3.13.** *If $\lambda = \cos(\theta) + \sin(\theta)\xi k \in \mathbb{H}$ with $\xi \in \mathbb{R}^3$ and $|\xi| = 1$, then $w \mapsto \lambda w \lambda^*$ fixes ξ (and hence any real multiple thereof).*

4 Euclidean Similarities

We now consider the problem of classifying $f = \begin{pmatrix} \lambda & \mu \\ 0 & \lambda^{*-1} \end{pmatrix} \in \mathrm{SL}(2, C_n)$. Clearly, f is loxodromic iff $|\lambda| \neq 1$. The problem is thus to recognize whether f is elliptic or parabolic when $|\lambda| = 1$ without explicitly checking fixed points. See also [Wa].

We recall from Greenberg [Gr]:

Lemma 4.1. *Any Euclidean isometry f has a decomposition:*

$$\begin{aligned} f &= \begin{pmatrix} 1 & \eta \\ 0 & 1 \end{pmatrix}\begin{pmatrix} \lambda & 0 \\ 0 & \lambda' \end{pmatrix}\begin{pmatrix} 1 & \xi \\ 0 & 1 \end{pmatrix}\begin{pmatrix} 1 & -\eta \\ 0 & 1 \end{pmatrix} \\ &= \begin{pmatrix} 1 & \eta \\ 0 & 1 \end{pmatrix}\begin{pmatrix} 1 & \xi \\ 0 & 1 \end{pmatrix}\begin{pmatrix} \lambda & 0 \\ 0 & \lambda' \end{pmatrix}\begin{pmatrix} 1 & -\eta \\ 0 & 1 \end{pmatrix} \\ &= \begin{pmatrix} \lambda & \lambda\xi - \lambda\eta + \eta\lambda' \\ 0 & \lambda' \end{pmatrix} \end{aligned}$$

with $\lambda\xi\lambda^* = \xi$ *and* $\eta \cdot \xi = 0$. *Further,* f *is elliptic iff* $\xi = 0$.

Lemma 4.2. *If* $\lambda \in \Gamma_n, \lambda \notin \mathbb{R}$ *and* $\eta, \xi \in \mathcal{V}$ *with* η, ξ *not both zero,* $\eta \cdot \xi = 0$, $\lambda\xi\lambda^* = \xi$ *and* $\mathcal{D}^+(\lambda) = 2m$ *then:*

$$\mathcal{D}^-(\lambda\xi - \lambda\eta + \eta\lambda') = \begin{cases} 2m+1 & \xi \neq 0 \\ 2m-1 & \xi = 0. \end{cases}$$

Proof. By replacing $\lambda\xi - \lambda\eta + \eta\lambda'$ with $\alpha(\lambda\xi - \lambda\eta + \eta\lambda')\alpha^*$ we may assume that:

$$\lambda = \lambda' \ , \ \xi \in \mathbb{R} \text{ and either } \eta = i_1 \text{ or } \eta = 0.$$

If $\eta = 0$ the result is immediate, so assume $\eta = i_1$ and let $\lambda = \lambda_0 + \lambda_1 i_1$ with $i_1 \notin \lambda_0$ and $i_1 \notin \lambda_1$.

We then have:

$$\begin{aligned} \lambda\xi - \lambda\eta + \eta\lambda' &= \lambda\xi - (\lambda_0 + \lambda_1 i_1)i_1 + i_1(\lambda_0 + \lambda_1 i_1)' \\ &= \lambda\xi + 2\lambda_1 \end{aligned}$$

and the result follows. □

Theorem 4.3. $\begin{pmatrix} \lambda & \mu \\ 0 & \lambda' \end{pmatrix} \in \mathrm{SL}(2, \mathcal{C}_n)$ *with* $\mathcal{D}^+(\lambda) = 2m$ *and* $\mu \neq 0$ *is:*

pure parabolic iff $\lambda \in \mathbb{R}$;
parabolic iff $\mathcal{D}^-(\mu) = 2m+1$;
elliptic iff $\mathcal{D}^-(\mu) = 2m-1$.

Proof. Follows directly from Lemma 4.1 and Lemma 4.2. □

For Euclidean isometries in three dimensions, the above simplifies to:

Theorem 4.3*. $\begin{pmatrix} \lambda & \mu \\ 0 & \lambda' \end{pmatrix} \in \mathrm{SL}(2, \mathbb{H})$ *is:*

pure parabolic if $\lambda \in \mathbb{R}$;
screw parabolic if $\mu \notin \mathbb{R}^3$;
elliptic otherwise.

5 Fixed points of Möbius Transformations.

In [A2] Ahlfors describes how to find the fixed points of an elliptic or parabolic element. A variant on his method enables us to find the fixed points of a general element of $\mathrm{SL}(2, \mathbb{H})$.

The following lemma is useful in obtaining normal forms:

Lemma 5.1. $(|\nu\eta|+\nu\bar{\eta})(|\nu|\eta)(|\nu\eta|+\bar{\nu}\eta)^{-1}=|\eta|\nu$ *provided* $|\nu\eta|+\nu\bar{\eta}\neq 0$.

Proof. Direct computation. □

Lemma 5.2. *If* $T\in \mathrm{SL}(2,C_n)$ *then it is affine conjugate to some matrix of the form:*

$$\begin{pmatrix} \lambda & -\lambda^{*-1} \\ \lambda & 0 \end{pmatrix}, \qquad T^2\infty \neq \infty,$$

$$\begin{pmatrix} 0 & -\lambda^{*-1} \\ \lambda & 0 \end{pmatrix}, \qquad T^2\infty = \infty,\ T\infty\neq\infty.$$

Proof. Suppose

$$T=\begin{pmatrix} a & b \\ c & d \end{pmatrix},\ S=\begin{pmatrix} \dfrac{|\nu|+\bar{\nu}}{|\nu|} & \dfrac{|\nu|+\bar{\nu}}{|\nu|}c^{-1}d \\ 0 & |\nu|+\nu \end{pmatrix},\ \text{where } \nu=ac^{-1}+c^{-1}d.$$

Then

$$STS^{-1}=\begin{pmatrix} \lambda & -\lambda^{*-1} \\ \lambda & 0 \end{pmatrix},\ \text{where } \lambda=|\nu|(|\nu|+\nu)\,c\,(|\nu|+\bar{\nu})^{-1}.$$

The above holds unless:

(1) $c=0$ in which case T fixes infinity.

(2) $\nu=0$ in which case T^2 fixes infinity.

(3) $\nu\in\mathbb{R}^-$ in which case putting $S=\begin{pmatrix} 1 & 0 \\ 0 & \nu \end{pmatrix}$ gives STS^{-1} of the required form, with $\lambda=\nu c$.

If $T^2\infty=\infty$, $T\infty\neq\infty$ then:

$$RTR^{-1}=\begin{pmatrix} 0 & -c^{*-1} \\ c & 0 \end{pmatrix},\ \text{where } R=\begin{pmatrix} 1 & -ac^{-1} \\ 0 & 1 \end{pmatrix}.$$ □

We now consider two notions of trace:

Definition 5.3.

$$\mathrm{tr}\left(\begin{pmatrix} a & b \\ c & d \end{pmatrix}\right)=a+c^{-1}dc\,,\qquad \hat{\mathrm{tr}}\left(\begin{pmatrix} a & b \\ c & d \end{pmatrix}\right)=a+d^*.$$

Remark 5.4. We recall from [W2, Lemma 15] that $\mathrm{Re}(\mathrm{tr})=\mathrm{Re}(\hat{\mathrm{tr}})$ is a conjugacy invariant. Note also that if $c=c^*\neq 0$ then $\mathrm{tr}=\hat{\mathrm{tr}}$.

Theorem 5.5. *If* $T = \begin{pmatrix} a & b \\ c & d \end{pmatrix} \in \mathrm{SL}(2, \mathcal{C}_n)$ *is conjugate to an element of* $\mathrm{SL}(2, \mathbb{R})$ *then* $b, c \in \mathbb{R}^{n+1}$ *and* $\mathrm{tr}(T) = \hat{\mathrm{tr}}(T) \in \mathbb{R}$.

Conversely, if $T = \begin{pmatrix} a & b \\ c & d \end{pmatrix} \in \mathrm{SL}(2, \mathcal{C}_n)$ *with* $c = c^* \neq 0$ *and* $\mathrm{tr}(T) \in \mathbb{R}\backslash\{0\}$, *then T is conjugate to an element of* $\mathrm{SL}(2, \mathbb{R})$. *If* $\mathrm{tr}(T) = 0$, *then T is conjugate to an element of* $\mathrm{SL}(2, \mathbb{R})$ *provided* $c \in \mathbb{R}^{n+1}$.

Proof. Any non-trivial element of $\mathrm{SL}(2, \mathbb{R})$ is $\mathrm{GL}(2, \mathbb{R})$ conjugate to one of the form $\begin{pmatrix} \tau & -1 \\ 1 & 0 \end{pmatrix}$.

Thus:

$$\begin{aligned} T &= \begin{pmatrix} \alpha & \beta \\ \gamma & \delta \end{pmatrix} \begin{pmatrix} \tau & -1 \\ 1 & 0 \end{pmatrix} \begin{pmatrix} \delta^* & -\beta^* \\ -\gamma^* & \alpha^* \end{pmatrix} \\ &= \begin{pmatrix} \tau\alpha\delta^* + \alpha\gamma^* + \beta\delta^* & -\alpha\alpha^* - \tau\alpha\beta^* - \beta\beta^* \\ \tau\gamma\delta^* + \gamma\gamma^* + \delta\delta^* & -\gamma\alpha^* - \tau\gamma\beta^* - \delta\beta^* \end{pmatrix} \end{aligned}$$

and the first part follows.

For the converse, conjugate T by a translation S so that $STS^{-1} = \begin{pmatrix} \tau & -c^{*-1} \\ c & 0 \end{pmatrix}$. Either $\tau \in \mathbb{R}\backslash\{0\}$ and hence $c \in \mathbb{R}^{n+1}$ or we assume that $c \in \mathbb{R}^{n+1}$. We may thus conjugate by some $R = \begin{pmatrix} \lambda & 0 \\ 0 & \lambda^{*-1} \end{pmatrix}$ so that $RSTS^{-1}R^{-1} = \begin{pmatrix} \tau & -1 \\ 1 & 0 \end{pmatrix}$. □

For an elliptic or parabolic transformation g, not fixing infinity, it is well known that the fixed point(s) lie on the intersection of the isometric spheres of g and its inverse. The following lemma is an algebraic version.

Lemma 5.6. *If* $T = \begin{pmatrix} \lambda & -\lambda^{*-1} \\ \lambda & 0 \end{pmatrix} \in \mathrm{SL}(2, \mathcal{C}_n)$ *is elliptic or parabolic then there exists* $\eta \in \mathbb{R}^{n+1}$ *such that* $\lambda\eta\lambda'^{-1} = \eta$ *and* $\mathrm{Re}(\eta) = \frac{1}{2}$.

Proof. Any elliptic or parabolic fixing $\eta \in \mathbb{R}^{n+2}$ is of the form:

$$\begin{aligned} T &= \begin{pmatrix} 1 & \xi \\ 0 & 1 \end{pmatrix} \begin{pmatrix} \alpha & 0 \\ \gamma & \alpha' \end{pmatrix} \begin{pmatrix} 1 & -\xi \\ 0 & 1 \end{pmatrix} \\ &= \begin{pmatrix} \alpha + \xi\gamma & -\alpha\xi + \xi(\alpha' - \gamma\xi) \\ \gamma & \alpha' - \gamma\xi \end{pmatrix} \end{aligned}$$

With the given normalization:

$$\alpha' = \gamma\xi \text{ and } \gamma = \alpha + \xi\gamma = \lambda.$$

It follows that $\lambda\xi\lambda'^{-1} = \xi$ and $\mathrm{Re}(\xi) = \frac{1}{2}$. If $\xi = \eta + t i_{n+1}$ with $t \in \mathbb{R}$ then noting that $\lambda i_{n+1}\lambda'^{-1} = i_{n+1}$ the desired η is obtained. □

Remark 5.7. Utilizing the map $T = \begin{pmatrix} 1 & -i_n \\ -i_n & 1 \end{pmatrix} : \mathbb{H}^{n+1} \to \mathbb{B}^{n+1}$, it follows that the orientation preserving Möbius transformations of $\mathbb{B}^n$ consist of those elements of $\mathrm{SL}(2, C_n)$ of the form: $\begin{pmatrix} a & b \\ b' & a' \end{pmatrix}$.

Lemma 5.8. *If* $T = \begin{pmatrix} \lambda & -\lambda^{*-1} \\ \lambda & 0 \end{pmatrix} \in \mathrm{SL}(2, C_n)$ *and there exists* $\eta \in \mathbb{R}^{n+1}$ *such that* $\lambda\eta\lambda'^{-1} = \eta$ *with* $\mathrm{Re}(\eta) = \frac{1}{2}$, *then either:*

(1) T *preserves the ball* $|w - \eta| \le r$ *with* $r^2 = |\eta|^2 - \dfrac{1}{|\lambda|^2}$.

(2) T *is elliptic fixing* $\eta + r i_n$ *with* $r^2 = \dfrac{1}{|\lambda|^2} - |\eta|^2$.

(3) $|\lambda\eta| = 1$ *and* T *is elliptic or parabolic fixing* η.

Proof. For η such that $\lambda\eta\lambda'^{-1} = \eta$ and $\mathrm{Re}(\eta) = \frac{1}{2}$:

$$\begin{pmatrix} 1 & -\eta \\ 0 & 1 \end{pmatrix}\begin{pmatrix} \lambda & -\lambda^{*-1} \\ \lambda & 0 \end{pmatrix}\begin{pmatrix} 1 & \eta \\ 0 & 1 \end{pmatrix} = \begin{pmatrix} \eta'\lambda & |\eta|^2(1 - 1/|\lambda|^2)\lambda' \\ \lambda & \eta\lambda' \end{pmatrix}$$

and the result follows. □

Lemma 5.9. *If* $T = \begin{pmatrix} \lambda & -\lambda^{*-1} \\ \lambda & 0 \end{pmatrix} \in \mathrm{SL}(2, C_n)$ *then either:*

(1) T *is elliptic and there exists* $\eta \in \mathbb{R}^{n+1}$ *such that:*
$\lambda\eta\lambda'^{-1} = \eta$ *with* $\mathrm{Re}(\eta) = \frac{1}{2}$ *and* $|\lambda\eta| < 1$.

(2) T *is parabolic and there exists a unique* $\eta \in \mathbb{R}^{n+1}$ *such that:*
$\lambda\eta\lambda'^{-1} = \eta$ *with* $\mathrm{Re}(\eta) = \frac{1}{2}$ *and* $|\lambda\eta| = 1$.

(3) T *is loxodromic otherwise.*

Proof. The result follows directly from Lemma 5.8. □

Lemma 5.10. *If* $T = \begin{pmatrix} \lambda & -\lambda^{*-1} \\ \lambda & 0 \end{pmatrix}$ *with* $\lambda \in \mathbb{H}$ *then either:*

(1) $\lambda \in \mathbb{R}^3$ *or*

(2) *There exists* η *such that* $\lambda\eta\lambda'^{-1} = \eta$ *with* $\mathrm{Re}(\eta) = \frac{1}{2}$ *and either:*

(a) T *preserves the ball* $|w - \eta| \le r$ *with* $r^2 = |\eta|^2 - \dfrac{1}{|\lambda|^2}$ *and for some* S, $STS^{-1} \in \mathrm{SL}(2, \mathbb{C})$.

(b) T *is elliptic fixing* $\eta + r i_n$ *with* $r^2 = \dfrac{1}{|\lambda|^2} - |\eta|^2$.

(c) $|\lambda\eta| = 1$ *and* T *is elliptic or parabolic fixing* η.

Proof. Follows from the above. □

Lemma 5.11. *If* $\eta = \eta_{\mathbb{R}} + \eta_{\mathcal{C}} \in \mathcal{V}$ *with* $|\eta| = 1$, *and* $w^2 = \eta$ *with* $w = w_{\mathbb{R}} + w_{\mathcal{C}} \in \mathcal{V}$ *then:*

$$w_{\mathbb{R}} = \pm\sqrt{\frac{1+\eta_{\mathbb{R}}}{2}}, \quad w_{\mathcal{C}} = \frac{\eta_{\mathcal{C}}}{2w_{\mathbb{R}}}, \quad \eta \neq -1.$$

If $\eta = -1$ *then any* w *with* $\mathrm{Re}(w) = 0$ *satisfies* $w^2 = \eta$.

Proof.

$$\begin{aligned} & w^2 = \eta \\ \text{iff} \quad & w_{\mathbb{R}}^2 + 2w_{\mathbb{R}}w_{\mathcal{C}} - |w_{\mathcal{C}}|^2 = \eta_{\mathbb{R}} + \eta_{\mathcal{C}} \\ \text{iff} \quad & w_{\mathbb{R}} = \pm\sqrt{\frac{1+\eta_{\mathbb{R}}}{2}}, \quad w_{\mathcal{C}} = \frac{\eta_{\mathcal{C}}}{2w_{\mathbb{R}}}, \quad \eta \neq -1. \end{aligned}$$ □

Lemma 5.12. *If* $T = \begin{pmatrix} \lambda & -\lambda^{*-1} \\ \lambda & 0 \end{pmatrix}$ *and* $\lambda\lambda^* = \lambda^*\lambda$, *then the fixed points satisfy:*

$$\eta = \frac{1}{2} + \sqrt{\frac{1}{4} - (\lambda\lambda^*)^{-1}}. \tag{5.1}$$

Proof. If w is a fixed point:

$$\lambda w - \lambda^{*-1} = w\lambda w.$$

Assuming that λ commutes with w:

$$w^2 - w + (\lambda\lambda^*)^{-1} = 0.$$

We now note that if $\lambda\lambda^* = \lambda^*\lambda$, then λ and η commute provided $\frac{1}{4}-(\lambda\lambda^*)^{-1}$ is not real and negative. When $\frac{1}{4}-(\lambda\lambda^*)^{-1}$ is real and negative T is elliptic preserving the real axis and the result follows from Lemma 5.6 and Lemma 5.8 □

Lemma 5.13.

$$T = \begin{pmatrix} a & b \\ -b' & a' \end{pmatrix} \in \mathrm{SU}(2,\mathbb{H}) \quad \textit{and} \quad \begin{pmatrix} \cos(\theta)e^{k\phi} & \sin(\theta)e^{k\phi} \\ -\sin(\theta)e^{k\phi} & \cos(\theta)e^{k\phi} \end{pmatrix}$$

are conjugate iff

$$\begin{aligned} \mathrm{Re}(a) &= \cos(\theta)\cos(\phi), \\ \mathrm{Re}(bk) &= -\sin(\theta)\sin(\phi). \end{aligned}$$

Proof. Follows from Lemma 3.13*. □

6 The Cross Ratio

Definition 6.1. For an ordered quadruple $\{z_1, z_2, z_3, z_4\}$ of points of $\mathbb{R}^n$, no three of which coincide, we define

$$[z_1, z_2, z_3, z_4] = (z_1 - z_3)(z_1 - z_2)^{-1}(z_2 - z_4)(z_3 - z_4)^{-1}$$

and call $[z_1, z_2, z_3, z_4]$ the cross ratio of z_1, z_2, z_3, z_4. We extend the definition by continuity to permit z_1, z_2 or z_4 to be infinite. If $z_3 = \infty$ then $\mathrm{Re}[\ ,\ ,\ ,\ ]$ and $|[\ ,\ ,\ ,\ ]|$ are well defined by continuity. It can be checked that in case two of the points coincide, the value of the cross ratio is independent of the points and is 0, 1, or ∞.

Lemma 6.2. *If* $f = \begin{pmatrix} a & b \\ c & d \end{pmatrix} \in \mathrm{SL}(2, \mathcal{C}_n)$, *then*

$$[fz_1, fz_2, fz_3, fz_4] = (cz_3 + d)^{*-1}[z_1, z_2, z_3, z_4](cz_3 + d)^*.$$

Proof. This follows from (see [A1], [W2])

$$fx - fy = (cy + d)^{*-1}(x - y)(cx + d)^{-1}. \qquad \square$$

Corollary 6.3.

$$\begin{aligned} \mathrm{Re}[fz_1, fz_2, fz_3, fz_4] &= \mathrm{Re}[z_1, z_2, z_3, z_4], \\ |[fz_1, fz_2, fz_3, fz_4]| &= |[z_1, z_2, z_3, z_4]|. \end{aligned}$$

Proof. For any x, $y \in \Gamma_n$, we have

$$\mathrm{Re}(xax^{-1}) = \mathrm{Re}(a), \quad |xy| = |x||y|. \qquad \square$$

Corollary 6.4. *Four points* z_1, z_2, z_3, z_4 *are on a circle if and only if* $[z_1, z_2, z_3, z_4]$ *is a real number.*

Proof. There exists a Möbius transformation $f \in \mathrm{M}(n)$ which maps z_1, z_2, z_3 to ∞, 0, 1. If z_1, z_2, z_3, z_4 are on a circle, then $[fz_1, fz_2, fz_3, fz_4]$ is real and hence $[z_1, z_2, z_3, z_4]$ real by Lemma 6.2. If $[z_1, z_2, z_3, z_4]$ is real, then $[fz_1, fz_2, fz_3, fz_4]$ is real and thus fz_4 real. So fz_1, fz_2, fz_3, fz_4 are on the real line and hence z_1, z_2, z_3, z_4 on a circle. $\square$

Lemma 6.5. *Given four distinct points* z_1, z_2, z_3, z_4 *in* $\overline{\mathbb{R}}^n$:

(1) $[z_1, z_2, z_3, z_4] + [z_2, z_1, z_3, z_4] = 1$,

(2) $[z_1, z_2, z_3, z_4][z_4, z_2, z_3, z_1] = 1$,

(3) $|[z_1, z_2, z_3, z_4]| = |[z_2, z_1, z_4, z_3]|$,

(4) $\|[z_1, z_2, z_3, z_4]\| = \|[z_3, z_4, z_1, z_2]\|$.

Proof. [1] By the definition of the cross ratio:

$$[x, \infty, 0, y] = xy^{-1}, \quad [\infty, x, 0, y] = (y - x)y^{-1}$$

for all distinct points x and y in $\mathbb{R}^n - \{0\}$. Thus,

$$[x, \infty, 0, y] + [\infty, x, 0, y] = 1.$$

We choose $f = \begin{pmatrix} a & b \\ c & d \end{pmatrix}$ so that $f\infty = z_2$, $f0 = z_3$. Then

$$\begin{aligned}
&[z_1, z_2, z_3, z_4] + [z_2, z_1, z_3, z_4] \\
&= [ff^{-1}z_1, f\infty, f0, ff^{-1}z_4] + [f\infty, ff^{-1}z_1, f0, ff^{-1}z_4] \\
&= d^{*-1}([f^{-1}z_1, \infty, 0, f^{-1}z_4] + [\infty, f^{-1}z_1, 0, f^{-1}z_4])d^* \\
&= 1.
\end{aligned}$$

[2 – 4] By computation. □

Our cross ratio is related to the complex cross ratio of Ahlfors [A3] by

Definition 6.6. The complex cross ratio of four points is

$$z = (z_1, z_2, z_3, z_4) = \mathrm{Re}[z_1, z_2, z_3, z_4] + |\mathrm{Im}[z_1, z_2, z_3, z_4]|i.$$

Lemma 6.7. *The complex cross ratio $z = (z_1, z_2, z_3, z_4)$ is invariant under Möbius transformations and*

$$\bar{z} + \hat{z} = 1$$

where $\hat{z} = (z_2, z_1, z_3, z_4)$.

Proof. Both the real part and the absolute value of the imaginary part of z are invariant under Möbius transformations by Corollary 6.3. The second statement follows from Lemma 6.5. □

Lemma 6.8. *Let σ denote the hyperbolic line in H^{n+1} which is orthogonal to the geodesics (z_1, z_2) and (z_3, z_4) where z_1, z_2, z_3, z_4 are distinct points in $\overline{\mathbb{R}}^n$. Then*

$$\sinh^2(\delta/2 + I\theta/2) = -[z_1, z_2, z_3, z_4] \tag{6.1}$$

where I is a Clifford number with $I^2 = -1$, δ the hyperbolic distance between (z_1, z_2) and (z_3, z_4), and θ one of the two angles between the spheres or hyperplanes which contain $(z_1, z_2) \cup \sigma$ and $(z_3, z_4) \cup \sigma$, respectively.

Proof. The formula for the distance between the geodesics is known, for example, see [GM3] for a form in H^3 or [A3] for a formula of $\cosh(\delta)$ in H^{n+1}.

We include a proof of (6.1) for the convenience of the reader. Normalize by a Möbius transformation so that $z_1 = -1$, $z_2 = 1$, $z_3 = -e^{\delta+i\theta}$ and $z_4 = e^{\delta+i\theta}$, the common orthogonal σ being the i_n axis. By calculation

$$[z_1, z_2, z_3, z_4] = -\sinh^2(\delta/2 + i\theta/2). \tag{6.2}$$

The result now follows from Lemma 6.2. □

Corollary 6.9. *Let δ denote the hyperbolic distance between geodesics (z_1, z_2) and (z_3, z_4) where z_1, z_2, z_3, z_4 are distinct points in $\overline{\mathbb{R}}^n$. Then*

$$\cosh\delta = |[z_1, z_2, z_3, z_4]| + |1 - [z_1, z_2, z_3, z_4]|. \tag{6.3}$$

Proof. Let θ be as in Lemma 6.8. Then by Lemma 6.5 and Lemma 6.8

$$\begin{aligned}
|[z_1, z_2, z_3, z_4]| &= |\sinh^2(\delta/2 + I\theta/2)| \\
&= \sinh^2(\delta/2)\cos^2(\theta/2) + \cosh^2(\delta/2)\sin^2(\theta/2), \\
|[z_2, z_1, z_3, z_4]| &= |\cosh^2(\delta/2 + I\theta/2)| \\
&= \cosh^2(\delta/2)\cos^2(\theta/2) + \sinh^2(\delta/2)\sin^2(\theta/2).
\end{aligned}$$

We obtain (6.3) by adding the second equation to the first. □

Definition 6.10. Suppose that f is loxodromic with fixed points u and v in $\overline{\mathbb{R}}^n$. For any g in $\mathrm{M}(n)$ which does not fix u and v, let

$$\begin{aligned}
\beta(f, g) &= [u, v, gu, gv]^{-1}[u, v, gfg^{-1}u, gfg^{-1}v][v, u, gu, gv]^{-1}, \\
\alpha(f) &= \inf|\beta(f, g)|, \\
\beta(f) &= \sup|\beta(f, g)|
\end{aligned}$$

where the infimum or supremum is taken over all Möbius transformations g in $\mathrm{M}(n)$ that do not fix u and v.

Remark 6.11. Note that, by Lemma 6.2, $\mathrm{Re}(\beta(f, g))$, $|\mathrm{Im}(\beta(f, g))|$ and hence $\alpha(f)$ and $\beta(f)$ are conjugacy invariant.

Lemma 6.12. *Suppose that f is loxodromic with fixed points u and v in $\overline{\mathbb{R}}^n$. Then*

$$\begin{aligned}
\alpha(f) &= \inf \frac{|[a, b, fa, fb]|}{|[a, b, u, v][b, a, u, v]|}, \\
\beta(f) &= \sup \frac{|[a, b, fa, fb]|}{|[a, b, u, v][b, a, u, v]|}
\end{aligned}$$

where the infimum or supremum is taken over all distinct points a and b in the complement of the set $\{u, v\}$ in $\overline{\mathbb{R}}^n$.

Proof. This lemma follows from

$$|\beta(f,g)| = \frac{|[g^{-1}u, g^{-1}v, fg^{-1}u, fg^{-1}v]|}{|[g^{-1}u, g^{-1}v, u, v][g^{-1}v, g^{-1}u, u, v]|}. \qquad \square$$

Lemma 6.13. *Suppose that f is loxodromic in H^{2k+1} with translation length τ and rotation angles $\theta_1, \ldots, \theta_k$ in $(-\pi, \pi]$. Then*

$$\begin{aligned} \alpha(f) &= 4\sinh^2(\tau/2) + 4\sin^2(\theta/2), \\ \beta(f) &= 4\sinh^2(\tau/2) + 4\sin^2(\Theta/2) \end{aligned}$$

where $\theta = \min_{1\le j\le k}\{|\theta_j|\}$ and $\Theta = \max_{1\le j\le k}\{|\theta_j|\}$.

Proof. If f is loxodromic, then f is conjugate to the map from $\mathbb{C}^k$ to $\mathbb{C}^k$: $x \to \lambda x \lambda^*$ which can be represented by

$$|\lambda|^2 \begin{pmatrix} e^{i\theta_1} & & & \\ & e^{i\theta_2} & & \\ & & \ddots & \\ & & & e^{i\theta_k} \end{pmatrix} x.$$

Thus, for distinct points a and b in the complement of the set $\{0, \infty\}$ in $\overline{\mathbb{R}}^{2k}$,

$$\begin{aligned} \beta(f) &= \sup_{a,b} \frac{1}{|\lambda|^2} \frac{|a - \lambda a \lambda^*|}{|a|} \frac{|b - \lambda b \lambda^*|}{|b|} \\ &= \sup_{|x|=1} \frac{|x - \lambda x \lambda^*|^2}{|\lambda|^2}. \end{aligned}$$

For unit $x = (z_1, z_2, \ldots, z_k)^T$ in $\mathbb{C}^k$,

$$\begin{aligned} |x - \lambda x \lambda^*|^2 &= ||\lambda|^2 e^{i\theta_1} z_1 - z_1|^2 + \ldots + ||\lambda|^2 e^{i\theta_k} z_k - z_k|^2 \\ &= |\lambda|^4 + 1 - 2|\lambda|^2(|z_1|^2 \cos\theta_1 + \ldots + |z_k|^2 \cos\theta_k) \\ &\le |\lambda|^4 + 1 - 2|\lambda|^2 \cos\Theta. \end{aligned}$$

Suppose $|\theta_{j_0}| = \Theta$. Set $z_i = 0$ if $i \neq j_0$ and $z_{j_0} = 1$. Then, for this choice of x

$$|x - \lambda x \lambda^*|^2 = |\lambda|^4 + 1 - 2|\lambda|^2 \cos\Theta.$$

Thus

$$\begin{aligned} \beta(f) &= \frac{1}{|\lambda|^2}(|\lambda|^4 + 1 - 2|\lambda|^2 \cos\Theta) \\ &= 4\sinh^2(\tau/2) + 4\sin^2(\Theta/2). \end{aligned}$$

Similarly

$$\begin{aligned}\alpha(f) &= \inf_{|x|=1} \frac{|x-\lambda x \lambda^*|}{|\lambda|^2} \\ &\geq \frac{1}{|\lambda|^2}(|\lambda|^4+1-2|\lambda|^2\cos\theta)\end{aligned}$$

and we get

$$\alpha(f) = 4\sinh^2(\tau/2) + 4\sin^2(\theta/2).$$ □

Remark 6.14. $\alpha(f)$ depends on the dimension in which f lives while $\beta(f)$ does not. $\alpha(f)$ becomes zero whenever f is extended to higher dimensions.

Definition 6.15. Suppose $f \in \mathrm{M}(n)$ is elliptic. Let u and v be any two fixed points of f in $\mathbb{R}^{n+1}$. For any g in $\mathrm{M}(n+1)$ which does not fix u and v, let

$$\beta(f,g,u,v) = [u,v,gu,gv]^{-1}[u,v,gfg^{-1}u,gfg^{-1}v][v,u,gu,gv]^{-1}.$$

We define

$$\begin{aligned}\alpha(f,u,v) &= \inf|\beta(f,g,u,v)|, \\ \beta(f,u,v) &= \sup|\beta(f,g,u,v)|\end{aligned}$$

where the infimum or supremum is taken over all Möbius transformations g in $\mathrm{M}(n+1)$ that do not fix u and v. We will show that $\alpha(f,u,v)$ and $\beta(f,u,v)$ are independent of the choice of the fixed points.

Remark 6.16. Similar to the proof of Lemma 6.12, we have

$$\begin{aligned}\alpha(f,u,v) &= \inf \frac{|[a,b,fa,fb]|}{|[a,b,u,v][b,a,u,v]|}, \\ \beta(f,u,v) &= \sup \frac{|[a,b,fa,fb]|}{|[a,b,u,v][b,a,u,v]|}\end{aligned}$$

where the infimum or supremum is taken over all distinct points a and b in the complement of the set $\{u,v\}$ in $\overline{\mathbb{R}}^{n+1}$.

Lemma 6.17. *Suppose that f is elliptic in H^{2k} with rotation angles θ_1, $\ldots$, θ_k in $(-\pi,\pi]$. Then*

$$\begin{aligned}\alpha(f,u,v) &= 4\sin^2(\theta/2), \\ \beta(f,u,v) &= 4\sin^2(\Theta/2)\end{aligned}$$

where $\theta = \min_{1\leq j\leq k}\{|\theta_j|\}$ and $\Theta = \max_{1\leq j\leq k}\{|\theta_j|\}$.

Proof. Similar to the proof of Lemma 6.13 with $|\lambda| = 1$. □

Remark 6.18. If f is elliptic, then $\alpha(f,u,v)$ and $\beta(f,u,v)$ do not depend on the choice of the fixed points u and v. We will denote them by $\alpha(f)$ and $\beta(f)$ respectively. From the above they are clearly invariant.

Definition 6.19. Suppose that f is parabolic in H^{2k+2} with rotation angles $\theta_1, \ldots, \theta_k$ in $(-\pi, \pi]$. We define

$$\begin{aligned} \alpha(f) &= 0, \\ \beta(f) &= 4\sin^2(\Theta/2) \end{aligned}$$

where $\Theta = \max_{1\le j\le k}\{|\theta_j|\}$.

Lemma 6.20. *If $f \in M(n)$ is conjugate to $\begin{pmatrix} \lambda & \mu \\ 0 & \lambda^{*-1} \end{pmatrix}$, then*

$$\beta(f) = \sup|\lambda - x\lambda^{*-1}x^{-1}|^2 \tag{6.4}$$

where the supremum is taken over all $x \in \mathbb{R}^m\backslash\{0\}$ ($m = n+1$ if f is elliptic and $m = n$ otherwise).

Proof. Similar to the proof of Lemma 6.13. □

Lemma 6.21. *Suppose that $f = \begin{pmatrix} a & b \\ c & d \end{pmatrix} \in SL(2, C_n)$ and the real part of $a + d^*$ is zero. Then $\beta(f) \geq 4$.*

Proof. Suppose that f has rotation angles $\theta_1, \ldots, \theta_k$ in $(-\pi, \pi]$ and multiplier λ. Then

$$(a + d^*)_{\mathbf{R}} = (\lambda + 1/\lambda)\cos(\theta_1/2)\cdots\cos(\theta_k/2)$$

by [W2, Lemma 15]. So $\theta_j = \pi$ for some j. Therefore

$$\beta(f) \geq 4\sin^2(\Theta/2) = 4.$$ □

Lemma 6.22.

$$\alpha(T) \leq |\mathrm{tr}(T)^2 - 4| \leq \beta(T) \text{ provided } T\infty \neq \infty.$$

Proof. Normalize by a translation so that T fixes 0 and the result follows easily. □

Definition 6.23. Suppose that f is represented by $A \in SL(2, C_n)$. We define

$$\begin{aligned} m(f)^2 &= 4||A||^2 - 2|\mathrm{Re}[\mathrm{tr}(A)]|^2, \\ \Omega(f)^2 &= \beta(f) + ||A||^2 - 2 \end{aligned}$$

where $||A||$ is the usual Schmidt norm of matrices.

Remark 6.24. Recall from [W2] that $||A||^2 = 2\cosh(\rho(i_n, fi_n))$.

Lemma 6.25. *Both $m(f)$ and $\Omega(f)$ are invariant under conjugacy by elements of* $\mathrm{SU}(2, C_n)$. *Each of them is zero if and only if f is the identity.*

Proof. Both $||A||$ and $\mathrm{Re}[\mathrm{tr}(A)]$ are invariant under $\mathrm{SU}(2, C_n)$ conjugacy by [W2, Lemma 12 and 15]. We note that $||A||^2 \geq 2$ with equality only when $f \in \mathrm{SU}(2, C_n)$. If $\Omega(f)$ is zero, then both $||A||^2 = 2$ and $\beta(f)$ are zero. Thus f is the identity or elliptic and the rotation angles of this elliptic are all zero. So f has to be the identity. Finally

$$m(f)^2 = 4|a_C|^2 + 4|b|^2 + 4|c|^2 + 4|d_C|^2 + 2(a_{\mathbb{R}} - d_{\mathbb{R}})^2$$

where $\begin{pmatrix} a & b \\ c & d \end{pmatrix}$ is a matrix representation of f. Thus $m(f) = 0$ if and only if f is the identity. □

Lemma 6.26. *Suppose that f is represented by $A \in \mathrm{SL}(2, C_n)$. Then*

$$2\beta(f) \leq m(f)^2 \leq 4||A - I||^2. \tag{6.5}$$

Proof. Suppose that f is represented by $A = \begin{pmatrix} a & b \\ c & d \end{pmatrix}$. Then

$$\begin{aligned} 2\sup_z |a - zdz^{-1}|^2 &\leq 2(a_{\mathbb{R}} - d_{\mathbb{R}})^2 + 4|a_C|^2 + 4|d_C|^2 \\ &\leq m(f)^2 \\ &= 4||A - I||^2 - 2(a_{\mathbb{R}} + d_{\mathbb{R}} - 2)^2 \end{aligned}$$

and we obtain (6.5) by conjugating f to $\begin{pmatrix} \lambda & \mu \\ 0 & \lambda^{*-1} \end{pmatrix}$. □

7 Conjugacy Invariant Inequalities

Lemma 7.1. *Suppose that f is not elliptic. Define the map $\theta : \mathrm{M}(n) \to \mathrm{M}(n)$ by $\theta(g) = gfg^{-1}$. If $\theta^n(g)(\mathrm{fix}(f)) = \mathrm{fix}(f)$ for some $n \geq 1$, then $g(\mathrm{fix}(f)) = \mathrm{fix}(f)$ and hence $\langle f, g \rangle$ is elementary.*

Proof. The result follows as for the corresponding statement for $\mathrm{SL}(2, \mathbb{C})$ [B, Theorem 5.1.4] and the fact that a parabolic or loxodromic element has at most two fixed points in $\overline{\mathbb{R}}^n$. □

Corollary 7.2. *Suppose that f is not elliptic and that $\langle f, g\rangle$ is discrete. If $\theta^n(g)$ and f have a common fixed point for some $n \geq 1$, then $g(\text{fix}(f)) = \text{fix}(f)$ and hence $\langle f, g\rangle$ is elementary.*

Proof. This follows from Lemma 7.1 if f is parabolic. If f is loxodromic, then f and $\theta^n(g)$ must have two common fixed points since $\langle f, \theta^n(g)\rangle$ is discrete [B]. So $g(\text{fix}(f)) = \text{fix}(f)$ by Lemma 7.1. □

Theorem 7.3. *Suppose that f is loxodromic with fixed points u and v and that $\{gu, gv\}$ is not equal to $\{u, v\}$. If $\langle f, g\rangle$ is discrete, then*

$$\beta(f)(1 + |[u, v, gu, gv]|) \geq 1.$$

Proof. The statement is invariant with respect to conjugation by Möbius transformations. Thus by means of such a conjugation we may assume that $\text{fix}(f) = \{0, \infty\}$. So f and g can be represented by

$$A = \begin{pmatrix} \lambda & 0 \\ 0 & \lambda^{*-1} \end{pmatrix}, \quad B = \begin{pmatrix} a & b \\ c & d \end{pmatrix}, \quad ad^* - bc^* = 1. \tag{7.1}$$

Consider the Shimizu-Leutbecher sequence

$$B_0 = B, \qquad B_{n+1} = B_n A {B_n}^{-1}. \tag{7.2}$$

The relation (7.2) yields

$$\begin{pmatrix} a_{n+1} & b_{n+1} \\ c_{n+1} & d_{n+1} \end{pmatrix} = \begin{pmatrix} a_n\lambda d_n^* - b_n\lambda^{*-1}c_n^* & -a_n\lambda b_n^* + b_n\lambda^{*-1}a_n^* \\ c_n\lambda d_n^* - d_n\lambda^{*-1}c_n^* & -c_n\lambda b_n^* + d_n\lambda^{*-1}a_n^* \end{pmatrix}. \tag{7.3}$$

By calculation

$$[0, \infty, g_n(0), g_n(\infty)] = -b_n c_n^*,$$
$$[\infty, 0, g_n(0), g_n(\infty)] = a_n d_n^*.$$

By Definition 6.10,

$$|[0, \infty, g_{n+1}(0), g_{n+1}(\infty)]| \leq \beta(f)|[0, \infty, g_n(0), g_n(\infty)][\infty, 0, g_n(0), g_n(\infty)]|$$

and we have

$$|b_{n+1}c_{n+1}| \leq \beta(f)|b_n c_n|(1 + |b_n c_n|).$$

Suppose that $\mu = \beta(f)(1 + |bc|) < 1$. We will obtain a contradiction. By induction,

$$|b_n c_n| \leq \mu^n |bc| \leq |bc|$$

and hence

$$b_n c_n \to 0, \qquad a_n d_n^* \to 1.$$

It follows from (7.3) that

$$|a_n| \to |\lambda|, \qquad |d_n| \to |\lambda|^{-1}.$$

Now

$$|b_{n+1}| \le |b_n a_n| \sqrt{\beta(f)}.$$

Thus, by induction, $|b_n|/|\lambda|^n \to 0$, and similarly, $|c_n||\lambda|^n \to 0$. So

$$A^{-n} B_{2n} A^n = \begin{pmatrix} \lambda^{-n} a_{2n} \lambda^n & \lambda^{-n} b_{2n} \lambda^{*-n} \\ \lambda^{*n} c_{2n} \lambda^n & \lambda^{*n} d_{2n} \lambda^{*-n} \end{pmatrix}$$

has a subsequence that converges to a diagonal matrix. Since $\langle f, g\rangle$ is discrete, $\lambda^{-n_k} b_{2n_k} \lambda^{*-n_k} = 0$ and $\lambda^{*n_k} c_{2n_k} \lambda^{n_k} = 0$ for sufficiently large k. Thus $b_n = c_n = 0$ for infinitely many n. Hence $g(\text{fix}(f)) = \text{fix}(f)$ by Lemma 7.1. This contradicts the assumption that $\{gu, gv\}$ is not equal to $\{u, v\}$. Therefore $\mu \ge 1$. □

Theorem 7.4. *Suppose that $\langle f, g\rangle$ is discrete and that $\{gu, gv\}$ is not equal to $\{u, v\}$ where u and v are two fixed points of f. If*

$$\beta(f)(1 + |[u, v, gu, gv]|) = 1, \tag{7.4}$$

then f is elliptic of order at least seven.

Proof. We suppose that f is loxodromic. Then after a conjugation, f and g can be represented by the matrices in (7.1). We consider the Shimizu-Leutbecher sequence (7.2). Let

$$\omega_n = [0, \infty, g_n 0, g_n \infty] = -b_n c_n^*.$$

It follows from the definition of $\beta(f)$ that

$$|\omega_{n+1}| \le |\omega_n (1 - \omega_n)| \beta(f). \tag{7.5}$$

By Theorem 7.3,

$$\beta(f)(1 + |\omega_n|) \ge 1 \tag{7.6}$$

for all $n \ge 0$. The equation (7.4) gives

$$\beta(f)(1 + |\omega_0|) = 1. \tag{7.7}$$

Thus by (7.5), (7.6), and (7.7),

$$1 \le \beta(f)(1 + |\omega_1|) \le \beta(f)(1 + |\omega_0||1 - \omega_0|\beta(f)) \le 1.$$

By induction, we have

$$\beta(f)(1 + |\omega_n|) = 1 \tag{7.8}$$

for all $n \geq 0$. Hence

$$1-\beta(f)=|\beta(f)\omega_{n+1}| \leq \beta(f)|\omega_n(1-\omega_n)|\beta(f)=(1-\beta(f))|1-\omega_n|\beta(f).$$

We note that $\beta(f)<1$ since otherwise $\omega_0=0$ and f and g have a common fixed point. By Corollary 7.2 $\{gu, gv\}$ equals $\{u, v\}$, a contradiction. So

$$\beta(f)|1-\omega_n|=1. \tag{7.9}$$

It follows from (7.8) and (7.9) that

$$\omega_n=1-\frac{1}{\beta(f)} \tag{7.10}$$

for all $n \geq 0$. The relation (7.3) yields

$$-b_{n+1}c_{n+1}^*=a_n(\lambda-a_n^{-1}b_n\lambda^{*-1}b_n^{-1}a_n)b_n^*c_n(c_n^{-1}d_n\lambda^* d_n^{-1}c_n-\lambda^{-1})d_n^*.$$

Thus

$$\beta(f)=(\lambda-a_n^{-1}b_n\lambda^{*-1}b_n^{-1}a_n)(\lambda^{-1}-c_n^{-1}d_n\lambda^* d_n^{-1}c_n). \tag{7.11}$$

Switching one of the terms (both in Γ_n) on the right hand side of (7.11) to the left hand side and taking the real part, we get

$$\frac{\beta(f)}{|\lambda-a_n^{-1}b_n\lambda^{*-1}b_n^{-1}a_n|^2}\lambda_{\mathbf{R}}\left(1-\frac{1}{|\lambda|^2}\right)=\lambda_{\mathbf{R}}\left(\frac{1}{|\lambda|^2}-1\right). \tag{7.12}$$

If $\lambda_{\mathbf{R}}=0$, then $(\lambda+\lambda^{-1})_{\mathbf{R}}=0$ and $\beta(f) \geq 4$ by Lemma 6.21. This contradicts the fact that $\beta(f)<1$. Thus $|\lambda|=1$ by (7.12). This contradicts the assumption that f is loxodromic. Therefore f is elliptic. Since $\beta(f)<1$, f is of order at least seven. □

Remark 7.5. If the equation (7.4) holds, then each of the rotation angles of f has to be of order at least seven.

Theorem 7.6. *Suppose that f is elliptic and $\langle f, g\rangle$ is a non-elementary discrete group. Then*

$$\beta(f)\left(\frac{1}{4\sin^2(\pi/10)}+|[u,v,gu,gv]|\right) \geq 1$$

where u and v are any two fixed points of f.

Proof. The proof is similar to that of Theorem 7.3. To get a contradiction we first observe that the condition

$$\beta(f)(1+|[u,v,gu,gv]|)<1$$

guarantees a convergent subsequence (see the proof of Theorem 7.3). Then the condition

$$\beta(f)\left(\frac{1}{4\sin^2(\pi/10)}+|[u,v,gu,gv]|\right)<1$$

is used to show the group $\langle f,g\rangle$ is elementary (see [W2, p105] for details). □

The next theorem is concerned with non-elliptic elements in M(n). We denote the translation length of f by τ_f, which is zero if f is not loxodromic.

Theorem 7.7. *Suppose that f is not elliptic and that*

$$\rho = 2\cosh(\tau_f/2)\sqrt{\beta(f)} < 1.$$

If $\langle f,g\rangle$ is discrete and $\mathrm{fix}(f)\cap\mathrm{fix}(g)$ *is empty, then for any Möbius transformation h that sends $u\in$* $\mathrm{fix}(f)$ *to ∞,*

$$|\mathrm{tr}^2(hgh^{-1})[fgu,fg^{-1}u,gu,g^{-1}u]| \geq \frac{1}{2}(1-\rho+\sqrt{(1-\rho)^2-4\beta(f)}). \quad (7.13)$$

Proof. Our proof is a modification of an argument given by Waterman [W2]. After a conjugacy by the Möbius transformation h, we assume $u=\infty$. Then f and g can be represented by

$$A=\begin{pmatrix}\lambda & \mu\\ 0 & \lambda^{*-1}\end{pmatrix},\qquad B=\begin{pmatrix}a & b\\ c & d\end{pmatrix},\qquad ad^*-bc^*=1.$$

Consider the Shimizu-Leutbecher sequence

$$B_0=B,\qquad B_{n+1}=B_nAB_n{}^{-1}. \quad (7.14)$$

The relation (7.14) yields

$$\begin{pmatrix}a_{n+1} & b_{n+1}\\ c_{n+1} & d_{n+1}\end{pmatrix} = \begin{pmatrix}a_n\lambda d_n^*-b_n\lambda^{*-1}c_n^*-a_n\mu c_n^* & -a_n\lambda b_n^*+b_n\lambda^{*-1}a_n^*+a_n\mu a_n^*\\ c_n\lambda d_n^*-d_n\lambda^{*-1}c_n^*-c_n\mu c_n^* & -c_n\lambda b_n^*+d_n\lambda^{*-1}a_n^*+c_n\mu a_n^*\end{pmatrix}.$$

For all $n\geq 0$, c_n is not zero by Corollary 7.2. Let

$$\begin{aligned}\sigma_n &= f(a_nc_n^{-1})-a_nc_n^{-1},\\ \tau_n &= f(-c_n^{-1}d_n)-(-c_n^{-1}d_n).\end{aligned}$$

By calculations,

$$\begin{aligned}c_{n+1} &= -c_n\tau_n\lambda^{*-1}c_n^*,\\ -c_{n+1}^{-1}d_{n+1} &= a_nc_n^{-1}-c_{n+1}^{-1}c_n\lambda c_n^{-1},\\ a_{n+1}c_{n+1}^{-1} &= a_nc_n^{-1}+c_n^{*\,-1}\lambda^{*-1}c_n^*c_{n+1}^{-1}.\end{aligned} \quad (7.15)$$

Since

$$\begin{aligned}|(fx-x)-(fy-y)| &= |(\lambda-(x-y)\lambda^{*-1}(x-y)^{-1})(x-y)\lambda^*| \\ &\le \sqrt{\beta(f)}|x-y||\lambda|,\end{aligned}$$

it follows from (7.15) that for $n \geq 0$,

$$\begin{aligned}|c_{n+1}||\tau_{n+1}-\sigma_n| &\le \sqrt{\beta(f)}|\lambda|^2, \\ |c_{n+1}||\sigma_{n+1}-\sigma_n| &\le \sqrt{\beta(f)}.\end{aligned}$$

Thus,

$$|c_{n+1}\tau_{n+1}| \le |\lambda|^{-1}|c_n\tau_n||c_n\sigma_n| + \sqrt{\beta(f)}|\lambda|^2, \tag{7.16}$$

$$|c_{n+1}\sigma_{n+1}| \le |\lambda|^{-1}|c_n\tau_n||c_n\sigma_n| + \sqrt{\beta(f)}. \tag{7.17}$$

For the quadratic iteration

$$p(z) = Az^2 + Bz + C, \quad (0 < A,\ 0 \le B < 1,\ 0 < C),$$

$p^n(z)$ converges to z_1 if $0 < z < z_2$ where $z_1 < z_2$ are two fixed points of $p(z)$. It follows from (7.16) and (7.17) that a subsequence of $|c_n\tau_n c_n\sigma_n|$ converges to a number that is less than or equal to

$$\frac{1}{2}|\lambda|^2(1-\rho-\sqrt{(1-\rho)^2-4\beta(f)})$$

provided

$$|c_0\tau_0 c_0\sigma_0| < \frac{1}{2}|\lambda|^2(1-\rho+\sqrt{(1-\rho)^2-4\beta(f)}). \tag{7.18}$$

Suppose that (7.18) holds. We will obtain a contradiction. It follows from the relations

$$\begin{aligned}c_{n+1} &= -c_n\tau_n\lambda^{*-1}c_n^*, \\ a_{n+1} &= -a_n\tau_n\lambda^{*-1}c_n^* + c_n^{*-1}\lambda^{*-1}c_n^*, \\ d_{n+1} &= c_n\tau_n\lambda^{*-1}a_n^* + c_n\lambda c_n^{-1}, \\ b_{n+1} &\le |a_n|^2|\mu| + \sqrt{\beta(f)}|a_n b_n|\end{aligned}$$

that there exists a convergent subsequence of a_n, b_n, c_n, d_n with c_n approaching zero. Thus c_n is eventually zero by the discreteness of $\langle f, g\rangle$. This contradicts Corollary 7.2. Therefore

$$\begin{aligned}&|\lambda|^{-2}|c|^2|f(ac^{-1})-ac^{-1}||f(-c^{-1}d)-(-c^{-1}d)| \\ &\ge \frac{1}{2}(1-\rho+\sqrt{(1-\rho)^2-4\beta(f)}).\end{aligned}$$

It can checked that the left hand side of the above inequality equals that of (7.13). □

Corollary 7.8. *Suppose that f is non-elliptic with fixed point u and that g is parabolic with fixed point w. If $\langle f, g\rangle$ is discrete and $u \neq w$, then*

$$|[fgu, fg^{-1}u, gu, g^{-1}u][gu, w, g^{-1}u, u]^2| \geq \frac{1}{2}(1-\rho+\sqrt{(1-\rho)^2-4\beta(f)}) \tag{7.19}$$

where $\rho = 2\cosh(\tau_f/2)\sqrt{\beta(f)}$.

Proof. We choose a Möbius transformation h that sends u to ∞, w to 0. Then

$$hfh^{-1} = \begin{pmatrix} \lambda & \mu \\ 0 & \lambda^{*-1} \end{pmatrix}, \qquad hgh^{-1} = \begin{pmatrix} a & 0 \\ c & a^{*-1} \end{pmatrix}.$$

We observe that

$$|[gu, w, g^{-1}u, u]| = |\mathrm{tr}(hgh^{-1})a^{-1}| = |\mathrm{tr}(hgh^{-1})|$$

and thus (7.19) follows from Theorem 7.7. □

8 Isolation in a discrete group

Theorem 8.1. *Suppose that $\langle f, g\rangle$ is discrete and neither f nor g is elliptic. If f and g do not have a common fixed point, then*

$$m(f)m(g) \geq \frac{1}{9}, \tag{8.1}$$

$$\Omega(f)\Omega(g) \geq \frac{1}{17}. \tag{8.2}$$

Proof. We assume that $m(f) \leq m(g)$. Since $m(f)$ is invariant under conjugations by elements in $\mathrm{SU}(2, \mathcal{C}_n)$, we may assume that f and g are represented by

$$A = \begin{pmatrix} \lambda & \mu \\ 0 & \lambda^{*-1} \end{pmatrix}, \quad B = \begin{pmatrix} a & b \\ c & d \end{pmatrix}, \quad ad^* - bc^* = 1.$$

It can be checked that

$$\begin{aligned} 2\beta(f) &\leq m(f)^2, \\ 2(|a|-|d|)^2 &\leq m(g)^2, \\ |ad| = |1+bc^*| &\leq 1+1/8m(g)^2, \\ 8\sinh^2(\tau_f/2) &\leq \min\{m(f)^2, \Omega(f)^2\}. \end{aligned}$$

If $\rho = 2\cosh(\tau_f/2)\sqrt{\beta(f)} \geq 1$, then $m(f) \geq \sqrt{2\sqrt{5}-4}$ and hence (8.1) holds. If $\rho < 1$, then by Theorem 7.7,

$$|c|^2|f(ac^{-1}) - ac^{-1}||f(-c^{-1}d) - (-c^{-1}d)|$$
$$\geq \frac{1}{2}|\lambda|^2(1-\rho+\sqrt{(1-\rho)^2-4\beta(f)}).$$

So,

$$\frac{|\lambda|^2}{2}\left(1-\frac{m(f)}{2}\sqrt{m(f)^2+8}+\sqrt{\left(1-\frac{m(f)}{2}\sqrt{m(f)^2+8}\right)^2-2m^2(f)}\right)$$
$$\leq \frac{|\lambda|^2}{2}\left(1-\rho+\sqrt{(1-\rho)^2-4\beta(f)}\right)$$
$$\leq |c|^2|f(ac^{-1})-ac^{-1}||f(-c^{-1}d)-(-c^{-1}d)|$$
$$\leq |c|^2(\sqrt{\beta(f)}|ac^{-1}||\lambda|+|\mu\lambda|)(\sqrt{\beta(f)}|c^{-1}d||\lambda|+|\mu\lambda|)$$
$$\leq \frac{|\lambda|^2m(f)^2}{4}\left(2|ad|+\sqrt{2}|c|(|a|+|d|)+|c|^2\right)$$
$$\leq \frac{|\lambda|^2m(f)^2}{4}\left(2+\frac{m(g)^2}{2}+\frac{m(g)}{\sqrt{2}}\sqrt{m(g)^2+4}\right).$$

The above inequality does not hold for $m(f) \leq m(g) \leq 0.35$. So (8.1) follows. To prove (8.2) we may assume that $\Omega(f) \leq \Omega(g)$. Similar to the proof of (8.1), we have:

$$\frac{|\lambda|^2}{2}\left(1-\frac{\Omega(f)}{2}\sqrt{2\Omega(f)^2+16}+\sqrt{\left(1-\frac{\Omega(f)}{2}\sqrt{2\Omega(f)^2+16}\right)^2-4\Omega^2(f)}\right)$$
$$\leq |\lambda|^2\Omega(f)^2(|a|+|c|)(|d|+|c|)$$
$$\leq 2|\lambda|^2\Omega(f)^2\left(2+\Omega(g)^2\right).$$

Thus $0.249 \leq \Omega(f) \leq \Omega(g)$. □

Corollary 8.2. *Suppose that $\langle f, g\rangle$ is discrete and neither f nor g is elliptic. If f and g do not have a common fixed point, then*

$$||f-I||\cdot||g-I|| \geq \frac{1}{36}.$$

Proof. This follows from $m(f) \leq 2||A-I||$ and Theorem 8.1. □

Theorem 8.3. *If G is a torsion-free non-elementary discrete subgroup of* $\mathrm{M}(n)$, *then there exists an h such that*

$$\Omega(f) \geq 0.249$$

for each non-identity f in hGh^{-1}.

Proof. The proof is similar to that in $\mathrm{SL}(2,\mathbb{C})$. See [GM2] for details. Also see [M2] and [W1]. □

Definition 8.4. Suppose that f is loxodromic. Let u and v be the two fixed points of f in $\overline{\mathbb{R}}^n$ and $\mathcal{A}_f$ the geodesic in H^{n+1} joining u and v. $\mathcal{A}_f$ is called the axis of f.

Theorem 8.5. *Suppose that f and g are loxodromic with disjoint fixed point sets. If $\langle f, g\rangle$ is discrete, then*

$$\left(1+\sqrt{1+\cosh^2\rho(\mathcal{A}_f,\mathcal{A}_g)}\right)^2\beta(f)\beta(g)\geq 4 \tag{8.3}$$

where $\rho(\mathcal{A}_f,\mathcal{A}_g)$ is the hyperbolic distance between the axes of f and g.

Proof. Suppose that f has fixed points u and v and that g has fixed points w_1 and w_2. By Lemma 6.12

$$\beta(f) \geq |[w_1,w_2,u,v]^{-1}[w_1,w_2,fw_1,fw_2][w_2,w_1,u,v]^{-1}|,$$

$$\beta(g) \geq |[u,v,w_1,w_2]^{-1}[u,v,gu,gv][v,u,w_1,w_2]^{-1}|$$

and we have

$$\begin{aligned}\beta(f)\beta(g)\cosh^2\rho(\mathcal{A}_f,\mathcal{A}_g) &\geq \beta(f)\beta(g)|\sinh^2(\rho(\mathcal{A}_f,\mathcal{A}_g)+I\theta)| \\ &= \beta(f)\beta(g)|-4[u,v,w_1,w_2][v,u,w_1,w_2]| \\ &\geq 4\beta(f)|[u,v,gu,gv]|.\end{aligned}$$

By Theorem 7.3

$$\beta(f)+\frac{1}{4}\beta(f)\beta(g)\cosh^2\rho(\mathcal{A}_f,\mathcal{A}_g)\geq 1. \tag{8.4}$$

Similarly,

$$\beta(g)+\frac{1}{4}\beta(f)\beta(g)\cosh^2\rho(\mathcal{A}_f,\mathcal{A}_g)\geq 1. \tag{8.5}$$

The inequality (8.3) follows from (8.4) and (8.5). □

9 Applications to hyperbolic n-space

We first obtain tubular neighborhoods around short geodesics by means of Meyerhoff's solid tube construction [Me]. A solid tube of radius r around a geodesic in a hyperbolic n-manifold $\mathcal{M}$ is the set of all points in $\mathcal{M}$ which are a distance less than or equal to r from the geodesic. Any geodesic will be the image of the axis $\mathcal{A}_f$ for some loxodromic f in a torsion-free non-elementary discrete subgroup G of $\mathrm{M}(n-1)$. We will thus denote the geodesic itself by f.

Lemma 9.1. *If f is a geodesic in a complete hyperbolic n-manifold $\mathcal{M}$ with*

$$\beta(f)\leq\sqrt{2}-1,$$

then there exists a solid tube around f with radius r given by

$$\sinh^2(r) = \frac{\sqrt{1-\beta(f)}}{\beta(f)} - \frac{1}{2}.$$

Proof. We follow the solid tube construction used by Meyerhoff in the 3-dimensional case [Me]. Construct infinitely long tubes of the same radius around $\mathcal{A}_f$ and all of its images under the action of G. The tube in $\mathcal{M}$ will fail to be embedded as soon as the tube around $\mathcal{A}_f$ in H^n intersects the tube around some image $\mathcal{A}_h$ of $\mathcal{A}_f$ in H^n. We note that f and h are conjugate under G and denote h by gfg^{-1}. The radius of the solid tube when it first bumps into itself is $\delta/2$ where δ is the hyperbolic distance between $\mathcal{A}_f$ and $\mathcal{A}_{gfg^{-1}}$.

By Corollary 6.9,

$$\cosh(\delta) = |[u,v,gu,gv]| + |1-[u,v,gu,gv]|$$

where u and v are the fixed points of f.

So,

$$\begin{aligned}
\sinh^2(r) &= \frac{1}{2}(\cosh(\delta)-1)\\
&= \frac{1}{2}(|[u,v,gu,gv]| + |[v,u,gu,gv]|) - \frac{1}{2}\\
&\geq \sqrt{|[u,v,gu,gv]||[v,u,gu,gv]|} - \frac{1}{2}\\
&\geq \sqrt{\frac{|[u,v,gfg^{-1}u,gfg^{-1}v]|}{\beta(f)}} - \frac{1}{2}\\
&= \frac{1}{\beta(f)}\sqrt{|[u,v,gfg^{-1}u,gfg^{-1}v]|\beta(f)} - \frac{1}{2}\\
&\geq \frac{\sqrt{1-\beta(f)}}{\beta(f)} - \frac{1}{2}. \qquad \text{[Theorem 7.3]}
\end{aligned}$$

□

We need the following result that gives an estimate of the lattice constant $\Lambda(D)$ of a subset D of $\beta(f) \leq 4a^2$. In what follows, we denote $\int_0^{\pi/2} \sin^p t\,dt$ by $\sigma(p)$. We have (integration by parts)

$$\sigma(p) = \begin{cases} \dfrac{p-1}{p}\dfrac{p-3}{p-2}\dfrac{p-5}{p-4}\cdots\dfrac{3}{4}\dfrac{1}{2}\dfrac{\pi}{2}, & p \text{ even},\\[2ex] \dfrac{p-1}{p}\dfrac{p-3}{p-2}\dfrac{p-5}{p-4}\cdots\dfrac{4}{5}\dfrac{2}{3}1, & p \text{ odd}.\end{cases}$$

Theorem 9.2. *For $0 < a < 1$, let $D = \bigcap_{i=1}^{k} D_i$ where D_i is the component of the set*

$$\{(x, y_1, \ldots, y_k) : \sinh^2(x/2) + \sin^2(y_i/2) \leq a^2\}$$

which contains 0. *Then*

$$\Lambda(D) \geq (2a)^{k+1}\sigma(k+1).$$

Proof. For each i, there exists $b = b(a) \in (1, 5/4)$ such that

$$E_i = \{(bx)^2 + (y_i/b)^2 \leq (2a)^2\} \subset D_i$$

by [CGM]. Set $S = \bigcap_{i=1}^{k} E_i$. By the Minkowski convex body theorem, a lower bound of the lattice constant of S is $V/2^{k+1}$ where V is the volume of S. By Fubini's theorem,

$$V = \int_S dx dy_1 dy_2 \cdots dy_k = 4ab^{-1}(4ab)^k \int_0^{\pi/2} \sin^{k+1} t \, dt.$$

Since b and k are both greater than or equal to 1, we have

$$\Lambda(D) \geq \Lambda(S) \geq V/2^{k+1} \geq (2a)^{k+1}\sigma(k+1). \qquad \square$$

Corollary 9.3. *Suppose that* $\theta_1, \ldots, \theta_k$ *are* k *real numbers and that* $|x|$ *is less than* $2\sigma(k+1)/\pi^k$. *Then there exists an integer* $m \neq 0$ *such that*

$$\sinh^2(mx/2) + \max_{1 \leq i \leq k} \sin^2(m\theta_i/2) \leq \left(\frac{\pi^k |x|}{2\sigma(k+1)}\right)^{\frac{2}{k+1}}. \qquad (9.1)$$

Proof. Let D be the region described in Theorem 9.2 and let Λ be the lattice

$$\Lambda = \{m_0(|x|, \theta_1, \ldots, \theta_k) + m_1(0, 2\pi, 0, \ldots, 0), + \ldots + m_k(0, \ldots, 0, 2\pi)\}.$$

Suppose that a is a number with

$$\nu = \left(\frac{\pi^k |x|}{2\sigma(k+1)}\right)^{\frac{1}{k+1}} < a < 1. \qquad (9.2)$$

Then

$$det(\Lambda) = (2\pi)^k |x| < (2a)^{k+1}\sigma(k+1) \leq \Lambda(D).$$

By definition of the lattice constants, there is a non-zero lattice point in D. So there exist integers m, m_1, ... m_k, not all zero, such that

$$\sinh^2\left(\frac{mx}{2}\right) + \sin^2\left(\frac{m\theta_i + m_i 2\pi}{2}\right) \leq a^2, \quad 1 \leq i \leq k. \qquad (9.3)$$

Note that m is not zero since otherwise $(0, m_1 2\pi, \ldots, m_k 2\pi)$ is inside $D \backslash \{0\}$. This is impossible since

$$D \subset \bigcap_{i=1}^{k} \{(x, y_1, \ldots, y_k) : |y_i| \leq 2\arcsin(a) < \pi\}.$$

The inequality (9.3) holds for each a which satisfies (9.2), and we obtain (9.1) by letting $a \to \nu$. □

Corollary 9.4. *Suppose that f has translation length $\tau(f)$ and rotation angles $\theta_1, \ldots, \theta_k$ in $(-\pi, \pi]$. If $\tau(f)$ is less than $2\sigma(k+1)/\pi^k$, then there exists an integer $m \neq 0$ such that*

$$\beta(f^m) \leq 4\left(\frac{\pi^k \tau(f)}{2\sigma(k+1)}\right)^{\frac{2}{k+1}}.$$

Proof. For the m in Corollary 9.3,

$$\begin{aligned} \beta(f^m) &= 4(\sinh^2(m\tau(f)/2) + \max_{1\leq i\leq k} \sin^2(m\theta_i/2)) \\ &\leq 4\left(\frac{\pi^k \tau(f)}{2\sigma(k+1)}\right)^{\frac{2}{k+1}}. \end{aligned}$$ □

Corollary 9.5. *Let f be a geodesic in a complete hyperbolic n-manifold. If*

$$\tau(f) \leq \frac{2\sigma(k+1)}{\pi^k}\left(\frac{\sqrt{2}-1}{4}\right)^{\frac{k+1}{2}}, \tag{9.4}$$

then there exists an embedded solid tube around f whose radius satisfies

$$\sinh^2(r) = \frac{1}{4}\left(\frac{\pi^k \tau(f)}{2\sigma(k+1)}\right)^{-\frac{2}{k+1}}\left(1 - 4\left(\frac{\pi^k \tau(f)}{2\sigma(k+1)}\right)^{\frac{2}{k+1}}\right)^{\frac{1}{2}} - \frac{1}{2}. \tag{9.5}$$

Proof. Since the axes of f^m and its conjugates are the same as the axes of f and its conjugates, we can apply Lemma 9.1 to get a lower bound of the radius of the solid tube as long as $\beta(f^m)$ is less than $\sqrt{2}-1$. By Corollary 9.4, the assumption (9.4) guarantees the existence of such an m. □

Remark 9.6. If the manifold has a geodesic of length less than

$$\tau_0 = \frac{2\sigma(k+1)}{\pi^k}\left(\frac{\sqrt{2}-1}{4}\right)^{\frac{k+1}{2}}, \quad k = \left[\frac{n-1}{2}\right]$$

then there exists a tubular neighborhood of radius r (see (9.5)) and a lower bound for the volume of H^{n+1}/G is obtained. If the translation length of each geodesic is greater than τ_0, then there exists a ball of radius τ_0 in the fundamental domain and a lower bound for the volume of H^{n+1}/G is again obtained.

In the finite volume case, the existence of a cusp implies the existence of a pure translation. Hersonsky has given a lower bound in this case [H].

Therefore a lower bound for complete hyperbolic n-manifolds can be obtained after one has considered these three cases.

The next theorem states that if the translation lengths of both f and g are small, then their axes are far apart.

Corollary 9.7. *Suppose that f and g are loxodromic with disjoint fixed point sets. If $\langle f, g\rangle$ is discrete and $\max\{\tau(f), \tau(g)\}$ is less than $2\sigma(k+1)/\pi^k$, where $k = [(n-1)/2]$, then*

$$\left(1+\sqrt{1+\cosh^2\delta(\mathcal{A}_f,\mathcal{A}_g)}\right)^2 \tau(f)^{\frac{2}{k+1}}\tau(g)^{\frac{2}{k+1}} \geq \frac{1}{4}\left(\frac{2\sigma(k+1)}{\pi^k}\right)^{\frac{4}{k+1}} \tag{9.6}$$

where $\delta(\mathcal{A}_f, \mathcal{A}_g)$ is the hyperbolic distance between the axes of f and g.

Proof. We apply Theorem 8.5 to $\langle f^m, g^n\rangle$ where m and n are chosen according to Corollary 9.4. □

We finally obtain a lower bound for the radius of the inscribed ball in the fundamental domain. For related results, see [M2] and [FH].

Theorem 9.8. *Suppose that G is a discrete torsion-free non-elementary subgroup of $\mathrm{M}(n)$. Then there exists a point x in H^{n+1} such that*

$$\rho(x, fx) \geq \frac{2\sigma(k+1)}{\pi^k}\left(\frac{1}{136}\right)^{\frac{k+1}{2}}$$

for every f in $G\backslash\{\mathrm{id}\}$ where $k = [(n-1)/2]$.

Proof. By Theorem 8.3, there exists an h such that

$$\Omega(g) \geq 0.249, \quad g = hfh^{-1}$$

for each non-identity f in G. After this conjugation, we may assume

$$\Omega(f) \geq 0.249$$

for each f in $G\backslash\{\mathrm{id}\}$. By Corollary 9.3, if

$$\rho(i_n, fi_n) < \frac{2\sigma(k+1)}{\pi^k},$$

then there is a non-zero integer m such that

$$\sinh^2(m\rho(i_n, fi_n)/2) + \max_{1\leq i\leq k}\sin^2(m\theta_i/2) \leq \left(\frac{\pi^k\rho(i_n, fi_n)}{2\sigma(k+1)}\right)^{\frac{2}{k+1}}.$$

So

$$\begin{aligned}
\frac{1}{17} &\leq \Omega^2(f^m) \\
&= 4\sinh^2(m\tau_f/2) + 4\max_{1\leq i\leq k}\sin^2(m\theta_i/2) + 4\sinh^2(m\rho(i_n, fi_n)/2) \\
&\leq 8\sinh^2(m\rho(i_n, fi_n)/2) + 4\max_{1\leq i\leq k}\sin^2(m\theta_i/2) \\
&\leq 8\left(\sinh^2(m\rho(i_n, fi_n)/2) + \max_{1\leq i\leq k}\sin^2(m\theta_i/2)\right) \\
&\leq 8\left(\frac{\pi^k\rho(i_n, fi_n)}{2\sigma(k+1)}\right)^{\frac{2}{k+1}}.
\end{aligned}$$

Therefore,

$$\rho(i_n, fi_n) \geq \frac{2\sigma(k+1)}{\pi^k}\left(\frac{1}{136}\right)^{\frac{k+1}{2}}. \qquad \square$$

Acknowledgment

We wish to thank Professor F.W. Gehring for many helpful conversations.

References

[A1] L.V. Ahlfors, *Möbius transformations and Clifford numbers*, Differential Geometry and Complex Analysis, H. E. Rauch memorial volume, Springer-Verlag, 1985.

[A2] L.V. Ahlfors, *On the fixed points of Möbius transformations in* $\mathbb{R}^n$, Ann. Acad. Sci. Fenn. Ser. A. I. Math. **10** (1985), 15–27.

[A3] L.V. Ahlfors, *Möbius transformations in several dimensions*, Lecture notes, University of Minnesota, 1981.

[B] A.F. Beardon, *The Geometry of Discrete Groups*, Springer-Verlag, 1983.

[CGM] C. Cao, F.W. Gehring, and G.J. Martin, *Lattice constants and a lemma of Zagier*, to appear.

[FH] S. Friedland and S. Hersonskyi, *Jørgensen's inequality for discrete groups in normed algebras*, Duke Math. J. **69** (1993), 593–614.

[GM1] F.W. Gehring and G.J. Martin, *Stability and extremality in Jørgensen's inequality*, Complex Variables Theory Appl. **12** (1989), 277–282.

[GM2] F.W. Gehring and G.J. Martin, *Inequalities for Möbius transformations and discrete groups*, J. Reine Angew. Math. **418** (1991), 31–76.

[GM3] F.W. Gehring and G.J. Martin, *Commutators, collars and the geometry of Möbius groups*, J. d'Analyse Math. **63** (1994), 175–219.

[Gr] L. Greenberg, *Discrete subgroups of the Lorentz group*, Math. Scand. **10** (1962), 85–107.

[H] S. Hersonsky, *Covolume estimates for discrete groups of hyperbolic isometries*, Michigan Math. J. **40** (1993), 467–475.

[J] T. Jørgensen, *On discrete groups of Möbius transformations*, Amer. J. Math. **98** (1976), 739–749.

[M1] G.J. Martin, *On discrete Möbius groups in all dimensions: A generalization of Jørgensen's inequality*, Acta Math. **163** (1989), 253–289.

[M2] G.J. Martin, *Balls in hyperbolic manifolds*, J. London Math. Soc. **40** (1989), 257–265.

[Me] R. Meyerhoff, *A lower bound for the volume of hyperbolic 3-manifolds*, Canad. J. Math. **39** (1987), 1038–1056.

[V] K. Th. Vahlen, *Ueber Bewegungen und complexe Zahlen*, Math. Ann. **55** (1902), 585–593.

[Wa] M. Wada, Conjugacy Invariants of Möbius Transformations, Complex Variables **15** (1990), 125–133.

[W1] P.L. Waterman, *An inscribed ball for Kleinian groups*, Bull. London Math. Soc. **16** (1984), 525–530.

[W2] P.L. Waterman, *Möbius transformations in several dimensions*, Advances in Math. **101** (1993), 87–113.

University of Michigan, Ann Arbor, MI 48109
ccao@math.lsa.umich.edu

Northern Illinois University, Dekalb, IL 60115
waterman@math.niu.edu

Extremal Quasiconformal Mappings in Plane Domains

Clifford J. Earle and Li Zhong

1 Introduction

This paper is an expanded version of the lecture given by the first author at the Ann Arbor conference celebrating Fred Gehring's seventieth birthday and his many contributions to mathematics. Our primary purpose is to give an exposition of some of the main results of our joint paper [EL]. In view of the special occasion, we have devoted Sections 2 through 8 to a survey of some classical results about extremal quasiconformal mappings. Sections 9 and 10 report some results from [EL], and Section 11 reports some results obtained by Nikola Lakic while this paper was in progress. The interested reader can learn more about extremal mappings from Kurt Strebel's survey article [St3] and Edgar Reich's paper [R3] in this volume, both of which we highly recommend.

We have chosen to confine our discussion in this paper to subregions of the plane rather than the general Riemann surfaces that we consider in [EL]. Plane regions already provide a rich variety of examples, encompassing both the open unit disk Δ and the complement of a Cantor set in $\mathbb{C}$.

2 Extremal quasiconformal mappings

Extremal mappings have been prominent in the theory of quasiconformal mappings since its earliest days (see [Gr]). In order to discuss them we need to review some familiar definitions. Recall that a complex-valued function $w = f(z)$ defined in a region V in the complex plane $\mathbb{C}$ is called a *quasiconformal mapping* if it is a homeomorphism of V onto its image and its

---

[1]1991 *Mathematics Subject Classification*. Primary 30C62; Secondary 30C75, 32G15

[2]Research of the second author was supported by the NSF Grant (Tian-yuan) of China.

complex distributional derivatives

$$w_z = \frac{1}{2}\left(\frac{\partial f}{\partial x} - i\frac{\partial f}{\partial y}\right) \quad \text{and} \quad w_{\bar{z}} = \frac{1}{2}\left(\frac{\partial f}{\partial x} + i\frac{\partial f}{\partial y}\right)$$

are Lebesgue measurable locally square integrable functions on V that satisfy the inequality $|w_{\bar{z}}| \leq k|w_z|$ almost everywhere in V, for some real number k with $0 \leq k < 1$. Like many authors, we shall use "qc mapping" as an abbreviation for "quasiconformal mapping".

It is well known that if $w = f(z)$ is a qc mapping defined on the region V then the function w_z is nonzero almost everywhere in V. The function

$$\mu_f = \frac{w_{\bar{z}}}{w_z}$$

is therefore a well defined bounded measurable function on V, called the *complex dilatation* or *Beltrami coefficient* of f. The L^∞ norm of every Beltrami coefficient is less than one. Conversely, every function μ in $L^\infty(V, \mathbb{C})$ with norm less than one is the Beltrami coefficient of some qc mapping whose domain is V. (See [A] or [LV].)

The positive number

$$K(f) = \frac{1 + \|\mu_f\|_\infty}{1 - \|\mu_f\|_\infty}$$

is called the *dilatation* of f. We say that f is K*-quasiconformal* or K*-qc* if f is a qc mapping and $K(f) \leq K$.

Definition. Let V be a plane region and $\mathcal{S}$ be a set of qc mappings whose domain is V. The map f in $\mathcal{S}$ is said to be *extremal in* $\mathcal{S}$ (or simply *extremal* when there is no danger of ambiguity) if $K(f) \leq K(g)$ for all g in $\mathcal{S}$.

3 Two examples

H. Grötzsch initiated the study of extremal qc mappings, and his techniques have inspired many of the subsequent developments. They are well illustrated in the following famous example from [Gr].

Example 1 [Grötzsch]. Let a, b, c, and d be four distinct points on the unit circle S^1, and let a', b', c', and d' be their images under some sense-preserving homeomorphism h of the closed unit disk onto itself. Let $\mathcal{S}$ be the set of all qc mappings of the open unit disk Δ onto itself whose homeomorphic extensions to the closed unit disk send a, b, c, and d to a', b', c', and d' respectively. Grötzsch proved that there is exactly one qc mapping f that is extremal in $\mathcal{S}$. This map f has a beautiful geometric description.

Choose a one-to-one conformal map F of Δ onto the interior of a rectangle R so that the homeomorphic extension of F to the closed disk sends a, b, c, and d to the corners of R. Similarly, choose a one-to-one conformal map G of Δ onto the interior of a rectangle R' so that its homeomorphic extension sends a', b', c', and d' to the corners of R'. Let A be the unique affine map of $\mathbb{R}^2$ onto itself such that $A(R) = R'$ and the composite map $f = G^{-1} \circ A \circ F$ belongs to $\mathcal{S}$. This f is the unique extremal mapping in $\mathcal{S}$. The proof of this theorem, which uses the length-area method, can be found in many books (see for example the first chapter of [A] or [LV]).

It is easy to compute the Beltrami coefficient of the extremal mapping f that we have just described. One finds that

$$\mu_f = t\frac{\overline{F'}}{F'} = t\frac{|(F')^2|}{(F')^2} \;, \tag{3.1}$$

where the complex number t is the constant Beltrami coefficient of the affine map A. Note that $|t| < 1$ and that the holomorphic function $(F')^2$ belongs to $L^1(\Delta)$ since its L^1 norm equals the area of the rectangle R.

Teichmüller discovered that many extremal qc mappings, including the extremal map in the set of qc mappings homotopic to a given sense-preserving homeomorphism between two compact Riemann surfaces of genus greater than one, have Beltrami coefficients whose form resembles (3.1). In recognition of the importance of that discovery, a Beltrami coefficient μ in a plane region V is called a *Teichmüller differential* if there are a complex number t in Δ and a holomorphic function $\varphi(z)$ in $L^1(V)$, not identically zero, such that

$$\mu(z) = t\frac{|\varphi(z)|}{\varphi(z)} \quad \text{almost everywhere in } V \;. \tag{3.2}$$

A qc mapping f whose Beltrami coefficient is a Teichmüller differential is called a *Teichmüller mapping*. A Teichmüller mapping f is extremal in an appropriately defined set of qc mappings. Our next example shows that this set must depend on more than just the boundary values of f.

Example 2. Consider the annulus $V = \{z \in \mathbb{C} : e^{-\pi/2} < |z| < e^{\pi/2}\}$ and the qc mapping $f(z) = -|z|^{-2i}z = -z\exp(-2i\log|z|)$. It is obvious that f maps V onto itself and equals the identity on the boundary of V. The identity map is therefore the unique extremal map in the set of qc self-mappings of V with the same boundary values as f.

However, f is a Teichmüller mapping. In fact $\log f(z) = \pi i + (1-i)\log z - i\log\bar{z}$, locally in V, so μ_f is the Teichmüller differential

$$\mu_f(z) = -\left(\frac{i}{1-i}\right)\frac{|1/z^2|}{1/z^2} \;, \qquad z \in V \;.$$

A set of qc maps in which f is extremal must somehow exclude the identity map. We shall define the appropriate set $\mathcal{S}_f$ in the next section.

Remark. The Teichmüller mappings in Examples 1 and 2 were written, at least locally, as compositions of conformal and affine maps. Any Teichmüller mapping can be written in that form in some neighborhood of each point z in V where the function $\varphi(z)$ in (3.2) is nonzero.

4 The group $\mathcal{QC}_0(V)$

By definition, if V is a plane region the group $\mathcal{QC}_0(V)$ consists of the qc mappings g of the plane onto itself that fix every point of $\mathbb{C} \setminus V$ and are homotopic to the identity by a homotopy g_t in which each g_t is a homeomorphism of $\mathbb{C}$ onto itself that fixes every point of $\mathbb{C} \setminus V$. We denote the restriction of g to V by $g|V$.

Definition. If V is a plane region and f is a qc mapping of V into $\mathbb{C}$, we define the equivalence class of f to be the set

$$\mathcal{S}_f = [f] = \{f \circ (g|V) : g \in \mathcal{QC}_0(V)\} .$$

Intuitively, $[f]$ consists of the qc mappings of V into the plane that have the same "boundary values" as f and can be deformed to f while preserving these boundary values.

Corollary 1.1 in §6 asserts that if f is a Teichmüller mapping of V into $\mathbb{C}$ then f is extremal in its equivalence class. In particular the Teichmüller mappings in Examples 1 and 2 are extremal in their respective equivalence classes, with $V = \Delta$ in Example 1 and $V = \{z \in \mathbb{C} : e^{-\pi/2} < |z| < e^{\pi/2}\}$ in Example 2.

Grötzsch's theorem tells us that the qc mapping f defined in Example 1 is extremal not only in $[f]$ but in the larger class $\mathcal{S}$ defined there. We can obtain Grötzsch's theorem from Corollary 1.1 by appropriately modifying and extending f. Let A and B be Möbius transformations that map the upper half plane $\mathcal{H}^+$ onto Δ and satisfy $A(\infty) = a$ and $B(\infty) = a'$. (Here a and a' have the same meaning as in §3.) For each g in Grötzsch's set $\mathcal{S}$ there is a unique qc mapping $\widetilde{g}$ of $\mathbb{C}$ onto itself that satisfies

$$\widetilde{g}(\bar{z}) = \overline{\widetilde{g}(z)} \quad \text{and} \quad g(A(z)) = B(\widetilde{g}(z)) \quad \text{for all} \quad z \in \mathcal{H}^+ .$$

Set $V = \mathbb{C} \setminus \{A^{-1}(b), A^{-1}(c), A^{-1}(d)\}$. If f is Grötzsch's extremal map, the restriction of $\widetilde{f}$ to V is a Teichmüller mapping. Corollary 1.1 therefore says that $\widetilde{f}$ is extremal in the class $[\widetilde{f}]$. Since $K(g) = K(\widetilde{g})$ and $\widetilde{g}$ belongs to $[\widetilde{f}]$ for all g in $\mathcal{S}$ it follows that f is extremal in $\mathcal{S}$.

Remarks. Let p be an arbitrary complex number distinct from 0 and 1, and let f be an arbitrary qc map of $V = \mathbb{C} \setminus \{0, 1, p\}$ into $\mathbb{C}$. Teichmüller proved that the extremal map in the class $[f]$ is unique and is a Teichmüller mapping (see [A]).

Strebel proved that a Teichmüller mapping is always the unique extremal mapping in its class. That fact cannot be deduced from Theorem 1 but can be derived from inequalities of Reich and Strebel that also are used in the proof of Theorem 1 (see [Ga2]).

5 Two more examples

Not all extremal qc mappings are Teichmüller mappings. The first counterexample occurs in the famous paper [BA] of Beurling and Ahlfors about the boundary values of qc mappings. Their example is the unique extremal mapping in its equivalence class. The first example of nonuniqueness was published by Strebel (see [St2]). We shall briefly discuss these two examples. We thank Edgar Reich for calling the Beurling-Ahlfors example to our attention.

Example 3. Let V be the upper half plane and let K be a real number greater than one. In §5 of [BA] it is shown that the qc mapping $f(z) = z|z|^{K-1}$ of V onto itself is extremal in its class $[f]$. A simple calculation, similar to that in Example 2, shows that

$$\mu_f(z) = \left(\frac{K-1}{K+1}\right) \frac{|1/z^2|}{1/z^2} , \qquad z \in V .$$

This has the same form as (3.2), but f is not a Teichmüller mapping because $z \mapsto |1/z^2|$ is not an integrable function on the upper half plane.

The question whether f is the unique extremal mapping in the class $[f]$ was not considered in [BA]. To study the uniqueness it is convenient to use the one-to-one conformal mapping $w = h(z) = \frac{1}{2} \log z - \frac{\pi}{4}$ of the upper half plane onto the horizontal strip

$$\{w \in \mathbb{C} : |w - \bar{w}| < \pi/2\} . \tag{5.1}$$

The qc mapping $h \circ f \circ h^{-1}$ of that strip onto itself is the horizontal stretching

$$u + iv \mapsto Ku + iv , \qquad u \text{ and } v \text{ in } \mathbb{R} \text{ and } |v| < \pi/4 . \tag{5.2}$$

Strebel proved in [St1] that this stretching is the unique extremal map in its equivalence class, so (equivalently) the Beurling-Ahlfors mapping f is the unique extremal map in $[f]$.

Example 4 (Strebel's chimney). Throughout this paragraph x and y will be real numbers. Fix a real number $K > 1$. Consider the plane region

$$V = \{x + iy : y < 0\} \cup \{x + iy : |x| < 1\}$$

and the qc mapping $f(x+iy) = x+iKy$ of V onto itself. The map f is K-qc and its Beltrami coefficient μ_f is a nonzero constant. Since V has infinite area f is not a Teichmüller mapping. However, it is extremal in its class $[f]$. This class $[f]$ contains an infinite number of distinct extremal mappings. For example, if L is any real number in the closed interval $[K^{-1}, K]$ the K-qc mapping

$$h_L(x+iy) = \begin{cases} x + iKy\,, & |x| < 1 \text{ and } y \geq 0 \\ x + iLy\,, & y < 0 \end{cases}$$

is extremal in $[f]$ (see [St2]).

6 The Hamilton-Krushkal′ condition

An important characterization of the Beltrami coefficients of extremal qc mappings was established through the work of several people. This Hamilton-Krushkal′ condition was first formulated by S. L. Krushkal′, who proved its necessity in some important cases (see [Kru1] and [Kru2]). R. S. Hamilton independently formulated the same condition and showed that it is satisfied by any f that is extremal in its class $[f]$ (see [H]). Reich and Strebel later gave another proof of that result and showed conversely that any f satisfying the Hamilton-Krushkal′ condition is extremal in its class (see [RS], [Ga2], and [R3]).

The work of Reich and Strebel has shown that it is convenient to formulate the Hamilton-Krushkal′ condition in terms of Hamilton sequences. Set $Q(V)$ equal to the closed subspace of $L^1(V)$ consisting of the L^1 functions φ that are holomorphic in the region V.

Definition. Let μ_f be the Beltrami coefficient of some qc mapping f in the plane region V. A *Hamilton sequence* for μ_f is a sequence $\{\varphi_n\}$ in $Q(V)$ such that $\|\varphi_n\| = 1$ for all n and $\lim_{n\to\infty} \iint_V \mu_f(z)\varphi_n(z)\,dx\,dy = \|\mu_f\|_\infty$.

Remark. In requiring all the φ_n in a Hamilton sequence to have norm one (rather than norm at most one) we are following well established practice. This practice has the minor drawback that no Hamilton sequences exist unless $Q(V)$ has positive dimension. For that reason we shall assume from now on that the complement $\mathbb{C} \setminus V$ of V contains at least three points. The dimension of $Q(V)$ is finite if and only if $\mathbb{C} \setminus V$ is a finite set.

Theorem 1 [Krushkal′, Hamilton, Reich and Strebel]. *Let V be a plane region whose complement $\mathbb{C} \setminus V$ contains at least three points. Let f*

be a qc mapping whose domain is V, and let μ_f be its Beltrami coefficient. Then f is extremal in $[f]$ if and only if μ_f has a Hamilton sequence.

Corollary 1.1. *Let V be as above. Every Teichmüller mapping with domain V is extremal in its equivalence class.*

Proof. If f is a Teichmüller mapping its Beltrami coefficient is a Teichmüller differential and, by (3.2), can be written in the form $\mu_f = k|\varphi|/\varphi$ with $0 \leq k < 1$, φ in $Q(V)$, and $\|\varphi\| = 1$. The sequence $\{\varphi_n\}$ with $\varphi_n = \varphi$ for all n is obviously a Hamilton sequence for μ_f. □

Corollary 1.2. *Let V be as above and let f and g be qc mappings whose domain is V. If f is extremal in $[f]$ and μ_g is a scalar multiple of μ_f then g is extremal in $[g]$.*

Proof. Let $\mu_g = \alpha\mu_f$. If $\alpha = 0$ there is nothing to prove. If $\alpha \neq 0$ and $\{\varphi_n\}$ is a Hamilton sequence for μ_f, then $\{|\alpha|\varphi_n/\alpha\}$ is a Hamilton sequence for μ_g.

Example 5. Let V be the horizontal strip (5.1). Since the horizontal stretching (5.2) is extremal in its equivalence class, Corollary 1.2 tells us that every affine qc mapping of V onto itself is extremal in its class. That fact provides another proof that the Teichmüller mapping $f(z)$ in Example 2 is extremal. Consider the holomorphic covering map $z = F(w) = e^{2iw}$ of the strip onto the annulus in Example 2. The affine qc mapping

$$\widehat{f}(u + iv) = u + 2v + \frac{\pi}{2} + iv\ , \qquad u \text{ and } v \text{ in } \mathbb{R} \text{ and } |v| < \pi/4\ ,$$

maps V onto itself and is a lift of f in the sense that $F \circ \widehat{f} = f \circ F$ in V. Every g in $[f]$ has a unique lift $\widehat{g}$ in $[\widehat{f}]$, and $K(\widehat{g}) = K(g)$. Therefore the extremality of $\widehat{f}$ in $[\widehat{f}]$ implies that f is extremal in $[f]$.

Corollary 1.2 says nothing about uniqueness of the extremal map, and more subtle methods are required to study uniqueness. We have already mentioned that Strebel proved the unique extremality of the horizontal stretching (5.2) in [St1]. The unique extremality of all affine maps of the horizontal strip V follows (after a one-to-one conformal mapping of V onto Δ) from Theorem 6 of Sethares's paper [Se] or Theorem 2 of the Hayman-Reich paper [HR]. The Hayman-Reich theorem relies in part on Theorem B of Reich's paper [R2], and Reich suggested to us that the unique extremality of the affine maps of V could be proved by direct application of the methods of [R2]. We shall give a proof here to illustrate these methods. Our proof will use the following very special case of Theorem B of [R2].

Proposition [Reich]. *Let V be any plane region such that $\mathbb{C} \setminus V$ contains at least three points. Suppose there are positive constants B and C and a sequence $\{\varphi_n\}$ in $Q(V)$ with the following properties:*

(a) *$\varphi_n(w) \to 1$ as $n \to \infty$ for all w in V,*

(b) *$|\varphi_n(w)| \le B$ for all positive integers n and all w in V,*

(c) *for every positive integer n we have*

$$\iint_V (|\varphi_n(w)| - \mathrm{Re}(\varphi_n(w)))\, du\, dv \le C\,. \tag{6.1}$$

Then every affine qc mapping f of V into $\mathbb{C}$ is uniquely extremal in its class $[f]$.

To apply the proposition we must define an appropriate sequence $\{\varphi_n\}$ in the horizontal strip V. Choose a strictly increasing sequence of positive real numbers r_n converging to 1 and set $\varphi_n(w) = [r_n/(\cosh^2(w) - r_n^2 \sinh^2(w))]^2$ for all n in $\mathbb{Z}^+$ and all w in V. Since $|\cosh(w)| > |\sinh(w)|$ for all w in V it is clear that each φ_n belongs to $Q(V)$ and that condition (a) is satisfied. It is also easy to verify that (b) holds with $B = 1$.

To verify (c), let a_n be the positive real number such that $\tanh(a_n) = r_n$ and set $V^+ = \{w \in V : \mathrm{Re}(w) > 0\}$. Then

$$\begin{aligned}
&|\varphi_n(w)| - \mathrm{Re}(\varphi_n(w)) \\
&\quad = \frac{2r_n^2(1 - r_n^2)^2[\mathrm{Im}(\cosh^2(w))]^2}{|\cosh^2(w) - r_n^2\sinh^2(w)|^4} \\
&\quad < 2\frac{(1 - r_n^2)^2}{|\cosh(w) - r_n\sinh(w)|^4} \times \frac{|\cosh(w)|^4}{|\cosh(w) + r_n\sinh(w)|^4} \\
&\quad < 8\frac{(1 - r_n^2)^2}{|\cosh(w) - r_n\sinh(w)|^4} = 8\frac{1}{|\cosh(w - a_n)|^4}
\end{aligned}$$

for all n in $\mathbb{Z}^+$ and w in V^+. Therefore

$$\begin{aligned}
&\iint_V (|\varphi_n(w)| - \mathrm{Re}(\varphi_n(w)))\, du\, dv \\
&\qquad = 2\iint_{V^+} (|\varphi_n(w)| - \mathrm{Re}(\varphi_n(w)))\, du\, dv \\
&\qquad < 16\iint_V \frac{du\, dv}{|\cosh(w - a_n)|^4}\,,
\end{aligned}$$

which is independent of n. This confirms that all affine qc mappings of the strip V into $\mathbb{C}$ are uniquely extremal. We thank Edgar Reich for informing us of this fact and that it can be deduced from Theorem B of [R2].

7 Degenerating Hamilton sequences

Teichmüller mappings are the extremal mappings with the simplest geometry, since they are compositions of conformal and affine mappings. One method for studying extremal mappings that are not Teichmüller mappings is to use degenerating Hamilton sequences.

Definition. Let μ_f be the Beltrami coefficient of the qc mapping f. The Hamilton sequence $\{\varphi_n\}$ for μ_f *degenerates* if $\varphi_n(z) \to 0$ uniformly on compact subsets of V as $n \to \infty$.

The connection between degenerating Hamilton sequences and Teichmüller mappings is given by

Lemma 1 [Reich and Strebel]. *If μ_f has a Hamilton sequence that does not degenerate, then f is a Teichmüller mapping.*

Proof. We include the standard proof of Lemma 1 in order to publicize Nikola Lakic's very simple derivation of formula (7.1) below.

Because of the mean value property of holomorphic functions, the closed unit ball of $Q(V)$ is a normal family. The hypothesis of the lemma therefore implies that μ_f has a Hamilton sequence $\{\varphi_n\}$ that converges uniformly on compact subsets of V to a holomorphic function φ that is not identically zero.

The crucial step is to verify the formula

$$\lim_{n\to\infty} \|\varphi_n - \varphi\| = 1 - \|\varphi\| \ . \tag{7.1}$$

Here is Lakic's proof. Since $0 \le |\varphi_n - \varphi| + |\varphi| - |\varphi_n| \le 2|\varphi|$, Lebesgue's dominated convergence theorem implies that

$$\begin{aligned}
&\lim_{n\to\infty} (\|\varphi_n - \varphi\| + \|\varphi\| - \|\varphi_n\|) \\
&\quad = \iint_V \lim_{n\to\infty} (|\varphi_n(z) - \varphi(z)| + |\varphi(z)| - |\varphi_n(z)|) dx\, dy = 0 \ .
\end{aligned}$$

That yields (7.1) because each φ_n has norm one.

By (7.1) (or Fatou's lemma), $0 < \|\varphi\| \le 1$. If $\|\varphi\| = 1$, then (7.1) implies that $\lim_{n\to\infty} \|\varphi_n - \varphi\| = 0$. Therefore

$$\begin{aligned}
\iint_V \mu_f(z)\varphi(z)\, dx\, dy &= \lim_{n\to\infty} \iint_V \mu_f(z)\varphi_n(z)\, dx\, dy \\
&= \iint_V \|\mu_f\|_\infty |\varphi(z)|\, dx\, dy \ ,
\end{aligned}$$

so $\mu_f\varphi = \|\mu_f\|_\infty |\varphi|$ almost everywhere in V and f is a Teichmüller mapping.

If $0 < \|\varphi\| < 1$ set $\psi_n = (\varphi_n - \varphi)/\|\varphi_n - \varphi\|$ for all n and consider the sequence $\{\psi_n\}$ in $Q(V)$. Again using (7.1) we find that

$$\|\mu_f\|_\infty \ \ge \ \lim_{n\to\infty} \left| \iint_V \mu_f(z)\psi_n(z)\, dx\, dy \right|$$

$$= \left| \frac{\|\mu_f\|_\infty - \iint_V \mu_f(z)\varphi(z)\,dx\,dy}{1 - \|\varphi\|} \right|$$

$$\geq \frac{\|\mu_f\|_\infty - \left|\iint_V \mu_f(z)\varphi(z)\,dx\,dy\right|}{1 - \|\varphi\|} \geq \|\mu_f\|_\infty \,.$$

Therefore $\iint_V \mu_f(z)\varphi(z)\,dx\,dy = \|\mu_f\|_\infty\|\varphi\| = \iint_V \|\mu_f\|_\infty|\varphi(z)|\,dx\,dy$. Again we conclude that $\mu_f\varphi = \|\mu_f\|_\infty|\varphi|$ almost everywhere in V and f is a Teichmüller mapping. □

Remark. Let f be a qc mapping whose domain V is a plane region such that $\mathbb{C} \setminus V$ contains at least three points. Combining Lemma 1 and the proof of Corollary 1.1 in §6, we see that f is a Teichmüller mapping if and only if μ_f has a Hamilton sequence that does not degenerate. It is, however, possible for the Beltrami coefficient of a Teichmüller mapping to have a degenerating Hamilton sequence. For an example see §5 of Reich's paper [R1].

8 Boundary dilatation and Strebel's Frame Mapping Criterion

Lemma 1 shows that it is desirable to find geometric conditions on an extremal mapping that will prevent its Beltrami coefficient from having a degenerating Hamilton sequence. Strebel's Frame Mapping Criterion provides such conditions in terms of the "boundary dilatation", which we shall now define.

Definition. Let f be a qc mapping whose domain is the plane region V. The number $H^*(f) = \inf\{K(f|(V \setminus E)) : E \text{ a compact subset of } V\}$ is called the *boundary dilatation* of f. The number $H([f]) = \inf\{H^*(f \circ g) : g \in \mathcal{QC}_0(V)\}$ is called the *boundary dilatation* of the class $[f]$.

Theorem 2 [Strebel's Frame Mapping Criterion]. *Let f be a qc mapping whose domain is the plane region V. If $H([f]) < K(f)$ and f is extremal in $[f]$, then μ_f has no degenerating Hamilton sequence.*

Corollary 2.1 [Strebel]. *Let f be a qc mapping whose domain is the plane region V. If $H([f]) < K(f)$ and f is extremal in $[f]$, then f is a Teichmüller mapping.*

See Chapter 6 of [Ga2] for a proof of Theorem 2.

Remark. In general it is not possible to apply Theorem 2 without some a priori knowledge about the extremal mapping. However, Theorem 2 does imply that the class $[f]$ contains a Teichmüller mapping if its boundary dilatation $H([f])$ equals one, for in that case any extremal map in $[f]$ will either be conformal or have dilatation greater than $H([f])$. Thus, for example, if f is a qc mapping of Δ onto itself and $H^*(f) = 1$, then the class $[f]$ contains a Teichmüller mapping. Any sense-preserving C^1 diffeomorphism of the unit circle onto itself can be extended to such an f (see §3 of [GS]), so it can also be extended to a Teichmüller mapping of Δ onto itself.

9 The converse of Strebel's Frame Mapping Criterion

In our study [EL] of the geometry of the Teichmüller metric we found a striking dichotomy. The geometry is simple at points corresponding to qc mappings f that satisfy the hypothesis of Strebel's Frame Mapping Criterion; it is complicated at points corresponding to mappings f such that μ_f has a degenerating Hamilton sequence (see §10 below). This dichotomy made us wonder whether the converse of the Frame Mapping Criterion holds. That is indeed the case.

Theorem 3 [EL]. *Let f be a qc mapping whose domain is the plane region V. Suppose the set $\mathbb{C} \setminus V$ is infinite. If f is extremal in $[f]$ and $H([f]) = K(f)$, then μ_f has a degenerating Hamilton sequence.*

Proof. The complete proof is given in §5 of [EL]. We shall sketch it here under the assumption that f is not conformal (so that μ_f has positive norm).

Exhaust V by an increasing sequence of nonempty compact subsets E_n, each contained in the interior of the next. For each n let ν_n be the L^∞ function on V that equals zero on E_n and equals μ_f on $V \setminus E_n$. Since $H^*(f) \geq H([f]) = K(f) \geq H^*(f)$, $\|\nu_n\|_\infty = \|\mu_f\|_\infty$.

Let f_n be a qc mapping of V into $\mathbb{C}$ whose Beltrami coefficient $\mu_{f_n} = \nu_n$. It is easy to verify that f_n is extremal in $[f_n]$ (see [EL]). By Theorem 1, there is a function φ_n in $Q(V)$ such that $\|\varphi_n\| = 1$ and

$$\begin{aligned}\frac{1}{n} &> \left| \iint_V \nu_n(z)\varphi_n(z)\,dx\,dy - \|\nu_n\|_\infty \right| \\ &= \left| \iint_{V\setminus E_n} \mu_f(z)\varphi_n(z)\,dx\,dy - \|\mu_f\|_\infty \right| .\end{aligned}$$

Since

$$\left| \iint_{V\setminus E_n} \mu_f(z)\varphi_n(z)\,dx\,dy \right| \leq \|\mu_f\|_\infty \iint_{V\setminus E_n} |\varphi_n(z)|\,dx\,dy \leq \|\mu_f\|_\infty ,$$

we conclude that $\iint_{V\setminus E_n} |\varphi_n(z)|\,dx\,dy \to 1$ and $\iint_{E_n} |\varphi_n(z)|\,dx\,dy \to 0$ as $n \to \infty$. It is therefore obvious that $\{\varphi_n\}$ is a degenerating Hamilton sequence for μ_f. □

Remark. Theorem 3 was already known when $V = \Delta$. In that case the assumption that f is extremal and $H([f]) = K(f)$ implies, by a theorem of Fehlmann (see [F]), that f has a substantial boundary point in the sense of Reich [R1]. The existence of that boundary point ζ implies that μ_f has a Hamilton sequence $\{\varphi_n\}$ that converges to zero uniformly in $\Delta\setminus N$ for every neighborhood N of ζ (see [R1]). That sequence is a degenerating Hamilton sequence for μ_f.

10 The metric geometry of the Teichmüller space of V

The Teichmüller pseudo-metric d on the set of classes of qc mappings with domain V is defined by setting

$$\begin{aligned} d([f_1],[f_2]) &= \min\left\{\log\left(K\left(\tilde{f}_2 \circ f_1{}^{-1}\right)\right) : \tilde{f}_2 \in [f_2]\right\} \\ &= \min\left\{\log\left(K\left(f_2 \circ g \circ f_1{}^{-1}\right)\right) : g \in \mathcal{QC}_0(V)\right\}. \end{aligned}$$

The Teichmüller space $Teich(V)$ is the quotient space obtained by identifying the classes $[f_1]$ and $[f_2]$ when $d([f_1],[f_2]) = 0$. For a more concrete description of $Teich(V)$, we define the Teichmüller class of the qc mapping f with domain V to be the set

$$\begin{aligned} [f]_T &= \{\tilde{f} : \tilde{f} \text{ is qc with domain } V \text{ and } d([f],[\tilde{f}]) = 0\} \\ &= \{\tilde{f} : \tilde{f} \text{ is qc with domain } V \text{ and } K(\tilde{f} \circ g \circ f^{-1}) = 1 \\ &\qquad \text{for some } g \in \mathcal{QC}_0(V)\} \\ &= \{\tilde{f} : \tilde{f} \text{ is qc with domain } V \text{ and } \mu_{\tilde{f}} = \mu_{\hat{f}} \text{ for some } \hat{f} \in [f]\}. \end{aligned}$$

$Teich(V)$ consists of the Teichmüller classes $[f]_T$, f qc with domain V. The Teichmüller pseudo-metric defined above induces the *Teichmüller metric* d_T on $Teich(V)$. Thus

$$d_T([f_1]_T,[f_2]_T) = d([f_1],[f_2]) = \min\{\log(K(f_2 \circ g \circ f_1{}^{-1})) : g \in \mathcal{QC}_0(V)\}$$

for all qc mappings f_1 and f_2 with domain V.

In our paper [EL] we study the geometry of the Teichmüller metric from the point of view of Busemann's geometry of geodesics (see [B]). We use the following basic definition from [B].

Definition. A *straight line* in a metric space X is the image of a distance preserving map from the real line $\mathbb{R}$ into X.

Observe that a straight line in X is necessarily a closed subset of X because $\mathbb{R}$ is a complete metric space.

We are interested in the straight lines in $Teich(V)$ that contain the basepoint $[id]_T$. (Here id is the identity map of V onto itself.) It is well known that for each point $[f]_T$ in $Teich(V)$ there is at least one straight line containing $[f]_T$ and the basepoint. To see why this is so, consider a qc mapping f of V into $\mathbb{C}$ such that f is extremal in the class $[f]$ and the Teichmüller class $[f]_T$ is not the basepoint. Let μ_f be the Beltrami coefficient of f. For each real number t in the open interval $(-1,1)$ let f_t be a qc mapping of V into $\mathbb{C}$ whose Beltrami coefficient is $t\mu_f/\|\mu_f\|_\infty$. It is easy to verify that the set of Teichmüller classes $[f_t]_T$, $-1 < t < 1$, is a straight line in $Teich(V)$ containing both $[f]_T$ and the basepoint.

The question arises whether there can be more than one straight line containing the basepoint and a given second point $[f]_T$. As Kravetz pointed out in the early and correct sections of his flawed paper [K], there is a unique straight line through any two given points of the Teichmüller space of a compact Riemann surface, and the same statement holds in the Teichmüller space of any Riemann surface obtained from a compact Riemann surface by removing a finite point set. However, the Teichmüller space of any other Riemann surface contains pairs of points that lie on infinitely many straight lines. That was first proved by the second author in [Li1] for $Teich(\Delta)$. The general case was proved later by H. Tanagawa in [T] and independently in [Li2] by the second author, who also proved the existence of simple closed geodesics (see [Li3]).

There are numerous metric spaces in which any two distinct points lie on infinitely many straight lines and simple closed geodesics. Perhaps the simplest is the plane $\mathbb{R}^2$ with the metric

$$d_\infty(\bar{x},\bar{y}) = \max\{|x_1-y_1|,|x_2-y_2|\}\,, \quad \bar{x}=(x_1,x_2) \text{ and } \bar{y}=(y_1,y_2) \text{ in } \mathbb{R}^2\,.$$

We refer the reader to §2 of [EL] for a discussion of simple closed geodesics in the metric space $(\mathbb{R}^2, d_\infty)$, contenting ourselves here with this elementary

Lemma 2 [EL]. *Any two given points in $(\mathbb{R}^2, d_\infty)$ lie on an infinite number of straight lines.*

Proof. We emphasize that we are using Busemann's metric definition of straight line, which ignores the linear space structure of $\mathbb{R}^2$. Translations preserve distances and map straight lines to straight lines, so we may assume one of the given points is $\bar{0}$. Let $\bar{x}(\neq \bar{0})$ be the other. Choose $j=1$ or 2 so that $|x_j| = \|\bar{x}\|_\infty$ and choose $\bar{y}$ in R^2 so that $\|\bar{y}\|_\infty = \|\bar{x}\|_\infty$, $y_j = x_j$, and $\bar{y} \neq \bar{x}$. For any $r \geq 1$ the map

$$f_r(t) = \begin{cases} t\bar{x}, & t \in \mathbb{R} \text{ and } t \leq r \\ r\bar{x} + (t-r)\bar{y}, & t \in \mathbb{R} \text{ and } t > r \end{cases}$$

from $\mathbb{R}$ to $(\mathbb{R}^2, d_\infty)$ preserves distances, so its image is a straight line in $(\mathbb{R}^2, d_\infty)$ containing $\bar{0}$ and $\bar{x}$. □

The theorems in [EL] show how a degenerating Hamilton sequence can be used to construct a distance preserving map from $(\mathbb{R}^2, d_\infty)$ into Teichmüller space (see the proof of Theorem 4 below). Any two points in the image of that map lie on infinitely many straight lines and closed geodesics. The constructions in [T] and [Li3] also involve degenerating Hamilton sequences. That is not surprising in view of the next theorem, which shows that degenerating Hamilton sequences are essential for any such constructions. Although we state the theorem only for plane regions, it remains valid if the region V is replaced by any Riemann surface X whose Teichmüller space is infinite dimensional.

Theorem 4. *Let V be a plane region whose complement $\mathbb{C}\backslash V$ is an infinite set and let f be a qc mapping with domain V. Suppose that f is extremal in its class $[f]$. The following statements are equivalent:*

(a) *there are infinitely many straight lines in $Teich(V)$ containing both $[f]_T$ and the basepoint,*

(b) *there are at least two straight lines in $Teich(V)$ containing both $[f]_T$ and the basepoint,*

(c) $H([f]) = K(f)$,

(d) *μ_f has a degenerating Hamilton sequence,*

(e) *$[f]_T$ and the basepoint are in the image of a distance preserving map from $(\mathbb{R}^2, d_\infty)$ to $Teich(V)$.*

Proof. The equivalence of (a), (b), (c), and (d) is part of Theorem 6 in [EL], and in fact the equivalence of (c) and (d) is simply a restatement of Theorems 2 and 3 in §§8 and 9 of the present paper. The implication (e)$\Longrightarrow$(a) follows immediately from Lemma 2.

To show that (d) implies (e) we need some results from [EL]. Let ℓ^∞ be the Banach space of bounded sequences of real numbers with its usual metric

$$d(\{x_n\}, \{y_n\}) = \sup\{|x_n - y_n| : n \geq 1\}, \quad \text{for all } \{x_n\} \text{ and } \{y_n\} \text{ in } \ell^\infty .$$

For any real number x, let $\widehat{x}$ be the sequence in ℓ^∞ all of whose terms equal x. Let r equal the Teichmüller distance from $[f]_T$ to the basepoint. Theorem 6 and Lemma 3.1 of [EL] together show that if (d) holds then there is a distance preserving map F from ℓ^∞ into $Teich(V)$ such that $F(\widehat{0})$ is the basepoint and $F(\widehat{r}) = [f]_T$. Let A be a distance preserving map from $(\mathbb{R}^2, d_\infty)$ to ℓ^∞ whose image contains both $\widehat{0}$ and $\widehat{r}$. The composition $F \circ A$ is the required map from $(\mathbb{R}^2, d_\infty)$ to $Teich(V)$. $\square$

11 Regularity of the Teichmüller metric

In §10 and in our paper [EL] we focussed on the closed subset of $Teich(V)$ consisting of the Teichmüller classes of the qc mappings f such that μ_f has a degenerating Hamilton sequence. As Theorem 4 indicates, the geometry of d_T on that set has some wild properties. Lakic has recently shown that the behavior of d_T on the complementary subset of $Teich(V)$ is much more regular (see [La]). We shall state some of his results in Theorem 5, after a brief discussion of differentiable functions on $Teich(V)$. Although we shall state Theorem 5 only for plane regions, it remains valid if the region V is replaced by any Riemann surface X whose Teichmüller space is infinite dimensional.

The Bers embedding (see [Ga2], [Le], or [N]) allows us to regard $Teich(V)$ as a bounded open set in a Banach space that is isomorphic to the dual space of $Q(V)$ (see §12 of [E]). Therefore it is sufficient for us to discuss functions defined on open subsets of Banach spaces. Let F be a real valued function defined on the open set X of the Banach space E, and let E^* be the (real) Banach space of bounded $\mathbb{R}$-linear functionals on E. We say that F is continuously differentiable (or C^1) on X if there is a continuous map $x \mapsto \ell_x$ from X to E^* such that

$$\lim_{v\to 0} \frac{F(x+v)-F(x)-\ell_x(v)}{\|v\|} = 0 \quad \text{for all } x \in X .$$

If $x \in X$, $v \in E$, and $\lim_{t\to 0} \frac{F(x+tv)-F(x)}{t}$ exists, we say that F has a directional derivative at x in the direction v.

Theorem 5 [Lakic]. *Let V be a plane region whose complement $\mathbb{C} \setminus V$ is an infinite set and let τ be a point of $Teich(V)$. The following statements are equivalent:*

(a) *there is a qc mapping f_0 with domain V such that $\tau = [f_0]_T$, f_0 is extremal in $[f_0]$, and $H([f_0]) < K(f_0)$,*

(b) *the function $[f]_T \mapsto d_T([f]_T, [id]_T)$ is C^1 in some neighborhood of τ,*

(c) *the function $[f]_T \mapsto d_T([f]_T, \tau)$ is C^1 in some neighborhood of the basepoint,*

(d) *the function $[f]_T \mapsto d_T([f]_T, \tau)$ has a directional derivative at the basepoint in every direction.*

The set of points τ at which (a), (b), (c), and (d) hold is open and dense.

The proof that (a) implies (b) and (c) will be given in Lakic's forthcoming paper [La]. He uses inequalities of Reich and Strebel much as Gardiner used them in his treatment [Ga1] of the finite dimensional case, but the infinite dimensional case presents some additional difficulties.

It is clear that (c) implies (d). Theorem 1 of [EL] and its Corollary 1.2 together show that (d) implies (a).

The openness of the set of points satisfying condition (a) is rather obvious, and a proof is indicated in §5 of [EL]. Lakic's proof that this set is dense will be given in [La]. It is based on a very natural idea. Let μ_f be the Beltrami coefficient of an extremal qc mapping f whose Teichmüller class $[f]_T$ is not the basepoint. Given a compact set E in V and a real number r with $0 < r < 1$, let $f_{E,r}$ be a qc mapping whose domain is V and whose Beltrami coefficient equals μ_f in E and equals $r\mu_f$ in $V \setminus E$. Lakic proves that there are sequences $\{E_n\}$ and $\{r_n\}$ such that $r_n \to 1$ and $\tau_n = [f_{E_n,r_n}]_T$ satisfies (a) for every n.

It is interesting that the hypothesis of Strebel's Frame Mapping Criterion is a necessary and sufficient condition for the regularity properties in Theorem 5, and it is satisfying that they hold on a dense open subset of Teichmüller space.

References

[A] L.V. Ahlfors, *Lectures on Quasiconformal Mappings*, Van Nostrand-Reinhold, Princeton, New Jersey, 1966.

[BA] A. Beurling and L. Ahlfors, The boundary correspondence under quasiconformal mappings, *Acta Math.*, **96** (1956), 125–142.

[B] H. Busemann, *The Geometry of Geodesics*, Academic Press, New York, 1955.

[E] C.J. Earle, Teichmüller theory, in *Discrete Groups and Automorphic Functions* (W. J. Harvey, ed.), Academic Press, London, New York, 1977, pp. 143–162.

[EL] C.J. Earle and Li Zhong, Isometrically Embedded Polydisks in Infinite Dimensional Teichmüller Spaces, *J. Geom. Anal.*, to appear.

[F] R. Fehlmann, Über extremale quasikonforme Abbildungen, *Comment. Math. Helv.*, **56** (1981), 558–580.

[Ga1] F.P. Gardiner, On the variation of Teichmüller's metric, *Proc. Roy. Soc. Edinburgh* Sect. A, **85** (1980), 143–152.

[Ga2] F.P. Gardiner, *Teichmüller Theory and Quadratic Differentials*, Wiley-Interscience, New York, 1987.

[GS] F.P. Gardiner and D. Sullivan, Symmetric structures on a closed curve, *Amer. J. Math.*, **114** (1992), 683–736.

[Gr] H. Grötzsch, Über die Verzerrung bei schlichten nicht konformen Abbildungen und über eine damit zusammenhängende Erweiterung des Picardschen Satzes, *Ber. Verh. Sächs. Akad. Wiss. Leipzig*, **80** (1928), 503–507.

[H] R.S. Hamilton, Extremal quasiconformal mappings with prescribed boundary values, *Trans. Amer. Math. Soc.*, **138** (1969), 399–406.

[HR] W.K. Hayman and E. Reich, On Teichmüller mappings of the disk, *Complex Variables Theory Appl.*, **1** (1982), 1–12.

[Kra] S. Kravetz, On the geometry of Teichmüller spaces and the structure of their modular groups, *Ann. Acad. Sci. Fenn.*, **278** (1959), 1–35.

[Kru1] S. Krushkal', Teichmüller's theorem on extremal quasi-conformal mappings, *Siberian Math. J.*, **8** (1967), 231–244.

[Kru2] S. Krushkal', Extremal quasiconformal mappings, *Siberian Math. J.*, **10** (1969), 411–418.

[La] N. Lakic, to appear.

[Le] O. Lehto, *Univalent Functions and Teichmüller Spaces*, Graduate Texts in Mathematics 109, Springer-Verlag, New York, Berlin, Heidelberg, 1987.

[LV] O. Lehto and K. I. Virtanen, *Quasiconformal Mappings in the Plane*, Springer-Verlag, Berlin, Heidelberg, New York, 1973.

[Li1] Li Zhong, Non-uniqueness of geodesics in infinite-dimensional Teichmüller spaces, *Complex Variables Theory Appl.*, **16** (1991), 261–272.

[Li2] Li Zhong, Non-uniqueness of geodesics in infinite-dimensional Teichmüller spaces (II), *Ann. Acad. Sci. Fenn.*, **18** (1993), 355–367.

[Li3] Li Zhong, Closed geodesics and non-differentiability of the metric in infinite-dimensional Teichmüller spaces, *Proc. Amer. Math. Soc.*, **124** (1996), 1459–1465.

[N] S. Nag, *The Complex Analytic Theory of Teichmüller Spaces*, John Wiley and Sons, New York, 1988.

[R1] E. Reich, On the relation between local and global properties of boundary values for extremal quasiconformal mappings in *Discontinuous Groups and Riemann Surfaces* (L. Greenberg, ed.), Ann. Math. Stud. 79, Princeton University Press, 1974, pp. 391–407.

[R2] E. Reich, On criteria for unique extremality of Teichmüller mappings, *Ann. Acad. Sci. Fenn.*, **6** (1981), 289–301.

[R3] E. Reich, Extremal extensions from the circle to the disk, in this volume.

[RS] E. Reich and K. Strebel, Extremal quasiconformal mappings with given boundary values, in *Contributions to Analysis* (L. Ahlfors, et al., eds.), Academic Press, New York, 1974, pp. 375–392.

[Se] G.C. Sethares, The extremal property of certain Teichmüller mappings, *Comment. Math. Helv.*, **43** (1968), 98–119.

[St1] K. Strebel, Eine Abschätzung de Länge gewisser Kurven bei quasikonformer Abbildung, *Ann. Acad. Sci. Fenn.*, **243** (1957), 1–10.

[St2] K. Strebel, Zur Frage der Eindeutigkeit extremaler quasikonformer Abbildungen des Einheitskreises, *Comment. Math. Helv.*, **36** (1962), 306–323.

[St3] K. Strebel, Extremal quasiconformal mappings, *Resultate Math.*, **10** (1986), 168–210.

[T] H. Tanigawa, Holomorphic families of geodesic discs in infinite dimensional Teichmüller spaces, *Nagoya Math. J.*, **127** (1992), 117–128.

Department of Mathematics, Cornell University, Ithaca, New York 14853
Department of Mathematics, Peking University, Beijing 100871, P.R. China

On the Asymptotic Behavior of Quasiconformal Mappings in Space

V.Ya. Gutlyanskiĭ
O. Martio
V.I. Ryazanov
M. Vuorinen

ABSTRACT Conformality at a point of a space quasiconformal mapping is studied without the assumption of differentiability. First, we introduce a notion of asymptotic linearity for general mappings at a prescribed point. Then we prove that the simultaneous asymptotic linearity and analyticity of a mapping f at a point of discreteness imply that f preserves the angles between rays emanating from this point and the moduli of infinitesimal annuli centered at it and thus we strengthen an earlier result of Caraman. Finally, we give sufficient conditions for quasiconformal mappings to be asymptotically linear and analytic in terms of the distortion coefficient and, as a consequence, we generalize the well-known Belinskii weak conformality result to the n-dimensional case.

1 Introduction

A result of Gehring and Lehto [GL] states that a quasiconformal mapping of the plane has a total differential almost everywhere. This result was extended to $\mathbb{R}^n$ by Gehring, Väisälä and Reshetnyak. The simple example of a K^{n-1}-quasiconformal self-mapping of the space $\mathbb{R}^n$

$$f(x) = x|x|^{\frac{1}{K}-1}\ , \qquad f(0) = 0\ , \tag{1.1}$$

shows that f may fail to have a total differential at some point if $K > 1$. A similar example is provided by [B, p.40]

$$f(x) = x(1 - \ln|x|)\ , \qquad f(0) = 0\ , \quad |x| < \delta < 1\ , \tag{1.2}$$

[1]1991 *Mathematics Subject Classification* 30C55,60

where the distortion coefficient $K_f(x) \to 1$ as $x \to 0$. The mapping f also fails to be differentiable at 0.

The first local conformality and regularity results for quasiconformal mappings in the plane are due to Teichmüller [T], Wittich [W], Belinskii [B] and Lehto [L] (see also [RW], [Sch], [BJ], [LV] p.234). For $n \geq 3$ the regularity theory has been studied, for instance, by Reshetnyak [Re$_2$], Ferrand [Fer], Iwaniec [Iw] and Iwaniec and Martin [IM].

The above results on the local behavior of quasiconformal mappings have already found applications to the Nevanlinna value-distribution theory (see [D]), to the theory of asymptotically conformal quasicircles due to Carleson [CA], Becker and Pommerenke [BP] (see, for instance, [ABL], [GR$_1$]), to the theory of asymptotically symmetric functions on the real axis $\mathbb{R}$ due to Fehlmann [F], Gardiner and Sullivan [GS] (see, for instance, [GR$_2$]) and to the theory of quasiconformal structures on manifolds due to Donaldson and Sullivan [DS] (see [IM]).

In 1978 Reshetnyak noted that if $f : U \to \mathbb{R}^n$ is quasiregular and nonconstant such that for some nondecreasing function $\omega : [0,\infty[\to \mathbb{R}$ with

$$\int_0 \frac{\omega(t)}{t} dt < \infty \tag{1.3}$$

and with $K_f(x)-1 \leq \omega(|x-x_0|)$ for almost all $x \in U$, then f is differentiable and conformal at x_0 (see, [Re$_2$], Th.1.1, p.204).

In this paper we study a weak conformality of a space quasiconformal mapping f at a point x_0 where f is not differentiable. This means that f preserves angles between rays emanating from x_0 and the moduli of infinitesimal annuli. Since we do not require differentiability at x_0, a satisfactory characterization of such points x_0 in terms of the distortion coefficient $K_f(x)$ is possible. In particular, we show that if, say $x_0 = 0$, $f(0) = 0$,

$$\lim_{r\to 0} \frac{1}{m(B(0,r))} \int_{B(0,r)} K_f(x)dx = 1 \ , \tag{1.4}$$

then f is weakly conformal at 0 and f has an asymptotic representation $f(x) \sim A(|x|)x$. Here the matrix function $A(t)$ may fail to have a limit as $t \to 0$. We study the behavior of $A(t)$ as $t \to 0$ in detail.

We complete the introduction by giving an equivalent analytic definition of quasiconformality in space due to Väisälä ([V$_2$], p.115).

An ACL^n homeomorphism $f : G \to \mathbb{R}^n$ of the domain $G \subseteq \mathbb{R}^n$ is called a quasiconformal mapping $(1 \leq K < \infty)$ in G if f has a total differential a.e. and

$$\frac{1}{K} \max |f'(x)|^n \leq J(x) \leq K \min |f'(x)|^n \tag{1.5}$$

a.e. in G. Here as usual $J(x)$ denotes the Jacobian of the mapping $f(x)$ and

$$\max|f'(x)| = \max_{|h|=1}|f'(x)h| \ , \quad \min|f'(x)| = \min_{|h|=1}|f'(x)h| \ . \tag{1.6}$$

Note that the left hand side of (1.6) gives the norm $\|f'(x)\|$ of the linear transformation $f'(x) : \mathbb{R}^n \to \mathbb{R}^n$. The distortion coefficient of f at x is defined as $K_f(x) = \|f'(x)\|^n/J(x)$; we let $K_f(x) = 1$ at any point of differentiability of f where $J(x) = 0$. Thus, $K_f(x) \leq K$ a.e. for a K-quasiconformal mapping f and $K_f(x)$ is defined at every point of differentiability of f.

2 On asymptotic linearity

We introduce some preliminary definitions and elementary results. Let U be a domain in $\mathbb{R}^n$ containing the origin 0 and let $v, w : U \to \mathbb{R}^m$ be mappings, not necessarily continuous. We say that $v(x) = o(w(x))$ if for each $\varepsilon > 0$ there is a neighborhood V of 0 such that $\|v(x)\| \leq \varepsilon \|w(x)\|$ for all $x \in V \setminus \{0\}$. Here $\|\cdot\|$ need not be the usual Euclidean norm.

Definition 2.1. *The functions v and w are said to be equivalent as $x \to 0$, denoted as*

$$v(x) \sim w(x) \ , \tag{2.2}$$

if

$$\|v(x) - w(x)\| = o(\|v(x)\| + \|w(x)\|) \ . \tag{2.3}$$

Lemma 2.4. *Condition (2.3) holds if and only if as $x \to 0$*

$$\|v(x) - w(x)\| = o(\min(\|v(x)\|, \|w(x)\|)) \ . \tag{2.5}$$

Proof. If (2.5) holds, the functions v and w are clearly equivalent.

For the converse let $\varepsilon \in (0, \frac{1}{2}]$ and pick a neighborhood V of 0 such that

$$\|v(x) - w(x)\| \leq \varepsilon(\|v(x)\| + \|w(x)\|) \leq \frac{1}{2}(\|v(x)\| + \|w(x)\|)$$

for all $x \in V \setminus \{0\}$. Fix $x \in V \setminus \{0\}$ and assume that $\min(\|v(x)\|, \|w(x)\|) = \|v(x)\|$. Now the last inequality yields

$$\|w(x)\| - \|v(x)\| \leq \frac{1}{2}(\|v(x)\| + \|w(x)\|)$$

and thus

$$\|w(x)\| \leq 3\|v(x)\| \ .$$

This shows that

$$\|v(x) - w(x)\| \leq \varepsilon(\|v(x)\| + \|w(x)\|) \leq 4\varepsilon\|v(x)\|$$

and by symmetry it follows that

$$\|v(x) - w(x)\| \leq 4\varepsilon \min(\|v(x)\|, \|w(x)\|) .$$

This inequality proves (2.5).

Corollary 2.6. *The relations $v(x) \sim w(x)$, $v(x) - w(x) = o(v(x))$, and $v(x) - w(x) = o(w(x))$ are all equivalent.*

Lemma 2.7. *The relation $v(x) \sim w(x)$ is an equivalence relation.*

Proof. Reflexivity and symmetry are immediate. To show the transitivity let $v(x) \sim w(x)$ and $w(x) \sim h(x)$. Then Corollary 2.6 yields

$$\begin{aligned} \|v(x) - h(x)\| &\leq \|v(x) - w(x)\| + \|w(x) - h(x)\| \\ &\leq o(\|v(x)\|) + o(\|h(x)\|) \leq o(\|v(x)\| + \|h(x)\|) \end{aligned}$$

and hence $v(x) \sim h(x)$ as required.

If $m = 1$ then we have the usual equivalence of the real quantities. The usual Euclidean norm and the usual inner product of $\mathbb{R}^m$ are denoted by $| \; |$ and $(,)$, respectively.

Lemma 2.8. **(a)** *If $v(x) \sim w(x)$, then*

$$|v(x)| \sim |w(x)| . \tag{2.9}$$

(b) *If $v(x) \sim w(x)$, then*

$$(v(x), w(x)) \sim |v(x)|\,|w(x)| . \tag{2.10}$$

(c) *If (2.9) and (2.10) hold, then $v(x) \sim w(x)$.*

Proof. (a) follows by the triangle inequality from Definition 2.1.
(b) is easily obtained from (a), (2.4) and the identity

$$2|v||w| - 2(v, w) = |v - w|^2 - (|v| - |w|)^2 .$$

(c) is also a simple consequence of the above identity.

Remark 2.11. By Lemma 2.8, $v(x) \sim w(x)$ is equivalent to two geometric conditions (2.9) and (2.10). Condition (2.10) means that if $v(x) \sim w(x)$, then the angle between $v(x)$ and $w(x)$ converges to zero as $x \to 0$ and (2.9) means the equivalence of lengths of the vectors $v(x)$ and $w(x)$. The example $v, w : \mathbb{R}^n \to \mathbb{R}^m$,

$$v(x) = |x|e_1 , \quad w(x) = x ,$$

shows that condition (2.9) alone does not imply the equivalence $v(x) \sim w(x)$.

Remark 2.12. Let $A = w^{-1}(0)$. Condition (2.9) is equivalent to the conditions $\lim_{x\to 0,\ x\notin A} |v(x)|/|w(x)| = 1$ and $v(x) = 0$ for $x \in U \cap A \setminus \{0\}$ where U is a neighborhood of 0.

Remark 2.13. If the coordinate functions $v_p \sim w_p$, $p = 1,\dots,m$, then $v \sim w$. The example $v, w : \mathbb{R}^n \to \mathbb{R}^m$,

$$v_1 = \dots = v_{m-1} = w_1 = \dots = w_{m-1} = |x| \,, \quad v_m = 0 \,, \quad w_m = |x|^2 \,,$$

shows that the converse conclusion is false.

We write $v(x) \approx w(x)$ and say that v and w have the same order of smallness at the origin if

$$|v(x)|/c \le |w(x)| \le c|v(x)|$$

for some $c \ge 1$ as $x \to 0$.

Definition 2.14. *A mapping $f : U \to \mathbb{R}^m$, $f(0) = 0$, is said to be asymptotically linear at 0 if for each $\alpha \in \mathbb{R} \setminus \{0\}$*

$$f(\alpha x) \sim \alpha f(x) \tag{2.15}$$

as $x \to 0$ and

$$f(x+y) \sim f(x) + f(y) \tag{2.16}$$

as $x \to 0$ whenever $x + y \approx x \approx y$. If (2.15) holds uniformly with respect to α, $c^{-1} \le |\alpha| \le c$, for each $1 \le c < \infty$, then we say that f is uniformly asymptotically linear at 0.

Note that, for instance, the mapping (1.2) is asymptotically linear at 0. Another example of such a mapping is [B, p.40]

$$f(z) = ze^{i\sqrt{-\log|z|}} \,, \qquad |z| < \delta < 1 \,, \ z \in \mathbb{C} \,,$$

where $\mathbb{R}^2$ is interpreted as $\mathbb{C}$. Under this mapping radial lines are transformed to spirals. This last example can also be easily extended to the n-dimensional case. In both examples the functions are quasiconformal mappings.

In Section 4 we shall prove that for quasiconformal mappings asymptotic linearity implies uniform asymptotic linearity.

2.17. Quasiconformal invariance. Here we prove that the equivalence relation $v(x) \sim w(x)$ is an invariant under quasiconformal mappings. We also investigate the invariance of some other concepts.

In view of the geometrical criterion for the equivalence, its conformal invariance is obvious. However, a much stronger invariance holds:

Theorem 2.18. *Let $v, w : U \to \mathbb{R}^m$ be such that $v(x), w(x) \to 0$, $v(x) \sim w(x)$ as $x \to 0$ and let $V(x) = f(v(x))$, $W(x) = f(w(x))$ where f is an arbitrary quasiconformal mapping defined in a neighborhood of the origin for $\mathbb{R}^m$ with $f(0) = 0$. Then $V(x) \sim W(x)$ as $x \to 0$.*

Proof. Indeed, without loss of generality we may assume that $f : \overline{\mathbb{R}^m} \to \overline{\mathbb{R}^m}$ (see, e.g., [V_2], p.139). Using a Möbius transformation and conformal invariance of the considered equivalence one can also suppose that $f(\infty) = \infty$. Then (see [AVV_1], p.15)

$$c_1\left(\frac{|\xi - z|}{|\xi| + |z|}\right)^\beta \le \frac{|f(\xi) - f(z)|}{|f(\xi)| + |f(z)|} \le c_2\left(\frac{|\xi - z|}{|\xi| + |z|}\right)^\alpha$$

for some c_1, c_2, α and $\beta > 0$. Thus at the points x where $|v(x)| + |w(x)| > 0$ we have

$$c_1\left(\frac{|v(x) - w(x)|}{|v(x)| + |w(x)|}\right)^\beta \le \frac{|V(x) - W(x)|}{|V(x)| + |W(x)|} \le c_2\left(\frac{|v(x) - w(x)|}{|v(x)| + |(w(x)|}\right)^\alpha$$

and this proves the statement.

Note that the proof could also be based on the corresponding result of Agard [A] about the distortion of angles under quasiconformal mappings in space. In the plane a similar result is due to Agard and Gehring [AG].

Lemma 2.19. *Let $v, w : U \to \mathbb{R}^m$ be such that $v(x), w(x) \to 0$, $v(x) \approx w(x)$ as $x \to 0$ and let $V(x) = f(v(x))$, $W(x) = f(w(x))$ where f is an arbitrary quasiconformal mapping in a neighborhood of the origin for $\mathbb{R}^m$, $f(0) = 0$. Then $V(x) \approx W(x)$ as $x \to 0$.*

Proof. Arguing as above we may assume additionally that $f : \overline{\mathbb{R}^m} \to \overline{\mathbb{R}^m}$, $f(\infty) = \infty$, is a K-quasiconformal mapping of the extended space. Then

$$1/\eta_{K,n}\left(\frac{|z|}{|y|}\right) \le \frac{|f(y)|}{|f(z)|} \le \eta_{K,n}\left(\frac{|y|}{|z|}\right) \tag{2.20}$$

where $\eta_{K,n} : [0,\infty) \to [0,\infty)$ is a homeomorphism depending only on n and K (see [AVV_2], [Vu_2]). Thus, if

$$c^{-1}|w(x)| \le |v(x)| \le c|w(x)|$$

then

$$C^{-1}|W(x)| \le |V(x)| \le C|W(x)|$$

where $C = \eta_{K,n}(c)$.

Corollary 2.21. *Let $f : U \to \mathbb{R}^m$, $f(0) = 0$, be a quasiconformal mapping which is asymptotically linear at the origin. Then*

$$f(x_1 e_1 + \ldots + x_m e_m) \sim \frac{x_1}{|x|} f(|x|e_1) + \ldots + \frac{x_m}{|x|} f(|x|e_m) \tag{2.22}$$

as $x \to 0$ *when the direction cosines* $\cos\alpha_1 = x_1/|x|, \ldots, \cos\alpha_m = x_m/|x|$ *are constants. Here* $e_1, \ldots, e_m$ *is an orthonormal basis of the space* $\mathbb{R}^m$.

Corollary 2.21 follows immediately from Lemma 2.19 and the following elementary remark.

Remark 2.23. Suppose that $a \sim b$, $c \sim d$ and $a+c \approx a \approx c$, $b+d \approx b \approx d$. Then $a+c \sim b+d$. Indeed, by the triangle inequality

$$\frac{|(a+c)-(b+d)|}{|a+c|+|b+d|} \le \frac{|a-b|+|c-d|}{|a+c|+|b+d|} \le$$

$$\le M\frac{|a-b|+|c-d|}{|a|+|c|+|b|+|d|} \le M\left(\frac{|a-b|}{|a|+|b|}+\frac{|c-d|}{|c|+|d|}\right) \to 0\,.$$

3 Spherical analyticity at a point

Let $U \subset \mathbb{R}^n$ be a domain and let $f : U \to \mathbb{R}^m$ be a mapping. Let $x_0 \in U$ and $0 < r < d(x_0, \partial U)$. We denote by $H(x_0, f, r)$ the least upper bound of the numbers $t \ge 1$ such that

$$\sup_{|x-x_0|=r} |f(x)-f(x_0)| \le t \inf_{|x-x_0|=r} |f(x)-f(x_0)|\,. \tag{3.1}$$

The point x_0 is called a *point* of *spherical analyticity* for f if

$$\lim_{r\to 0} H(x_0, f, r) = 1\,. \tag{3.2}$$

We recall that for $m = n$, if f has a total differential $f'(x_0)$ at x_0 and $J(x_0) = \det f'(x_0) \ne 0$, then f is said to be regular at x_0. P. Caraman [C, Th.18, p.332] proved that a homeomorphism f preserves angles between rays emanating from a regular point of spherical analyticity. In other words, at such a point x_0

$$\frac{(f'(x_0)a, f'(x_0)b)}{|f'(x_0)a||f'(x_0)b|} = \frac{(a,b)}{|a||b|} \tag{3.3}$$

for all $a, b \in \mathbb{R}^n \setminus \{0\}$. At a regular point $x_0 = 0 = f(x_0)$ condition (3.3) is the same as condition (3.6) below. Looking at example (1.2) we find that $x_0 = 0$ is a point of spherical analyticity and asymptotic linearity for f. Moreover, f preserves angles in the sense below, although f does not satisfy the assumptions of the Caraman theorem because f is not differentiable at 0. The following result explains the situation and extends the aforementioned Caraman theorem. Without loss of generality we may assume that $x_0 = 0$ and that $f(0) = 0$. We recall that a mapping $f : U \to \mathbb{R}^m$ is said to be *discrete* at $x_0 \in U$ if there is a neighborhood V of x_0 such that $V \cap f^{-1}(f(x_0)) = \{x_0\}$. Note that if $x_0 = 0$, $f(0) = 0$ and f is discrete at 0, then $|f(x)| > 0$ for x close to 0.

Theorem 3.4. *Suppose that $x_0 = 0$ is a point of discreteness, spherical analyticity and uniform asymptotic linearity for a mapping $f : U \to \mathbb{R}^m$, $f(0) = 0$. Then the following conditions hold: 1) For each $c \geq 1$ and $c^{-1} \leq |x|/|y| \leq c$, as $x \to 0$,*

$$|f(x)| \sim \frac{|x|}{|y|}|f(y)| \,. \tag{3.5}$$

2) f preserves the angles between rays emanating from the origin, i.e.

$$\lim_{t\to 0} \frac{(f(ta), f(tb))}{|f(ta)||f(tb)|} = \frac{(a,b)}{|a||b|} \tag{3.6}$$

for all $a, b \in \mathbb{R}^n \setminus \{0\}$.

Remark 3.7. (a) We recall that condition (3.5) means that for each $\varepsilon > 0$ there is a neighborhood V of 0 such that

$$\left| |f(x)| - \frac{|x|}{|y|}|f(y)| \right| \leq \varepsilon \left(|f(x)| + \frac{|x|}{|y|}|f(y)| \right)$$

for all $x, y \in V$ with $c^{-1} \leq |x|/|y| \leq c$. (b) Note that in (3.6) the inner product on the left hand side is in $\mathbb{R}^m$ and on the right hand side in $\mathbb{R}^n$, respectively.

Proof for Theorem 3.4. First we show (3.5). Fix $c > 1$ and let $c^{-1} \leq |y|/|x| = \varrho \leq c$. The uniform asymptotic linearity of f at 0 yields $|f(\varrho x)| \sim \varrho|f(x)|$. On the other hand, since 0 is a point of spherical analyticity of f, we have $|f(y)| \sim |f(\varrho x)|$. Indeed, for each $\varepsilon > 0$ there exists $r > 0$ such that

$$1 - \varepsilon < \frac{|f(\rho x)|}{|f(y)|} < 1 + \varepsilon$$

provided that $|y| < r$. Hence

$$\left| |f(\rho x)| - |f(y)| \right| \leq \varepsilon |f(y)|$$

and the claim follows, see Corollary 2.6. For (3.6) we consider two possibilities. First suppose that $b = -a \neq 0$. By the asymptotic linearity $f(-ta) \sim -f(ta)$ as $t \to 0$ and this means by Lemma 2.4 that

$$|f(ta) + f(-ta)| = o(\min(|f(-ta)|, |f(ta)|) \,. \tag{3.8}$$

Now the parallelogram law yields

$$(f(ta)), f(-ta)) = \frac{1}{2}(|f(ta) + f(-ta)|^2 - |f(ta)|^2 - |f(-ta)|^2) \tag{3.9}$$

and simple computation based on (3.8), the discreteness and the spherical analyticity of f at 0 give

$$\lim_{t\to 0} \frac{(f(ta), f(-ta))}{|f(ta)||f(-ta)|} = -1 . \tag{3.10}$$

Next let $b \neq -a$. Setting $x = ta$ and $y = tb$ we rewrite (3.5) in the form

$$\frac{|f(tb)|}{|f(ta)|} \sim \frac{|b|}{|a|} \tag{3.11}$$

as $t \to 0$; here the discreteness of f at 0 has also been used. Now we can use the asymptotic linearity and (3.5) to conclude

$$\frac{|f(ta) + f(tb)|}{|f(ta)|} \sim \frac{|f(t(a+b))|}{|f(ta)|} \sim \frac{|a+b|}{|a|} .$$

The parallelogram law now yields

$$1 + \frac{|f(tb)|^2}{|f(ta)|^2} + 2\frac{(f(ta), f(tb))}{|f(ta)|^2} \sim 1 + \frac{|b|^2}{|a|^2} + 2\frac{(a,b)}{|a|^2}$$

and we obtain (3.6) from this and (3.11). Theorem 3.4 follows.

4 Convergence and compactness theorems

First we recall some terminology. Let $A : \mathbb{R}^n \to \mathbb{R}^n$ be a nonsingular linear mapping. The symmetric linear map A^*A has n eigenvectors $l_1, \dots, l_n$, which form an orthonormal basis of $\mathbb{R}^n$. We let $E = E(A)$ denote the ellipsoid $A(B(0,1))$ and write $0 < \lambda_1 \leq \lambda_2 \leq \dots \leq \lambda_n$ for the lengths of semi-axes of E. Now each λ_i is the positive square root of an eigenvalue of A^*A. We arrange the eigenvectors $l_1, \dots, l_n$ so that l_i is an eigenvector corresponding to the eigenvalue λ_i^2, $i = 1, 2, \dots, n$.

Let $f : G \to \mathbb{R}^n$ be a mapping which is differentiable at $x \in G$ with a nonsingular derivative $f'(x)$, i.e., f is regular at x. We let $C(x; h)$ denote the n-dimensional open cube centered at x with edge length $h > 0$ and oriented along the vectors $l_1, \dots, l_n$ corresponding to $A = f'(x)$ in the above discussion.

We begin with the following result.

Lemma 4.1. *Let $f, f_j : G \to \mathbb{R}^n$ be sense-preserving ACL^n homeomorphisms. If $f_j \to f$ as $j \to \infty$ uniformly on compact subsets of G, then*

$$K_f(x) \leq \liminf_{h\to 0} \liminf_{j\to\infty} \frac{1}{h^n} \int\limits_{C(x;h)} K_{f_j}(y)\, dy \tag{4.2}$$

at every regular point $x \in G$ of f.

The next corollary gives a counterpart of the well-known Strebel result for plane quasiconformal mappings (see [St]).

Corollary 4.3. *Let $f, f_j : G \to \mathbb{R}^n$ be K-quasiconformal mappings and suppose that $f_j \to f$ uniformly on compact subsets of G. Then*

$$K_f(x) \le \limsup_{j\to\infty} K_{f_j}(x) \tag{4.4}$$

for a.e. $x \in G$.

Proof for Corollary 4.3. Since

$$\int_{C(x;h)} K_{f_j}(y)\,dy \le \int_{C(x;h)} \sup_{i\ge j}\{K_{f_i}(y)\}\,dy$$

for $j = 1, 2, \ldots$, we obtain

$$\begin{aligned}\liminf_{j\to\infty} \int_{C(x;h)} K_{f_j}(y)\,dy &\le \lim_{j\to\infty} \int_{C(x;h)} \sup_{i\ge j}\{K_{f_i}(y)\}\,dy \\ &\le \int_{C(x;h)} \limsup_{j\to\infty} K_{f_j}(y)\,dy\end{aligned}$$

where the last conclusion follows from the Lebesgue convergence theorem because $K_{f_j} \le K$ a.e. Now this and (4.2) together with the differentiation of an integral (see [S], p.118) yield (4.4).

Proof for Lemma 4.1. We may assume that $x = 0$, $f(0) = 0$, and $f_j(0) = 0$, $j = 1, 2, \ldots$. As above we let $e_1, \ldots, e_n$ be an orthonormal basis in $\mathbb{R}^n$ formed by the eigenvectors of $f'(0)^* f'(0)$. We let λ_i, $i = 1, 2, \ldots, n$, be the lengths of the corresponding semi-axes of $E = E(f'(0))$. We abbreviate $C(h) = C(0; h)$. For $\varepsilon > 0$ we can choose $\delta = \delta(\varepsilon) > 0$ such that for $h \in (0, \delta)$ and for all $y \in C(h)$

$$|f(y) - f'(0)y| < h\varepsilon \tag{4.5}$$

since f is differentiable at 0. Since $f_j \to f$ locally uniformly, we have for all $y \in C(h)$

$$|f_j(y) - f'(0)y| < h\varepsilon \tag{4.6}$$

for $j > j_0$. The set $f'(0)C(h)$ is the rectangular parallelopiped

$$(-\lambda_1 h/2, \lambda_1 h/2) \times \cdots \times (-\lambda_n h/2, \lambda_n h/2)$$

where edges are along the basis vectors $\tilde{e}_1, \ldots, \tilde{e}_n$,

$$\tilde{e}_i = \frac{f'(0)e_i}{|f'(0)e_i|}, \qquad i = 1, 2, \ldots, n,$$

of $\mathbb{R}^n$. Inequality (4.6) yields that the points $f_j(y)$, $y \in C(h)$, all lie in the parallelopiped

$$(-(\frac{\lambda_1}{2}+\varepsilon)h,\ (\frac{\lambda_1}{2}+\varepsilon)h) \times \cdots \times (-(\frac{\lambda_n}{2}+\varepsilon)h,\ (\frac{\lambda_n}{2}+\varepsilon)h)\ .$$

Here $\mathbb{R}^n$ is again equipped with the basis $\tilde{e}_1, \ldots, \tilde{e}_n$. Thus

$$m(f_j(C(h))) \leq h^n(\lambda_1+2\varepsilon)(\lambda_2+2\varepsilon)\ldots(\lambda_n+2\varepsilon) \tag{4.7}$$

and since f_j is an ACL^n homeomorphism, see [Re$_3$, Corollary 1, p.182, and Theorem 2.2, p.99],

$$\int\limits_{C(h)} J_{f_j}(y)\,dy = m(f_j(C(h))) \leq h^n[J_f(0)+\Delta(\varepsilon)] \tag{4.8}$$

by (4.6) because $J_f(0) = \lambda_1\lambda_2\ldots\lambda_n$; here $\Delta(\varepsilon) \to 0$ as $\varepsilon \to 0$.

Next consider the $(n-1)$-dimensional cube $C^*(h)$ with center at $x=0$ and edges (length h) oriented along $e_1, \ldots, e_{n-1}$. Consider a segment $l(z)$, $z \in C^*(h)$, perpendicular to $C^*(h)$ inside $C(h)$ and write $l_j(z)$ for the length of the curve $f_j(l(z))$. Since f is ACL, we have

$$l_j(z) = \int\limits_{-h/2}^{h/2} |f_j'(z,y_n)e_n|\,dy_n \tag{4.9}$$

for almost every $z \in C^*(h)$ with respect to the $(n-1)$-dimensional Lebesgue measure. On the other hand, (4.6) implies

$$l_j(z) \geq (|f'(0)e_n| - 2\varepsilon)h = (\lambda_n - 2\varepsilon)h\ .$$

Hence (4.9) yields

$$\int\limits_{-1/2}^{1/2} |f_j'(z,th)e_n|dt \geq \lambda_n - 2\varepsilon$$

and by the Hölder inequality

$$\int\limits_{-1/2}^{1/2} |f_j'(z,th)e_n|^{n/2}dt \geq (\lambda_n - 2\varepsilon)^{n/2}$$

and thus

$$\int\limits_{-h/2}^{h/2} |f_j'(z,y_n)e_n|^{n/2}\,dy_n \geq h(\lambda_n - 2\varepsilon)^{n/2}$$

for a.e. $z \in C^*(h)$. Integrating over $C^*(h)$ and using the Fubini theorem we obtain

$$\int\limits_{C(h)} |f_j'(y)e_n|^{n/2}\, dy \geq h^n(\lambda_n - 2\varepsilon)^{n/2} . \tag{4.10}$$

Next the Hölder inequality gives

$$\begin{aligned} \int\limits_{C(h)} |f_j'(y)e_n|^{n/2}\, dy &\leq \int\limits_{C(h)} \|f_j'(y)\|^{n/2}\, dy \leq \int\limits_{C(h)} K_{f_j}(y)^{1/2} J_{f_j}(y)^{1/2}\, dy \\ &\leq \Big(\int\limits_{C(h)} K_{f_j}(y)\, dy \Big)^{1/2} \Big(\int\limits_{C(h)} J_{f_j}(y)\, dy \Big)^{1/2} . \end{aligned} \tag{4.11}$$

Here $\|f_j'(y)\| = \sup\limits_{|h|\leq 1} |f_j'(y)h|$ and the inequality $\|f_j'(y)\|^n \leq K_{f_j}(y) J_{f_j}(y)$ has also been used. Now (4.11) together with (4.8) and (4.10) yield

$$\frac{(\lambda_n - 2\varepsilon)^n}{J_f(0) + \Delta(\varepsilon)} \leq \frac{1}{h^n} \int\limits_{C(h)} K_{f_j}(y)\, dy .$$

Letting first $j \to \infty$ and then $h \to \infty$ and finally $\varepsilon \to 0$ we complete the proof.

The local behavior of a quasiconformal mapping can be transformed to the global behavior with help of the following lemma. We first introduce some notation.

Let $f : \mathbb{R}^n \to \mathbb{R}^n$, $f(0) = 0$, be a K-quasiconformal mapping. For each $t > 0$ let

$$F_t(y) = \frac{\omega_t(f(ty))}{\lambda_t} , \qquad y \in \mathbb{R}^n , \ \ \lambda_t = |f(te_1)| ,$$

where ω_t is a rotation of $\mathbb{R}^n$ sending $f(te_1)/\lambda_t$ to the vector e_1 and $e_1, \ldots, e_n$ is a fixed orthonormal basis in $\mathbb{R}^n$.

We say that a family of mappings $g_t : \mathbb{R}^n \to \mathbb{R}^n$, $t > 0$, is a normal family if every sequence g_{t_j}, $j = 1, 2, \ldots$, $t_j \to 0$, has a subsequence which converges locally uniformly.

Lemma 4.12. **(1)** *The mapping family F_t, $t > 0$, is a normal family.* **(2)** *If $F_{t_j} \to F_0$ as $t_j \to 0$, $j \to \infty$, locally uniformly, then $F_0 : \mathbb{R}^n \to \mathbb{R}^n$ is K-quasiconformal mapping normalized as $F_0(0) = 0$ and $F_0(e_1) = e_1$.* **(3)** *If in addition*

$$\lim_{r \to 0} \frac{1}{\Omega_n r^n} \int\limits_{B(0,r)} K_f(x) dx = 1 ,$$

then the mapping F_0 *in (2) is an orthogonal mapping; here* $\Omega_n = m(B(0,1))$.

Proof. Every quasiconformal mapping $f : \mathbb{R}^n \to \mathbb{R}^n$ has an extension to a homeomorphism $f : \overline{\mathbb{R}}^n \to \overline{\mathbb{R}}^n$, $f(\infty) = \infty$. Since $K_{F_t}(y) = K_f(ty)$, $t > 0$, the claim (1) follows from [V$_2$], Theorem 20.5. The claim (2) follows from [V$_2$], Corollary 21.3 and 37.4.

To prove (3) let $\rho > 0$. Then

$$\frac{1}{\Omega_n \rho^n} \int\limits_{B(0,\rho)} K_{F_t}(y)\, dy = \frac{1}{\Omega_n t^n \rho^n} \int\limits_{B(0,t\rho)} K_f(x) dx \to 1$$

as $t \to 0$. Since $K_{F_t}(y) \geq 1$, this shows that $K_{F_t}(y) \to 1$ in measure as $t \to 0$. Without loss of generality we may then assume that $K_{F_{t_j}}(y) \to 1$ a.e. By Corollary 4.3 we have $K_{F_0}(y) = 1$ a.e. and by the special Liouville theorem (see Remark 4.17) and the normalization the mapping F_0 is an orthogonal mapping.

The asymptotic linearity of f at 0 reflects in the limit mapping F_0 in the following way.

Lemma 4.13. *Let* $f : U \to \mathbb{R}^n$, $f(0) = 0$, *be a quasiconformal mapping, asymptotically linear at* 0. *If* $t_j > 0$, $t_j \to 0$ *and* $F_j = F_{t_j}$, *then the sequence* F_j *has a subsequence which converges locally uniformly to a linear quasiconformal mapping* $F_0 : \mathbb{R}^n \to \mathbb{R}^n$, $F_0(e_1) = e_1$.

Proof. By Lemma 4.12 we can choose a subsequence, denoted again by F_j, such that $F_j \to F_0$ locally uniformly and $F_0 : \mathbb{R}^n \to \mathbb{R}^n$ is a quasiconformal mapping satisfying $F_0(0) = 0$ and $F_0(e_1) = e_1$. It remains to show that F_0 is linear. Let $\alpha \in \mathbb{R} \setminus \{0\}$ and $z \in \mathbb{R}^n$. Now

$$F_0(\alpha z) = \lim_{j\to\infty} \frac{\omega_{t_j}(f(t_j \alpha z))}{|f(t_j e_1)|},$$

$$\alpha F_0(z) = \lim_{j\to\infty} \alpha \frac{\omega_{t_j}(f(t_j z))}{|f(t_j e_1)|} = \lim_{j\to\infty} \frac{\omega_{t_j}(\alpha f(t_j z))}{|f(t_j e_1)|}$$

and hence, by (2.15),

$$\begin{aligned} |F_0(\alpha z) - \alpha F_0(z)| &= \lim_{j\to\infty} \frac{|f(t_j \alpha z) - \alpha f(t_j z)|}{|f(t_j e_1)|} \\ &\leq \lim_{j\to\infty} \varepsilon_j \frac{|f(t_j \alpha z)| + |\alpha||f(t_j z)|}{|f(t_j e_1)|} = 0 \end{aligned}$$

where $\varepsilon_j \to 0$ as $j \to \infty$. Note that the expression

$$\frac{|f(t_j \alpha z)| + |\alpha||f(t_j z)|}{|f(t_j e_1)|}$$

remains bounded as $j \to \infty$ due to the quasiconformality of f; this is a simple application of the boundedness of $H(0,f) = \limsup_{r\to 0} H(0,f,r)$ at 0 and a standard modulus estimate. Thus $F_0(\alpha z) = \alpha F_0(z)$.

Similarly, if $x, y \in \mathbb{R}^n \setminus \{0\}$, $x + y \neq 0$, then using (2.16) we can show $F_0(x+y) = F_0(x) + F_0(y)$. Thus F_0 is a linear mapping as required.

Lemma 4.14. *Let $f : U \to \mathbb{R}^n$, $f(0) = 0$, be a quasiconformal mapping. Then the asymptotic linearity of f at the origin implies the uniform asymptotic linearity there.*

Proof. Arguing as above we may assume that $f : \overline{\mathbb{R}^n} \to \overline{\mathbb{R}^n}, f(\infty) = \infty$. Suppose that there exist sequences $\alpha_j \in \mathbb{R}$, $c^{-1} \le |\alpha_j| \le c$, and $x_j \in \mathbb{R}^n \setminus \{0\}$, $x_j \to 0$ as $j \to \infty$ such that

$$\frac{|f(\alpha_j x_j) - \alpha_j f(x_j)|}{|f(\alpha_j x_j)| + |\alpha_j f(x_j)|} \ge \varepsilon \tag{4.15}$$

for some $\varepsilon > 0$. Without loss of generality we may suppose that $\eta_j = x_j/|x_j| \to \eta_0 \in \mathbb{R}^n$, $|\eta_0| = 1$, $\alpha_j \to \alpha_0$, $c^{-1} \le |\alpha_0| \le c$, and also that $F_{t_j} \to F_0$, $t_j = |x_j|$, as $j \to \infty$. Here the auxiliary functions F_t have been defined in Lemma 4.12.

By Lemma 4.12 the limit mapping F_0 is a quasiconformal mapping and from (2.15) we observe that

$$F_0(\alpha z) = \alpha F_0(z) \tag{4.16}$$

for all $\alpha \in \mathbb{R}$ and all $z \in \mathbb{R}^n$. Moreover, by Lemma 4.12

$$F_{t_j}(\eta_j) \to F_0(\eta_0) \neq 0, \qquad F_{t_j}(\alpha_j \eta_j) \to F_0(\alpha_0 \eta_0) = \alpha_0 F_0(\eta_0) \neq 0\ .$$

Consequently, from this and (4.16) we obtain

$$\lim_{j\to\infty} \frac{|F_j(\alpha_j \eta_j) - \alpha_j F_j(\eta_j)|}{|F_j(\alpha_j \eta_j)| + |\alpha_j F_j(\eta_j)|} = \frac{|F_0(\alpha_0 \eta_0) - \alpha_0 F_0(\eta_0)|}{|F_0(\alpha_0 \eta_0)| + |\alpha_0 F_0(\eta_0)|} = 0\ .$$

The last statement contradicts (4.15).

Remark 4.17. In the proof of Lemma 4.12 we used the special Liouville theorem: Every 1-quasiconformal mapping of $\mathbb{R}^n$, $n \ge 2$, is a similarity. For a simple proof of this result see [TV]. Note that the general Liouville theorem ([G1], [Re1], [BI]), valid for $n \ge 3$, is not needed here.

5 On weak asymptotic conformality

We have proved in the preceding sections, in particular, that simultaneous spherical analyticity and asymptotic linearity for a quasiconformal homeomorphism $f : U \to \mathbb{R}^n$, at a prescribed point imply the local geometric properties, typical for a conformal mapping. The mapping f need not be differentiable at that point. The next result gives a sufficient condition for the above properties in terms of the distortion coefficient $K_f(x)$.

Theorem 5.1. *Let $f : U \to \mathbb{R}^n$, $f(0) = 0$, be a quasiconformal mapping and*

$$\lim_{r\to 0} \frac{1}{\Omega_n r^n} \int\limits_{|x|<r} K_f(x)dx = 1 \tag{5.2}$$

Then $x = 0$ is a point of asymptotic linearity and spherical analyticity for the mapping f. Here Ω_n is the volume of the unit ball in $\mathbb{R}^n$.

Combining Lemma 4.13 with assertions of Theorem 3.4 and 5.1 we obtain the following corollaries.

Corollary 5.3. *Suppose that a quasiconformal mapping $f : U \to \mathbb{R}^n$ satisfies (5.2). Then the origin is a point of uniform asymptotic linearity for the mapping f.*

Corollary 5.4. *Suppose that a quasiconformal mapping $f : U \to \mathbb{R}^n$ satisfies (5.2). Then f preserves the moduli of infinitesimal spherical rings centered at the origin.*

This statement was first proved in [K] by methods of Reshetnyak's theory of mappings with bounded distortion [Re$_2$].

Corollary 5.5. *Suppose that a quasiconformal mapping $f : U \to \mathbb{R}^n$ satisfies (5.2). Then f preserves the angles between rays emanating from the origin.*

Proof of Theorem 5.1. Without loss of generality we may assume that $f : \overline{\mathbb{R}^n} \to \overline{\mathbb{R}^n}$ (see [V$_2$], p.139). Applying a Möbius transformation and conformal invariance of asymptotic linearity and spherical analyticity at a point we can also suppose that $f(\infty) = \infty$. In what follows, we will use the notation of Lemma 4.12. By this lemma, any limit function F_0 is an normalized orthogonal mapping of the space $\mathbb{R}^n$.

To prove that $x = 0$ is a point of spherical analyticity for f let us suppose the converse. Then there exist sequences x_j and $z_j \in \mathbb{R}^n\setminus\{0\}$, $|x_j| = |z_j| \to 0$ as $j \to \infty$ such that

$$\lim_{j\to\infty} \frac{|f(z_j)|}{|f(x_j)|} = 1 + \varepsilon \tag{5.6}$$

for some $\varepsilon > 0$. Without loss of generality we can assume that $y_j = z_j/|z_j| \to y_0 \in \mathbb{R}^n$, $\eta_j = x_j/|x_j| \to \eta_0 \in \mathbb{R}^n$, $|\eta_0| = |y_0| = 1$, as $j \to \infty$. Moreover, by Lemma 4.12 we can assume also that $F_{t_j} \to F_0$, locally uniformly as $j \to \infty$, where we set $t_j = |x_j|$. Consequently, $F_{t_j}(y_j) \to F_0(y_0)$,

$F_{t_j}(\eta_j) \to F_0(\eta_0)$ as $j \to \infty$. By Lemma 4.12, under the condition (5.2), F_0 is an orthogonal mapping. Hence

$$\lim_{j\to\infty} \frac{|F_{t_j}(y_j)|}{|F_{t_j}(\eta_j)|} = 1$$

and this contradicts (5.6).

To prove that $x = 0$ is a point of asymptotic linearity for f we will proceed as follows: First we show that

$$f(\rho x) \sim \rho f(x)\,, \qquad \forall \rho \in \mathbb{R} \setminus \{0\} \tag{5.7}$$

as $x \to 0$. Then we will prove that

$$f(x+z) \sim f(x) + f(z) \tag{5.8}$$

as $x \to 0$ where $x \approx z \approx x + z$.

Assume that $f(\rho x) \not\sim \rho f(x)$. Then there exist $\rho \in \mathbb{R} \setminus \{0\}$ and a sequence $x_j \to 0$, $x_j \in \mathbb{R}^n \setminus \{0\}$, such that

$$\frac{|f(\rho x_j) - \rho f(x_j)|}{|f(\rho x_j)| + |\rho||f(x_j)|} \geq \varepsilon \tag{5.9}$$

for some $\varepsilon > 0$. Without loss of generality we may assume that $\eta_j = x_j/|x_j| \to \eta_0 \in \mathbb{R}^n$, $|\eta_0| = 1$. Setting $t_j = |x_j|$ and arguing as above we obtain

$$\lim_{j\to\infty} \frac{|f(\rho x_j) - \rho f(x_j)|}{|f(\rho x_j)| + |\rho f(x_j)|} = \lim_{j\to\infty} \frac{|F_{t_j}(\rho\eta_j) - \rho F_{t_j}(\eta_j)|}{|F_{t_j}(\rho\eta_j)| + |\rho F_{t_j}(\eta_j)|} =$$

$$\frac{|F_0(\rho\eta_0) - \rho F_0(\eta_0)|}{|F_0(\rho\eta_0)| + |\rho F_0(\eta_0)|} = 0$$

and this contradicts (5.9).

Next, let us suppose that $f(x+z) \not\sim f(x) + f(z)$. Then there exist sequences $x_j, z_j \in \mathbb{R}^n \setminus \{0\}$, $x_j, z_j \to 0$ as $j \to \infty$ such that $x_j + z_j \approx x_j \approx z_j$ and

$$\frac{|f(x_j + z_j) - f(x_j) - f(z_j)|}{|f(x_j + z_j)| + |f(x_j) + f(z_j)|} \geq \varepsilon \tag{5.10}$$

for some $\varepsilon > 0$.

Let $y_j = (x_j + z_j)/|x_j + z_j|$, $\eta_j = x_j/|x_j + z_j|$, $\zeta_j = z_j/|x_j + z_j|$ and assume that $y_j \to y_0$, $\eta_j \to \eta_0$, $\zeta_j \to \zeta_0$, $y_0, \eta_0, \zeta_0 \in \mathbb{R}^n$ as $j \to \infty$ and $|y_0| = 1$. Then, setting $t_j = |x_j + z_j|$ and arguing once more as above we deduce, applying Lemma 4.12, that

$$\lim_{j\to\infty} \frac{|F_{t_j}(y_j) - F_{t_j}(\eta_j) - F_{t_j}(\zeta_j)|}{|F_{t_j}(y_j)| + |F_{t_j}(\eta_j) + F_{t_j}(\zeta_j)|} = \frac{|F_0(y_0) - F_0(\eta_0) - F_0(\zeta_0)|}{|F_0(y_0)| + |F_0(\eta_0) + F_0(\zeta_0)|} = 0$$

and this contradicts (5.10). The proof for Theorem 5.1 is complete.

Remark 5.11. Example (1.2) shows that under condition (5.2) the mapping f may still fail to be differentiable at $x = 0$. Hence the coefficient $K_f(x)$ is not defined at $x = 0$. If we complete the definition of $K_f(x)$ for the mapping (1.2) by letting $K_f(0) = 1$ then, condition (5.2) means that the origin is the Lebesgue point for $K_f(x)$. Now the point $x = 0$ is also a point of the approximative continuity for $K_f(x)$ because $K_f \in L^\infty$.

6 Asymptotic linearity and differentiability

We have already noted that the mapping (1.2) is asymptotically linear at the origin. However, it does not have a total differential at this point. We will show that the asymptotic linearity may be interpreted as a kind of differentiability in the sense of Belinskii ([B], p.41), see also [GR$_1$].

Theorem 6.1. *Let $f : U \to \mathbb{R}^n$, $f(0) = 0$, be a quasiconformal mapping. Then the following assertions are equivalent:*

1) *f is asymptotically linear at $x = 0$;*

2) *For $t > 0$ there is a linear mapping $A(t) : \mathbb{R}^n \to \mathbb{R}^n$, such that*

$$f(x) \sim A(|x|)x \tag{6.2}$$

as $x \to 0$ and

$$A(t) \sim A(\tau) \tag{6.3}$$

as $t \to 0$ whenever $\tau \approx t$.

Here linear mappings are interpreted as $n \times n$ matrices and the equivalence in (6.3) is understood with respect to the so-called matrix norm. The norm of the linear operator $A : \mathbb{R}^n \to \mathbb{R}^n$ (see [H], p.178) can be defined in any of the following equivalent ways:

$$\|A\| = \sup_{x \in \mathbb{R}^n \setminus \{0\}} \frac{|Ax|}{|x|}, \tag{6.4}$$

$$\|A\| = \sup_{e \in S^{n-1}} |Ae|, \tag{6.5}$$

$$\|A\| = \sup_{x,y \in \mathbb{R}^n \setminus \{0\}} \frac{(Ax, y)}{|x||y|}, \tag{6.6}$$

$$\|A\| = \sup_{e,y \in S^{n-1}} (Ae, y). \tag{6.7}$$

Corollary 6.8. *The linear mapping* $A(t)$ *in 6.1 2) satisfies*

$$\lim_{\substack{t,\tau\to 0\\ \tau\approx t}} \frac{\|A(t)\|}{\|A(\tau)\|} = 1\ .$$

Remark 6.9. The matrix function $A(t)$ in Theorem 6.1 is not unique. However, as $A(t)$ the matrix function

$$\begin{aligned} A_f(t) &= (f(te_1)/t \cdots f(te_n)/t) \\ &= \begin{pmatrix} f_1(te_1)/t & f_1(te_2)/t & \cdots & f_1(te_n)/t \\ f_2(te_1)/t & f_2(te_2)/t & \cdots & f_2(te_n)/t \\ \vdots & \vdots & \ddots & \vdots \\ f_n(te_1)/t & f_n(te_2)/t & \cdots & f_n(te_n)/t \end{pmatrix} \end{aligned} \tag{6.10}$$

can always be used.

The proof of Theorem 6.1 will be based on the following lemma that gives a number of conditions equivalent to (6.3).

Lemma 6.11. *Let* $f : U \to \mathbb{R}^n$, $f(0) = 0$, *be a quasiconformal mapping satisfying (6.2). Then the following relations are equivalent:*

$$\lim_{\substack{\varrho,\tau\to 0\\ \varrho\approx\tau}} \frac{\|A(\varrho) - A(\tau)\|}{\|A(\varrho)\| + \|A(\tau)\|} = 0\ , \tag{6.12}$$

$$\lim_{\substack{\varrho,\tau\to 0\\ \varrho\approx\tau}} \frac{|A(\varrho)x - A(\tau)x|}{|A(\varrho)x| + |A(\tau)x|} = 0\ , \qquad x \in \mathbb{R}^n \setminus \{0\}\ , \tag{6.13}$$

$$\lim_{\substack{\varrho,\tau\to 0\\ \varrho\approx\tau}} \frac{|A(\varrho)y - A(\tau)y|}{|A(\varrho)y| + |A(\tau)y|} = 0\ , \qquad y \in S^{n-1}\ , \tag{6.14}$$

$$\lim_{\substack{\varrho,\tau\to 0\\ \varrho\approx\tau}} \frac{|A(\varrho)e_p - A(\tau)e_p|}{|A(\varrho)e_p| + |A(\tau)e_p|} = 0\ , \qquad p = 1, \ldots, n\ , \tag{6.15}$$

where $e_1, \ldots, e_n$ *is any orthonormal basis in the space* $\mathbb{R}^n$. *Moreover, the limits (6.13) and (6.14) are uniform with respect to* $x \in C$ *and* $y \in S^{n-1}$, *respectively, provided that* $C \subset \mathbb{R}^n \setminus \{0\}$ *is compact. The matrix function* $A(t)$ *in Theorem 6.1 is unique up to the equivalence* $A_1(t) \sim A_2(t)$ *as* $t \to 0$ *in the above sense.*

Proof. The equivalence of (6.13) and (6.14) is obvious. Condition (6.14) immediately implies (6.15). Conversely, since the mappings $A(t)$ are linear, (6.15) implies (6.13). Next we show that a pointwise limit in (6.14) implies that the limit is uniform on S^{n-1}. Indeed, let us suppose that there exist sequences $y_j \in S^{n-1}$, $\varrho_j, \tau_j \in \mathbb{R} \setminus \{0\}$, $c^{-1} \leq \tau_j/\varrho_j \leq c$, $\tau_j, \varrho_j \to 0$ as $j \to \infty$ such that

$$\frac{|A(\varrho_j)y_j - A(\tau_j)y_j|}{|A(\varrho_j)y_j| + |A(\tau_j)y_j|} \geq \varepsilon \tag{6.16}$$

for some $\varepsilon > 0$. Without loss of generality we can additionally assume that $y_j \to y_0 \in S^{n-1}$ and $t_j = \tau_j/\varrho_j \to t_0 > 0$ as $j \to \infty$ and, moreover, by Lemma 4.12 that $F_{\tau_j}(y) \to F_0(y)$ as $j \to \infty$ locally uniformly with respect to $y \in \mathbb{R}^n$ where F_0 is a quasiconformal mapping normalized by $F_0(0) = 0$. Furthermore, for $t > 0$

$$F_0(ty) = tF_0(y)$$

in view of (6.2) and (6.14) for every fixed $y \in S^{n-1}$. And finally,

$$F_{\tau_j}(y_j) \to F_0(y_0) \neq 0\ , \qquad F_{\tau_j}(t_j y_j) \to F_0(t_0 y_0) \neq 0$$

as $j \to \infty$. Thus

$$\lim_{j\to\infty} \frac{|F_{\tau_j}(t_j y_j) - t_j F_{\tau_j}(y_j)|}{|F_{\tau_j}(t_j y_j)| + |t_j F_{\tau_j}(y_j)|} = \frac{|F_0(t_0 y_0) - t_0 F_0(y_0)|}{|F_0(t_0 y_0)| + |t_0 F_0(y_0)|} = 0$$

and this contradicts (6.16). Therefore, conditions (6.13)–(6.15) are equivalent where the limits in (6.13) and (6.14) can be pointwise or uniform with respect to the parameters $x \in C$ and $y \in S^{n-1}$, respectively, where $C \subset \mathbb{R}^n \setminus \{0\}$ is compact. Finally, if $\rho \approx \tau$ and $x, y \in S^{n-1}$, then

$$A(\rho)y = \frac{1}{\rho} A(\rho)(\rho y) \sim \frac{1}{\rho} f(\rho y)\ , \qquad A(\tau)x = \frac{1}{\tau} A(\tau)(\tau x) \sim \frac{1}{\tau} f(\tau y)$$

and since $\frac{1}{\rho} f(\rho y) \approx \frac{1}{\tau} f(\tau y)$, see Lemma 2.19, we obtain $|A(\rho)y| \approx |A(\tau)x|$. Now

$$\|A(\varrho) - A(\tau)\| = \sup_{y\in S^{n-1}} |A(\varrho)y - A(\tau)y|$$

and consequently, (6.12) is equivalent to (6.14). Since the last statement of Lemma 6.11 is obvious, the lemma follows.

Proof of Theorem 6.1. 1) $\Rightarrow$ 2). By Corollary 2.21 condition (6.2) holds as $x \to 0$ for every fixed $y = x/|x|$ and for $A_f(t)$ given by (6.10). Equivalence (6.3) follows from the (6.15) of Lemma 6.11 because $f(\varrho e_p) \sim f(\tau e_p)\varrho/\tau$, $p = 1, \ldots, n$, by Lemma 4.13. It remains to show that (6.2) holds as $|x| \to 0$ independently of $y = x/|x|$, that is uniformly with respect to $y \in S^{n-1}$. To this end, let us suppose that there exists a sequence $x_j \to 0$ such that

$$\frac{|f(x_j) - A_f(|x_j|)x_j|}{|f(x_j)| + |A_f(|x_j|)x_j|} \geq \varepsilon \tag{6.17}$$

for some $\varepsilon > 0$. Without loss of generality we can additionally assume that $y_j = x_j/|x_j| \to y_0 \in \mathbb{R}^n$, $|y_0| = 1$, as $j \to \infty$. Using the terminology of Lemma 4.12 we may also suppose that $F_{t_j}(y) \to F_0(y)$, $t_j = |x_j|$, as $j \to \infty$ locally uniformly with respect to $y \in \mathbb{R}^n$ where F_0 is a linear quasiconformal mapping. From this convergence we have, in particular, that the matrix function

$$\begin{aligned} B_f(t_j) &= \frac{t_j \omega_{t_j} A_f(t_j)}{\lambda_{t_j}} \\ &= (F_{t_j}(e_1) \ldots F_{t_j}(e_n)) \to (F_0(e_1) \ldots F_0(e_n)) = B_f(0) \end{aligned}$$

as $j \to \infty$ where $F_t(e_p)$, $p = 1, 2, \ldots, n$, are interpreted as columns of the corresponding matrices. Consequently,

$$B_f(t_j) y_j \to B_f(0) y_0 \ .$$

Thus, in view of the linearity of F_0 we obtain that

$$\lim_{j \to \infty} \frac{|F_{t_j}(y_j) - B_f(t_j) y_j|}{|F_{t_j}(y_j)| + |B_f(t_j) y_j|} = \frac{|F_0(y_0) - B_f(0) y_0|}{|F_0(y_0)| + |B_f(0) y_0|} = 0 \ ,$$

and this contradicts (6.17). 2) $\Rightarrow$ 1). By Lemma 6.11 in view of uniformity of the limit (6.13) with respect to the parameter $x \in \mathbb{R}^n \setminus \{0\}$ we have from (6.2), (6.3) that

$$f(\alpha x) \sim \alpha A(|\alpha x|) x \sim \alpha A(|x|) x \sim \alpha f(x)$$

as $x \to 0$ for all $\alpha \in \mathbb{R} \setminus \{0\}$ and similarly

$$f(x+y) \sim A(|x+y|) x + A(|x+y|) y \sim f(x) + f(y)$$

as $x \to 0$ with $x \approx y \approx x + y$.

Acknowledgments

This research was made possible in part by the grants U96000 and U96200 from the International Science Foundation (ISF).

References

[A] Agard S., Angles and quasiconformal mappings in space, *J. Analyse Math.*, **22** (1969), 177–200.

[AG] Agard S., Gehring F.W., Angles and quasiconformal mappings, *Proc. London Math. Soc.* **14A** (1965), no.3, 1–21.

[ABL] Anderson J.M., Becker J, Lesley F.D., On the boundary correspondence of asymptotically conformal automorphisms, *J. London Math. Soc.*, **38** (1988), no.3, 453–462.

[AVV$_1$] Anderson G.D., Vamanamurthy M.K., Vuorinen M., Conformal invariants, quasiconformal maps and special functions, *Lecture Notes Math.*, **1508** (1992), 1–19.

[AVV$_2$] Anderson G.D., Vamanamurthy M.K., Vuorinen M., Dimension-free quasiconformal distortion in n-space, *Trans. Amer. Math. Soc.*, **297** (1986), no.2, 687–706.

[BP] Becker J., Pommerenke Ch., Über die quasikonforme Fortsetzung schlichten Funktionen, *Math. Zeit.*, **161** (1978), 69–80.

[B] Belinskii P.P., *General properties of quasiconformal mappings*, Nauka, Novosibirsk, 1974, Russian.

[BI] Bojarski B., Iwaniec T., Another approach to Liouville theorem, *Math. Nachr.*, **107** (1982), 253–262.

[BJ] Brakalova M., Jenkins J.A., On the local behavior of certain homeomorphisms, *Kodai Math. J.*, **17** (1994), 201–213.

[C] Caraman P., *n-dimensional quasiconformal mappings*, Bucuresti-Kent, 1974.

[CA] Carleson L., On mappings conformal at the boundary, *J. Analyse Math.*, **19** (1967), 1–13.

[DS] Donaldson S., Sullivan D., Quasiconformal 4-manifolds, *Acta Math.*, **163** (1989), 181–252.

[D] Drasin D., On the Teichmüller-Wittich-Belinskii theorem, *Results in Mathematics*, **10** (1986), 54–65.

[DU] Dugundji J., *Topology*, Allyn Bacon, Inc., Boston, 1974.

[F] Fehlmann R., Über extremale quasikonforme Abbildungen, *Comment. Math. Helv.*, **56** (1981), 558–580.

[Fer] Ferrand J., Regularity of conformal mappings of Riemannian manifolds, *Lect. Notes in Math.*, **743** (1979), 197–203.

[GS] Gardiner F.P., Sullivan D.P., Symmetric structures on a closed curve, *Amer. J. Math.*, **114** (1992), 683–736.

[G$_1$] Gehring F.W., Rings and quasiconformal mappings in space, *Trans. Amer. Math. Soc.*, **103** (1962), 353–393.

[G$_2$] Gehring F.W., Quasiconformal mappings, in *Complex analysis and its applications*, 2, International Atomic Energy Agency, Vienna, 1976, pp. 213–268.

[GL] Gehring F.W., Lehto O., On the total differentiability of functions of a complex variable, *Ann. Acad. Sci. Fenn. Ser. A.I.*, **272** (1960), 1–9.

[GR$_1$] Gutlyanskiĭ, V.Ya., Ryazanov, V.I., On asymptotically conformal curves, *Complex Variables J.*, **25** (1994), no.4, 357–366.

[GR$_2$] Gutlyanskiĭ, V.Ya., Ryazanov, V.I., On boundary correspondence under quasiconformal mappings, *Ann. Acad. Sci. Fenn. Ser. Math.*, **21** (1996), 167–178.

[H] Halmos P.R., *Finite-dimensional vector spaces*, 2nd. ed., Van Nostrand Reinhold, New York, etc., 1958.

[HKM] Heinonen J., Kilpelinen, T., Martio, O., *Nonlinear Potential Theory of Degenerate Elliptic Equations*, Clarendon Press, Oxford-New York-Tokyo, 1993.

[Iw] Iwaniec T., Regularity theorems for solutions of partial differential equations related to quasiregular mappings in several variables, *Dissertationes Mathematicae*, **CXCVIII** (1982), 1–48.

[IM] Iwaniec T., Martin G., Quasiregular mappings in even dimensions, *Acta Math.*, **170** (1993), 29–81.

[K] Kudryavtseva N.A., On a condition of weak conformality, *Sibirskii Mat. Zh.*, **35** (1994), no.2, 377–379.

[L] Lehto O., On the differentiability of quasiconformal mappings with prescribed complex dilatation, *Ann. Acad. Sci. Fenn. A I*, **275** (1960), 1–28.

[LV] Lehto O., Virtanen K., *Quasikonforme Abbildungen*, Springer-Verlag, Berlin, etc., 1965.

[MRV] Martio O., Rickman S., Väisälä J., Distortion and singularities of quasiregular mappings, *Ann. Acad. Sci. Fenn. Ser. A1*, **465** (1970), 1–13.

[M] Mostow G.D., Quasiconformal mappings in n-space and rigidity of hyperbolic space forms, *Inst. Hautes Études Sci. Publ. Math.*, **34** (1968), 53–104.

[Re_1] Reshetnyak Yu.G., Stability of conformal mappings in multi-dimensional spaces, *Sibir. Mat. Zh.*, **8** (1967), 91–114.

[Re_2] Reshetnyak Yu.G., *Stability theorems in geometry and analysis*, Nauka, Sibirskoe otdelenie, Novosibirsk, 1982, Russian.

[Re_3] Reshetnyak Yu.G, Space Mappings with Bounded Distortion, *Transl. of Math. Monographs, AMS*, **73** (1989).

[RW] Reich E., Walczak R.K., On the behavior of quasiconformal mappings at a point, *Trans. Amer. Math. Soc.*, **117** (1965), 338–351.

[S] Saks S., *Theory of the integral*, Dover Publ. Inc., New York, 1937.

[Sch] Schatz A., On the local behavior of homeomorphic solutions of Beltrami's equations, *Duke Math. J.*, **35** (1968), 289–306.

[St] Strebel K., Ein Konvergenzsatz für Folgen quasikonformen Abbildungen, *Comment. Math. Helv.*, **44** (1969), 469–475.

[T] Teichmüller O., Untersuchungen über konforme und quasikonforme Abbildung, *Deutsche Math.*, **3** (1938), 621–678.

[TV] Tukia P., Väisälä J., A remark on 1-quasiconformal maps, *Ann. Acad. Sci. Fenn. Ser. A*, **10** (1985), 561–562.

[V_1] Väisälä J., On quasiconformal mappings in space, *Ann. Acad. Sci. Fenn. Ser. A*, **298** (1961), 1–36.

[V_2] Väisälä J., Lectures on n-Dimensional Quasiconformal Mappings, *Lecture Notes in Math.*, **229**, Springer-Verlag, Berlin-New York, 1971.

[Vu_1] Vuorinen M., Conformal Geometry and Quasiregular Mappings, *Lecture Notes in Math.*, **1319**, Springer-Verlag, Berlin-New York, 1988.

[Vu_2] Vuorinen M., Quadruples and spatial quasiconformal mappings, *Math. Z.*, **205** (1990), 617–628.

[W] Wittich H., Zum Beweis eines Satzes über quasikonforme Abbildungen, *Math. Z.*, **51** (1948), 275–288.

Institute of Applied Mathematics and Mechanics, NAS of Ukraine, ul. Roze Luxemburg 74, 340114, Donetsk, Ukraine
Department of Mathematics, University of Helsinki P.O. Box 4 (Yliopistonk. 5), FIN–00014, Helsinki, Finland

The Gehring Lemma

Tadeusz Iwaniec

ABSTRACT I am honored and pleased to comment on the work of Professor Frederick W. Gehring on the occasion of his 70th birthday. Because of his pioneering ideas, the theory of quasiconformal mappings has enjoyed an extensive development. Reverse Hölder inequalities represent one of Fred's supreme accomplishments, with enormous repercussions for nonlinear PDEs. I am pleased to acknowledge the immense influence Fred has had on me and my work. His wonderful lemma, in particular, has shaped my own views on the L^p-theory of mappings of finite distortion and their governing equations. It is to Fred and this lemma that my article is dedicated.

1 Introduction

The Lemma is often phrased as a self-improving property of reverse Hölder inequalities. It states:

Gehring's Lemma. *Let Ω be an open subset of $\mathbb{R}^n$ and let $1 < p < \infty$. Suppose a nonnegative measurable function h on Ω satisfies*

$$\fint_Q h^p \le A_p \left(\fint_Q h \right)^p \tag{1.1}$$

for all cubes $Q \subset \Omega$ with a constant A_p independent of the cube. Then there exist a new exponent $s > p$ and a constant A_s depending only on n, p and A_p such that

$$\fint_Q h^s \le A_s \left(\fint_Q h \right)^s \tag{1.2}$$

Here and subsequently, the symbol $\fint_Q h$, also denoted at times by h_Q, stands for the L^1-mean of h over the cube Q. That is,

$$h_Q = \fint_Q h = \frac{1}{|Q|} \int_Q h$$

[1]Research supported by NSF Grant DMS-9401104.

In what follows Q may expand to the entire space $\mathbb{R}^n$ and, as a convention, we take h_Q to mean zero in such cases.

There are a few key lemmas in real-variable theory which comprise to the arsenal of all analysts. To my mind, Gehring's Lemma is one of those elegant and powerful tools which will be appreciated forever. Its fundamental character is attested to by its numerous generalizations and applications. My only regret is that I won't be able to include many of these contributions here.

The origin of reverse Hölder inequalities can be traced back to work of B. Muckenhoupt [M]. While the early contributors were concerned with L^p-weights, the proper founder of the theory, Fred Gehring, made a decisive step to a firm foundation of the L^p-theory of quasiregular mappings and nonlinear PDEs. For the sake of brevity, I will need to rephrase some of the well-known definitions and results.

Definition 1.1. A homeomorphism $f : \Omega \to \mathbb{R}^n$ of Sobolev class $W^{1,n}_{\mathrm{loc}}(\Omega, \mathbb{R}^n)$ is said to be K-quasiconformal, $1 \leq K < \infty$, if its differential matrix $Df(x)$ and the Jacobian determinant $J(x, f) = \det Df(x)$ satisfy

$$|Df(x)|^n \leq KJ(x, f). \tag{1.3}$$

The definition of a K-quasiregular mapping just amounts to dropping the injectivity of f.

Our account begins with the Acta paper [G1], in which Fred derives inequalities relating L^1- and L^n-means of the differential of a K-quasiconfomal mapping:

$$⨍_Q |Df|^n \leq A_n(K) \left(⨍_Q |Df| \right)^n \tag{1.4}$$

for all cubes $Q \subset \Omega$ which are sufficiently far from the boundary that diam $f(Q) \leq$ dist $[f(Q), \mathbb{R}^n \setminus f(\Omega)]$ (see also O. Martio [Ma]). The implications of this inequality were unprecedented in the theory of multidimensional quasiconformal mappings.

Integrability Theorem. *Every K-quasiconformal mapping $f : \Omega \to \mathbb{R}^n$ belongs to the Sobolev space $W^{1,p}_{\mathrm{loc}}(\Omega, \mathbb{R}^n)$ with an exponent $p = p(n, K)$ greater than the dimension.*

This fundamental theorem not only extended an earlier result of B. Bojarski [B1] but also opened the most direct way to the analytic foundation of quasiregular mappings in $\mathbb{R}^n$ (see [BI], [I3,5] and the references given there).

For quasiregular mappings the arguments establishing inequality (1.4) no longer apply. Nevertheless, as shown by A. Elcrat and N.G. Meyers [EM], slightly weaker estimates still remain valid; namely,

$$\fint_Q |Df|^n \le C(n)K\left(\fint_{2Q} |Df|^{\frac{n^2}{n+1}}\right)^{\frac{n+1}{n}}, \tag{1.5}$$

where $2Q \subset \Omega$ is the cube with the same center as Q but enlarged to twice $Q's$ size. The so-called *weak reverse Hölder inequalities* emerged:

$$\fint_Q h^p \le A_p\left(\fint_{2Q} h\right)^p \tag{1.6}$$

for all cubes $Q \subset 2Q \subset \Omega$ and some $1 < p < \infty$. The derivation of (1.5) and its counterparts for nonlinear PDEs essentially relied upon Caccioppoli type estimates. The classical references are: [EM], [GM], [BI], [Gi], [GG] and [HKM]. For the Jacobian of a mapping $f \in W^{1,n}_{\text{loc}}(\Omega, \mathbb{R}^n)$ the Caccioppoli estimate takes the form:

$$\int_\Omega \varphi(x)J(x,f)dx \le \int_\Omega |\nabla\varphi|\ |f-c|\ |Df|^{n-1}, \tag{1.7}$$

where $\varphi \in C_0^\infty(\Omega)$ and $c \in \mathbb{R}^n$. By virtue of the Poincaré-Sobolev Lemma this yields

$$\fint_Q J(x,f)dx \le C(n)\left(\fint_{2Q} |Df|^{\frac{n^2}{n+1}}\right)^{\frac{n+1}{n}} \tag{1.8}$$

provided $J(x,f) \ge 0$. Hence (1.5) is a straightforward consequence of (1.3) (see [BI] for more general estimates and a thorough treatment). In more recent years new interest in estimates (1.7) and (1.8) has arisen in connection with integrability properties of the Jacobian determinant. But we shall not develop this point here. We refer the reader to [GIM] and the references given there.

Gehring's Lemma for the weak reverse Hölder inequalities (1.6) required only minor modification of the original proof. In Section 3, we shall examine Gehring's arguments quite carefully and in a setting of greater generality. Some minor hypothesis, accidentally omitted in the statement of a lemma but mentioned in an alternative proof of his main result, has inspired others to improve upon and extend his lemma. Sometimes I find it difficult to give credit where it is properly due for these innovations. They are too numerous to discuss them all, but some will be recognized here in various places.

The above mentioned oversight in Fred's paper raises an interesting question. Inequalities (1.1) clearly imply a pointwise estimate between the maximal function of h^p and that of h:

$$(Mh^p)(x) \le A_p[Mh(x)]^p. \tag{1.9}$$

Is the latter estimate sufficient to imply inequality (1.2)? The answer is indeed in the affirmative, but it requires a priori that $h \in L^s(\Omega)$ (see Proposition 5.2). The core of the proof is to examine carefully the dependence on

p of the constants occuring in maximal inequalities like (2.4). Surprisingly a routine approximation argument shows that one can dispense with this a priori integrability requirement if h satisfies the reverse Hölder inequalities. Within this framework, generalizations to the weak reverse Hölder inequalities become straightforward and it is our goal to present this technique in detail.

As a matter of fact, Fred's first attempt depended strongly on maximal functions, yielding the $L \log L$-integrability of the Jacobian of a quasiconformal mapping but not the full result, that is the Integrability Theorem. It is important to recognize, in this connection, the amazing contribution of Stefan Müller [Mu], who showed that the $L \log L$-integrability of a non-negative Jacobian $J(x, f)$ has nothing to do with quasiconformality on the part of f. Undoubtedly, Stefan Müller is the one who deserves much of the credit for raising this subject to its present level. Today the integrability theory of Jacobians has a life of its own. But we cannot enter into a discussion of this topic here.

I have tried to elaborate basic techniques and to present them in a manner accessible to students as material worthy of further exploration. Partly for this purpose, and partly to reserve this article as a testimony to the value of Gehring's Lemma itself, I have deliberately not discussed here questions concerning estimates with large exponents (see [I1,2]) or those below the natural exponent p (see [I4] and [IS]). These results, as evidenced by current literature, form the most important complement to Gehring's Lemma. I shall refer to them only by way of digression. I have attempted to recall all contributions to the development of Gehring's Lemma, but I have undoubtedly overlooked a number of them. Some not mentioned explicitely are included in the references.

2 Retrospection

We shall take a little time to trace the early investigations that led Fred to his lemma. The subject was linked with maximal inequalities from the very beginning. At the time the utility of maximal functions was already well recognized due to basic results of N. Wiener [W] and E. Stein [S1]. In order to get quickly to the substance of this approach we formulate, as a starting point, the most general and precise estimates for maximal functions.

Throughtout, Ω will be a cube in $\mathbb{R}^n$, which may expand to the entire space $\mathbb{R}^n$. For $h \in L^1(\Omega)$, the Hardy-Littlewood maximal function on Ω is defined by

$$Mh(x) = \sup \left\{ \fint_Q |h|; \quad x \in Q \subset \Omega \right\}$$

the supremum being taken over all cubes $Q \subset \Omega$ containing x. Sometimes, to indicate the dependence of M on Ω, we write $M_\Omega h$. Also, we use $M_p h$ as an abbreviation for $(Mh^p)^{\frac{1}{p}}$.

We shall consider two increasing functions

$$\Phi(t) = tF(t) \quad \text{and} \quad \Psi(t) = tF(t) - \int_0^t F(s)ds \tag{2.1}$$

where $F : \mathbb{R}_+ \to \mathbb{R}_+$ is a differentiable function with $F'(t) \geq 0$ for all $t \geq 0$.

Maximal Theorem. *Let h be a nonnegative measurable function on the cube Ω. Then*

$$\int_\Omega \Psi(Mh) \leq 3^n \int_\Omega \Phi(2h) \tag{2.2}$$

Conversely,

$$\int_\Omega \Phi(h) \leq F(h_\Omega) \int_\Omega h + 2^n \int_\Omega \Psi(Mh) \tag{2.3}$$

Here we recall that $h_\Omega = 0$ in case $\Omega = \mathbb{R}^n$ (see [GIM] for a proof and more inequalities). Maximal inequalities in Orlicz spaces had previously been established by R. Bagby and D. Parson [BP], but they did not give clear evidence that the constants depended only on the dimension. And this fact is vital for our applications.

Letting $F(t) = t^{p-1}, p > 1$, we recover Wiener's L^p-bound, as well as a reverse bound which exhibits an asymptotically correct constant, as p approaches 1:

$$\fint_\Omega h^p - \left(\fint_\Omega h\right)^p \leq 2^n \frac{p-1}{p} \fint_\Omega (Mh)^p \leq 6^n 2^p \fint_\Omega h^p. \tag{2.4}$$

For $\Omega = \mathbb{R}^n$ this translates to

$$2^{-n} \int_{\mathbb{R}^n} h^p \leq \frac{p-1}{p} \int_{\mathbb{R}^n} (Mh)^p \leq 3^n \cdot 2^p \int_{\mathbb{R}^n} h^p. \tag{2.5}$$

Setting $F(t) = \log(e+t)$, we find that $\Psi(t) = e + t - e\log(e+t)$. Thus Mh is integrable if and only if h belongs to the Zygmund class $L\log L$. This fact was already known to E. Stein in 1969 [S1]. In order to arrive at precise bounds we apply the Maximal Theorem to h/h_Ω in place of h. This yields

$$3^{-n-2} \fint_\Omega Mh \leq \fint_\Omega h \log\left(e + \frac{h}{h_\Omega}\right) \leq 2^n \fint_\Omega Mh \tag{2.6}$$

provided $h_\Omega \neq 0$, which excludes the degenerate case $\Omega = \mathbb{R}^n$. Note that the $L\log L$-mean in the middle of (2.6) defines a norm in $L\log L(\Omega)$ and, therefore, can be used to define a maximal operator $M_{L\log L}$ (see [GI] for the definition and results). Further, one has to be a little careful in taking the limit as Ω shrinks to a point. The definition of the maximal function depends on the cube Ω, so no point wise estimate follows from (2.6). We now recall the reverse Hölder inequalities

$$\fint_Q h^p \le A_p \left(\fint_Q h\right)^p \tag{2.7}$$

for all cubes $Q \subset \Omega$, with some $1 < p < \infty$, and a constant A_p independent of Q. This obviously implies a pointwise inequality between maximal funtions of h^p and h:

$$Mh^p \le A_p(Mh)^p$$

From Wiener's inequality we infer that $Mh^p \in L^1(\Omega)$ and

$$\int_\Omega Mh^p \le C(n,p) \int_\Omega h^p$$

Then, in view of Stein's result, it follows that h^p actually belongs to the Zygmund class $L\log L(\Omega)$. A precise estimate is obtained from (2.6) and (2.7); namely,

$$\fint_\Omega h^p \log(e+h) \le A_p(1)(h_\Omega)^p \log(e+h_\Omega) \tag{2.8}$$

Summarizing, the reverse Hölder inequalities (2.7) imply (2.8), the so-called reverse Jensen's inequality.

For a K-quasiconformal mapping this fact results, by way of (1.4), precisely in the $L\log L$-integrability of the Jacobian. Note that we did not really have to use (1.4); we could have instead applied (1.8) and so obtained Müller's result [Mu].

This first step toward higher integrability of h can actually be carried further and it is our aim next to demonstrate how. We shall proceed by induction to establish the following generalization of (2.8):

$$\int_\Omega h^p \log^k(e+h) \le A_p(k)(h_\Omega)^p \log^k(e+h_\Omega) \tag{2.9}$$

for $k = 0, 1, 2, \ldots$. The point is that inequalities (2.8) and (2.9) remain valid when we replace Ω by any subcube $Q \subset \Omega$. Suppose (2.9) has been proven for some $k \ge 1$. Using convexity of the function $t \longmapsto t\log^{\frac{k}{p}}(e+t)$ and Jensen's inequality, we derive the somewhat weaker estimates

$$\fint_Q h^p \log^k(e+h) \le A_p(k)\left[\fint_Q h\log^{\frac{k}{p}}(e+h)\right]^p$$

for every cube $Q \subset \Omega$. With the notation $H = h\log^{\frac{k}{p}}(e+h)$, we thus arrive at the reverse Hölder inequalities for H:

$$\fint_Q H^p \le A_p(k)\left(\fint_Q H\right)^p.$$

This makes it legitimate to apply (2.8) to H. Accordingly, there is a constant C depending on n, p and $A_p(k)$ such that

$$\fint_\Omega H^p \log(e+H) \le C(H_\Omega)^p \log(e+H_\Omega) \tag{2.10}$$

Since $h \le H$, we obtain

$$\fint_\Omega h^p \log^{k+1}(e+h) \le C(H_\Omega)^p \log(e+H_\Omega) \tag{2.11}$$

What is left is to estimate H_Ω in terms of h_Ω. This is done with the aid of the induction hypothesis (2.9) as follows

$$(H_\Omega)^p \le \fint_\Omega H^p = \fint_\Omega h^p \log^k(e+h) \le A_p(k)(h_\Omega)^p \log^k(e+h_\Omega)$$

The proof of the induction step is completed by substituting this into (2.11), the detailed verification being left to the reader. Although very simple and elegant, these arguments give no information on the constant $A_p(k)$. Nevertheless, more scrupulous application of the maximal inequalities reveals that $A_p(k) \le (Ck)^k$, $k = 1, 2, \dots$, where C depends only on n, p and the initial constant A_p in (2.7). We now apply (2.9), with h replaced by h/h_Ω, to obtain

$$\fint_\Omega h^p \log^k(e + \frac{h}{h_\Omega}) \le (2Ck)^k A_p (h_\Omega)^p \qquad k = 0, 1, \dots \tag{2.12}$$

This is justified, for the reverse Hölder inequalities are homogeneous. For sufficiently small $\varepsilon > 0$ we may multiply both sides of (2.12) by $\frac{\varepsilon^k}{k!}$ and sum over all integers $k = 0, 1, \dots$. This gives

$$\fint_\Omega h^p \left(e + \frac{h}{h_\Omega}\right)^\varepsilon \le A(n,p)(h_\Omega)^p,$$

whence

$$\fint_\Omega h^{p+\varepsilon} \le A(n,p)(h_\Omega)^{p+\varepsilon},$$

which is the desired estimate.

Although the actual proof of Gehring's Lemma makes no appeal to maximal functions it resembles the arguments commonly used for the maximal inequalities. The key tools are of course Vitali's Lemma and the Calderón-Zygmund Decomposition. In the next section we mimic Fred's original idea in somewhat greater generality.

3 Gehring's Lemma in Orlicz Spaces

Let $\Phi : \mathbb{R}_+ \to \mathbb{R}_+$ be a convex function of the form $\Phi(t) = tF(t)$, where $F(t)$ increases from zero to infinity as t runs over the interval $[0, \infty)$. We impose the doubling condition:

$$F(2t) \leq M\, F(t) \tag{3.1}$$

for some $M \geq 1$ and all $t \geq 0$. This implies that

$$F(\lambda t) \leq M^{1+\log_2 \lambda} F(t) \tag{3.2}$$

for all $\lambda \geq 1$ and $t \geq 0$.

The Orlicz space $L^{\Phi}(\mathbb{R}^n)$ consists of all measurable functions g such that $\Phi(|g|)$ is integrable. $L^{\Phi}(\mathbb{R}^n)$ is a Banach space equipped with the Luxemburg norm

$$\|g\|_{\Phi} = \inf\{\lambda > 0;\ \int \Phi\left(\frac{|g|}{\lambda}\right) < \Phi(1)\} \tag{3.3}$$

The best general reference for Orlicz spaces is [RR]. We shall present here what is perhaps the most general version of Gehring's Lemma.

Lemma 3.1. *Let $g, h \in L^{\Phi}(\mathbb{R}^n)$ be nonnegative measurable functions satisfying:*

$$\Phi^{-1}\left[\fint_Q \Phi(g)\right] \leq K \fint_{2Q} g + \Phi^{-1}\left[\fint_{2Q} \Phi(h)\right] \tag{3.4}$$

for every cube $Q \subset \mathbb{R}^n$, where the constant $K > 1$ is independent of the cube. Then there exists a positive exponent $\varepsilon = \varepsilon(n, K, M)$ and a constant $C = C(n, K, M)$ such that

$$\int_{\mathbb{R}^n} gF^{1+\varepsilon}(g) \leq C \int_{\mathbb{R}^n} hF^{1+\varepsilon}(h) \tag{3.5}$$

Proof. We point out that the functional $g \longmapsto \Phi^{-1}\left[\fint_Q \Phi(|g|)\right]$ may not grow at the same rate as the Φ-norm of g. This occurs typically when Φ has nearly linear growth. Examples; $\Phi(t) = t\log^{\alpha}(1+ t)$, $\alpha > 0$, and $\Phi(t) = t\log\log(e+t)$. There is no loss of generality in assuming that the integral of the right hand side of (3.5) is converging. Given any positive number t we put $s = (2K+2)t$. We shall make use of the familiar Calderón-Zygmund decomposition:

$$\mathbb{R}^n = \left(\bigcup_{j=1}^{\infty} Q_j\right) \cup \left(\mathbb{R}^n \setminus \bigcup_{j=1}^{\infty} Q_j\right)$$

where Q_j are pairwise disjoint cubes such that

$$\Phi(s) \leq \fint_{Q_j} \Phi(g) \leq 2^n \Phi(s) \tag{3.6}$$

for $j = 1, 2, \ldots,$ and

$$g(x) \leq s \quad \text{for almost every } x \in \mathbb{R}^n \setminus \bigcup_{j=1}^{\infty} Q_j \tag{3.7}$$

Consider the distribution sets of g and h

$$G_s = \{x \in \mathbb{R}^n;\ g(x) > s\} \quad \text{and} \quad H_s = \{x \in \mathbb{R}^n;\ h(x) > s\}$$

Thus (3.7) reads as $G_s \subset \cup Q_j$, while (3.6) yields

$$\int_{G_s} \Phi(g) \leq \sum_{j=1}^{\infty} \int_{Q_j} \Phi(g) \leq 2^n \Phi(s) \sum_{j=1}^{\infty} |Q_j| \tag{3.8}$$

Our nearest goal is to estimate the volume of $2Q_j$ by combining inequality (3.4) with(3.6).

First we observe that

$$s = \Phi^{-1}[\Phi(s)] \leq \Phi^{-1}\left[\fint_{Q_j} \Phi(g)\right] \leq K \fint_{2Q_j} g + \Phi^{-1}\left[\fint_{2Q_j} \Phi(h)\right] \tag{3.9}$$

Next we estimate the integrals in the right hand side as follows

$$\fint_{2Q_j} g \leq t + \frac{1}{|2Q_j|} \int_{2Q_j \cap G_t} g$$

Similarly, with the aid of the inequalities:

$$\Phi^{-1}(x+y) \leq \Phi^{-1}(x) + \Phi^{-1}(y) \quad \text{and} \quad \Phi^{-1}(x) \leq t + \frac{x}{F(t)},$$

for all $x, y \in \mathbb{R}_+$ and $t > 0$, we obtain

$$\begin{aligned}
\Phi^{-1}\left[\fint_{2Q_j} \Phi(h)\right] &\leq \Phi^{-1}\left[\frac{1}{|2Q_j|} \int_{2Q_j \cap H_t} \Phi(h) + \Phi(t)\right] \\
&\leq \Phi^{-1}\left[\frac{1}{|2Q_j|} \int_{2Q_j \cap H_t} \Phi(h)\right] + t \\
&\leq t + \frac{1}{F(t)|2Q_j|} \int_{2Q_j \cap H_t} \Phi(h) + t
\end{aligned}$$

Combining these letter two estimates with (3.9) yields

$$s \leq (K+2)t + \frac{K}{|2Q_j|}\int_{2Q_j \cap G_t} g + \frac{1}{F(t)|2Q_j|}\int_{2Q_j \cap H_t} \Phi(h)$$

This, in view of the definition of s, implies the desired estimate of the volume of $2Q_j$;

$$|2Q_j| \leq \frac{1}{t}\int_{2Q_j \cap G_t} g + \frac{1}{KtF(t)}\int_{2Q_j \cap H_t} \Phi(h) \tag{3.10}$$

We shall make use of this estimate only for selected (mutually disjoint) cubes. According to Vitali's Covering Lemma we can extract from $\{2Q_j\}_{j\in N}$ a disjoint family $\{2Q_{j'}\}_{j'\in N'\subset N}$ such that $|\bigcup_{j=1}^{\infty} 2Q_j| \leq 3^n \sum_{j'} |2Q_{j'}|$. Then we can add inequalities (3.10) with respect to the indices j' to obtain

$$\begin{aligned}
\sum_{j=1}^{\infty} |Q_j| &\leq |\bigcup_{j=1}^{\infty} 2Q_j| \leq 3^n \sum_{j'} |2Q_{j'}| \\
&\leq \frac{3^n}{t}\sum_{j'}\int_{2Q_{j'} \cap G_t} g + \frac{3^n}{K\Phi(t)}\sum_{j'}\int_{2Q_{j'} \cap H_t} \Phi(h) \\
&\leq \frac{3^n}{t}\int_{G_t} g + \frac{3^n}{K\Phi(t)}\int_{H_t} \Phi(h)
\end{aligned}$$

Combining with (3.8) yields

$$\int_{G_s} \Phi(g) \leq \frac{6^n\Phi(s)}{t}\int_{G_t} g + \frac{6^n\Phi(s)}{K\Phi(t)}\int_{H_t} \Phi(h)$$

where we recall that $\Phi(s) = sF(s) \leq 4KtF(s)$. We also have a trivial estimate

$$\int_{G_t \setminus G_s} \Phi(g) = \int_{G_t \setminus G_s} gF(g) \leq F(s)\int_{G_t} g$$

These two latter estimates add up to the following

$$\int_{G_t} \Phi(g) \leq (4\cdot 6^n K + 1)F(s)\int_{G_t} g + 4\cdot 6^n \frac{F(s)}{F(t)}\int_{H_t} \Phi(h)$$

Observe that in view of (3.2) we have $F(s) \leq M^{3+\log_2 K} F(t)$. We then arrive at the inequality

$$\int_{g>t} gF(g) \leq \alpha F(t)\int_{g>t} g + \beta\int_{h>t} hF(h) \tag{3.11}$$

for every $t > 0$, where $\alpha = \alpha(n, K, M) > 1$ and $\beta = \beta(n, K, M) > 0$. Following [G1] we shall transform (3.11) into an inequality for Stieltjes integrals. For this we need to introduce a continuously decreasing function

$$\mu(t) = \int_{G_t} g \leq \frac{1}{F(t)} \int_{\mathbb{R}^n} \Phi(g) < \infty \tag{3.12}$$

Now, with the aid of the fomula: $\int_{G_t} gF(g) = -\int_t^\infty F(s)\, d\mu(s)$, we can rewrite (3.11) as

$$-\int_t^\infty F(s) d\mu(s) \leq \alpha F(t)\mu(t) + \beta \int_{H_t} hF(h) \tag{3.13}$$

Our next objective is to estimate the integral $-\int_0^\infty F^{1+\varepsilon}(s) d\mu(s)$, for $0 < \varepsilon < \frac{1}{\alpha - 1}$. In order to avoid delicate questions about convergence of improper integrals it will be necessary to truncate μ. Fix any $T > 0$ and define $\mu_T(t) = \mu(t)$ if $0 < t \leq T$ and $\mu_T(t) = 0$ if $t > T$. Inequality (3.13) is easily seen to be preserved, that is

$$-\int_t^\infty F(s) d\mu_T(s) \leq \alpha F(t)\mu_T(t) + \beta \int_{H_t} hF(h)$$

for all $t > 0$. Now we can integrate by parts. We leave it to the reader to verify that integration by parts will produce no boundary terms and only remark that to see this one should take into account that $F(0) = 0$ and $g, h \in L^\Phi(\mathbb{R}^n)$.

$$\begin{aligned}
&-\int_0^\infty F^{1+\varepsilon}(s) d\mu_T(s) \\
&= \int_0^\infty F^\varepsilon(t) \left[\int_t^\infty F(s) d\mu_T(s)\right]' dt \\
&= \varepsilon \int_0^\infty F^{\varepsilon-1}(t) F'(t) \left[-\int_t^\infty F(s) d\mu_T(s)\right] dt \\
&\leq \varepsilon\alpha \int_0^\infty F^\varepsilon(t) F'(t) \mu_T(t) dt + \varepsilon\beta \int_0^\infty F^{\varepsilon-1}(t) F'(t) \left[\int_{H_t} hF(h)\right] dt \\
&= \frac{\varepsilon\alpha}{\varepsilon+1} \int_0^\infty \mu_T(t) dF^{1+\varepsilon}(t) + \beta \int_0^\infty \left[\int_{H_t} hF(h)\right] dF^\varepsilon(t) \\
&= -\frac{\varepsilon\alpha}{\varepsilon+1} \int_0^\infty F^{1+\varepsilon}(s) d\mu_T(s) - \beta \int_0^\infty F^\varepsilon(t) \left[\int_{H_t} hF(h)\right]' dt
\end{aligned}$$

The first integral can be brought to the left hand side

$$\begin{aligned}
-\int_0^\infty F^{1+\varepsilon}(s) d\mu_T(s) \;&\leq\; \frac{\beta + \varepsilon\beta}{1 + \varepsilon - \varepsilon\alpha} \int_0^\alpha F^\varepsilon(t) \left[-\int_{H_t} hF(h)\right]' dt \\
&=\; \frac{\beta + \varepsilon\beta}{1 + \varepsilon - \varepsilon\alpha} \int_{\mathbb{R}^n} hF^{1+\varepsilon}(h)
\end{aligned}$$

We may now let T go to infinity to conclude with the inequality

$$-\int_0^\infty F^{1+\varepsilon}(s)d\mu(s) \le \frac{\beta+\varepsilon\beta}{1+\varepsilon-\varepsilon\alpha}\int_{\mathbb{R}^n} hF^{1+\varepsilon}(h)$$

In view of the definition of μ, the integral in the left hand side is equal to $\int gF^{1+\varepsilon}(g)$. Hence the lemma follows.

Gehring's Lemma in Orlicz spaces proves very useful in the study of the minima of variational convex functionals of the form

$$\mathcal{E}[u] = \int_\Omega E(x, \nabla u)dx$$

where the integrand $E : \Omega \times \mathbb{R}^n \to \mathbb{R}_+$ grows with the rate of $\Phi(|\xi|)$. In the coming section we shall discuss the case when Φ has power growth, $\Phi(t) = t^p$ with $p > 1$. Before embarking on that discussion, however, let us state the L^p-variant of Lemma 3.1.

Lemma 3.2. *Suppose g and h are nonnegative functions of class $L^p(\mathbb{R}^n)$, with $1 < p < \infty$, and satisfy*

$$\left(\fint_Q g^p\right)^{\frac{1}{p}} \le K\fint_{2Q} g + \left(\fint_{2Q} h^p\right)^{\frac{1}{p}} \tag{3.14}$$

for all cubes $Q \subset \mathbb{R}^n$. Then there exist a new exponent $s = s(n,p,K) > p$ and a constant $C = C(n,p,K)$ such that

$$\int_{\mathbb{R}^n} g^s \le C\int_{\mathbb{R}^n} h^s \tag{3.15}$$

In particular, a function $g \in L^p(\mathbb{R}^n)$ which satisfies the homogeneous weak reverse Hölder inequalities

$$\left(\fint_Q g^p\right)^{\frac{1}{p}} \le K\fint_{2Q} g \tag{3.16}$$

for all cubes $Q \subset \mathbb{R}^n$ must necessarily be the zero function. We want to emphasize that Lemma 3.2 is sufficient for major applications. In particular, there is no need for local variants of it, such as Proposition 6.1 later on.

4 Applications to PDEs

Without knowing the applications of Gehring's Lemma to partial differential equations, it is impossible to appreciate its fundamental character. Indeed, Gehring's Lemma attracted the attention of many prominent mathematicians in the field of nonlinear PDEs. For the most significant early contributions, see [EM], [GM], [St1,2], [GG] and [BI]. However, the study of higher integrability properties of solutions of linear equations originated much earlier by using singular integrals. It is rewarding to review briefly some of these classical lines. For this, we shall begin with a nonhomogeneous Laplace equation in $\mathbb{R}^n$.

$$\operatorname{div}\,(\nabla u - h) = f \tag{4.1}$$

where h is a given vector field of class $L^s(\mathbb{R}^n, \mathbb{R}^n)$, while f is a scalar function of class $L^{\frac{ns}{n+s}}(\mathbb{R}^n)$, $s \geq \frac{n}{n-1}$. The gradient of the solution is found explicitely by using singular integrals. Specifically, we can write

$$\nabla u = \mathcal{R}\langle \mathcal{R}|h\rangle + \mathcal{P}f \tag{4.2}$$

Here $\mathcal{R}$ stands for the Riesz transform

$$\mathcal{R}g(x) = c_n \int \frac{(x-y)g(y)}{|x-y|^{n+1}}dy$$

and $\mathcal{P}$ is the Riesz potential

$$\mathcal{P}f(x) = c_n \int \frac{(x-y)f(y)}{|x-y|^{n}}dy$$

By virtue of boundedness of these operators in L^s- and $L^{\frac{ns}{n+s}}$- spaces, respectively, we have a fundamental estimate

$$\|\nabla u\|_s \leq C_1\|h\|_s + C_2\|f\|_{\frac{ns}{n+s}} \tag{4.3}$$

This is a starting point for the derivation of local estimates for solutions of the homogeneous equation

$$\operatorname{div}\,\nabla u = 0 \quad \text{in } \Omega \subset \mathbb{R}^n \tag{4.4}$$

Multiplying by a function $\varphi \in C_0^\infty(\Omega)$, we obtain

$$\operatorname{div}\,[\nabla(\varphi u) - u\nabla\varphi] = \langle \nabla\varphi|\nabla u\rangle.$$

Hence (4.3) reads as

$$\|\varphi\nabla u\|_s \leq (1+C_1)\|u\nabla\varphi\|_s + C_2\,\big\|\,|\nabla\varphi|\,|\nabla u|\,\big\|_{\frac{ns}{n+s}} \tag{4.5}$$

Note that for $s = 2$ one easily obtains even stronger result. Indeed, multiplying (4.4) by $\varphi^2 u$ and integrating by parts yields

$$\|\varphi\nabla u\|_2 \leq 2\|u\nabla\varphi\|_2 \tag{4.6}$$

Thus the second term in (4.5) is not present in the estimate with natural exponent $s = 2$. We now proceed to the example representing a large class of nonlinear elliptic PDEs. This is the A-harmonic equation

$$\operatorname{div}\,A(x, \nabla u) = 0 \quad \text{in } \Omega \subset \mathbb{R}^n \tag{4.7}$$

where $A:\Omega\times\mathbb{R}^n\to\mathbb{R}^n$ is a given measurable mapping. Following the lead of the familiar p-harmonic operator we impose the following conditions

$$|A(x,\xi)-A(x,\zeta)| \le M(|\xi|+|\zeta|)^{p-2}|\xi-\zeta| \tag{4.8}$$

$$\langle \xi-\zeta \mid A(x,\xi)-A(x,\zeta)\rangle \ge m(|\xi|+|\zeta|)^{p-2}|\xi-\zeta|^2 \tag{4.9}$$

$$A(x,t\xi) = t^{p-1}A(x,\xi) \tag{4.10}$$

for all $\xi,\zeta\in\mathbb{R}^n$ and $t\ge 0$. Moreover, to make the forthcoming arguments a little easier, we assume that $2\le p<n$. The natural class to seek solutions from is, of course, the Sobolev space $W^{1,p}_{\text{loc}}(\Omega)$. A function $u\in W^{1,p}_{\text{loc}}(\Omega)$ is said to be A-harmonic if

$$\int_\Omega \langle A(x,\nabla u)|\nabla\eta(x)\rangle dx = 0 \tag{4.11}$$

for every $\eta\in W^{1,p}(\Omega)$ with compact support. Fix an arbitrary test function $\varphi\in C_0^\infty(\Omega)$. Substituting $\eta=|\varphi|^p u$ into (4.11), we obtain the Caccioppoli estimate

$$\|\varphi\nabla u\|_p \le \frac{pM}{m}\|u\nabla\varphi\|_p \tag{4.12}$$

This estimate paves the way to higher integrability of the gradient. Indeed, when combined with Poincaré inequality (see (4.19)) it leads to weak reverse Hölder inequalities:

$$\left(\fint_Q |\nabla u|^p\right)^{\frac{1}{p}} \le \left(\fint_{2Q} |\nabla u|^{\frac{np}{n+p}}\right)^{\frac{n+p}{np}} \tag{4.13}$$

for every cube $2Q\subset\Omega$. Let us observe in advance that Gehring's Lemma for weak reverse Hölder inequalities (see Propositon 6.1) certainly implies that $u\in W^{1,s}_{\text{loc}}(\Omega)$ with some $s>p$. But at this point we want to illustrate the advantage in using Lemma 3.2. For this, we first consider a nonhomogeneous A-harmonic equation in $\mathbb{R}^n$:

$$\text{div } A(x,\nabla u-h)=f \tag{4.14}$$

for $\nabla u\in L^p(\mathbb{R}^n,\mathbb{R}^n)$ with given $h\in L^p(\mathbb{R}^n,\mathbb{R}^n)$ and $f\in L^{\frac{nq}{n+q}}(\mathbb{R}^n)$, $\frac{1}{p}+\frac{1}{q}=1$. The existence and uniqueness of the solution are established by using Browder-Minty theory of monotone operators. We multiply (4.14) by u and integrate the left hand side by parts. The right hand side can be estimated with the aid of Sobolev's inequality

$$\left|\int fu\right| \le \|f\|_{\frac{nq}{n+q}}\|u\|_{\frac{np}{n-p}} \le C\|f\|_{\frac{nq}{n+q}}\|\nabla u\|_p$$

After lengthy but rather standard calculation we then conclude with the following basic estimate for the gradient

$$\int|\nabla u|^p \le C_1\int|h|^p + C_3\left(\int|f|^{\frac{nq}{n+q}}\right)^{\frac{n+q}{n}} \tag{4.15}$$

Inequality (4.15) suggests to examine a bounded nonlinear operator

$$\mathcal{A} : L^p(\mathbb{R}^n, \mathbb{R}^n) \times L^{\frac{nq}{n+q}}(\mathbb{R}^n) \to L^p(\mathbb{R}^n, \mathbb{R}^n) \tag{4.16}$$

defined by the rule $\mathcal{A}(h, f) = \nabla u$. This operator plays the same crucial role here as the Riesz transform did in the harmonic case. We may always write $f =$ div F, for some $F = \nabla\vartheta \in L^q(\mathbb{R}^n, \mathbb{R}^n)$. This can be done by solving nonhomogeneous Laplace equation (4.1) with $u = \vartheta$ and $h \equiv 0$. Precisely, we have $F = \mathcal{P}f$. Thus (4.14) takes the form

$$\text{div } A(x, \nabla u - h) = \text{div } F \tag{4.17}$$

Fix a cube $Q \subset \mathbb{R}^n$ and a cut-off function $\chi \in C_0^\infty(2Q)$. The latter means that $0 \le \chi \le 1$, $\chi \equiv 1$ on Q and $|\nabla\chi| \le C(n)/$ diam Q. Then multiply (4.17) by $\eta = \chi^p(u - u_{2Q})$ and integrate by parts. We will not bother the reader with additional routine and lengthy details. The final product of this calculation is the nonhomogeneous reverse Hölder inequality

$$\left(\fint_Q |\nabla u|^p\right)^{\frac{1}{p}} \le K\left(\fint_{2Q} |\nabla u|^{\frac{np}{n+p}}\right)^{\frac{n+p}{np}} + C\left[\fint_{2Q} (|h| + |F|^{q-1})^p\right]^{\frac{1}{p}} \tag{4.18}$$

Let us point out only that in deriving (4.18) we have made use of Poincaré's inequality:

$$\left(\fint_Q |u - u_{2Q}|^p\right)^{\frac{1}{p}} \le C(n,p) \text{ diam } Q\left(\fint_{2Q} |\nabla u|^{\frac{np}{n+p}}\right)^{\frac{n+p}{np}} \tag{4.19}$$

It is at this point that Gehring's Lemma comes to the play. Assuming that $h \in L^s(\mathbb{R}^n, \mathbb{R}^n)$ and $F \in L^{s(q-1)}(\mathbb{R}^n, \mathbb{R}^n)$ with some $s > p$, we can use Lemma 3.2 to infer that $\nabla u \in L^s(\mathbb{R}^n, \mathbb{R}^n)$, and we have a uniform bound

$$\int_{\mathbb{R}^n} |\nabla u|^s \le C_1 \int_{\mathbb{R}^n} |h|^s + C_2 \int_{\mathbb{R}^n} |F|^{s(q-1)} \tag{4.20}$$

provided s is sufficiently close to p. One may estimate the latter integral in terms of the function $f =$ div F. Indeed, $F = \mathcal{P}f$ and the standard inequality for Riesz potential yields $\|F\|_{s(q-1)} \le C\|f\|_{\frac{sn}{np-n+s}}$. This brings us to the following generalization of (4.3).

Proposition 4.1. *The nonlinear operator $\mathcal{A}$, originally defined by (4.16), actually extends to a bounded operator*

$$\mathcal{A} : L^s(\mathbb{R}^n, \mathbb{R}^n) \times L^{\frac{ns}{np-n+s}}(\mathbb{R}^n) \to L^s(\mathbb{R}^n, \mathbb{R}^n) \tag{4.21}$$

for all $p \le s < b(n, m, M, p)$.

In other words, the solution of (4.14) satisfies

$$\int_{\mathbb{R}^n} |\nabla u|^s \le C_1 \int_{\mathbb{R}^n} |h|^s + C_2 \left(\int_{\mathbb{R}^n} |f|^{\frac{ns}{np-n+s}}\right)^{\frac{np-n+s}{np-n}} \tag{4.22}$$

This result can be viewed as a nonlinear analogue of the L^s-bound of the Riesz transform. Yet another method shows that $\mathcal{A}$ is still bounded for exponents s slightly smaller than p, $a(n,m,M,p) < s \le p$. This can be found in [IS] [ISS], but the idea goes back as far as [I4].

We shall now proceed to the homegeneous equation (4.7) in an arbitrary open region $\Omega = \mathbb{R}^n$. As in the linear case, for each nonnegative test function $\varphi \in C_0^\infty(\Omega)$ we may write

$$\operatorname{div} A(x, \nabla\varphi u - u\nabla\varphi) = \langle \nabla\varphi^{p-1} | A(x, \nabla u)\rangle$$

This can be viewed as an equation in the entire space $\mathbb{R}^n$. Inequality (4.22) yields

$$\int_\Omega |\varphi\nabla u|^s \le C_1 \int_\Omega |u\nabla\varphi|^s + C_2 \left[\int_\Omega \left(|\nabla\varphi^{p-1}|\; |\nabla u|^{p-1}\right)^{\frac{ns}{np-n+s}}\right]^{\frac{np-n+s}{np-n}} \tag{4.23}$$

One may simplify (4.23) by applying Young's inequality to the latter integral. This will break it into two terms. One of the terms will be absorbed by the left hand side and the final result will read as follows:

Theorem 4.1. *Every A-harmonic function u belongs to $W^{1,s}_{\mathrm{loc}}(\Omega)$ and satisfies the estimate*

$$\|\varphi\nabla u\|_s \le C_1\|u\nabla\varphi\|_s + C_2 \left\|\; |\nabla\varphi||\nabla u| \;\right\|_{\frac{ns}{n+s}} \tag{4.24}$$

for every $\varphi \in C_0^\infty(\Omega)$, where $p \le s < b(n,m,M,p)$.

Note that for $s = p$ the second term in the right hand side of (4.24) is not present (see Caccioppoli's inequality (4.12)). The addition of this term is the new element in estimates beyond the natural exponent p for second order equations. This does not happen for the first order PDEs such as (1.3). For such examples we refer the reader to [ISS]. Note too that (4.24) implies weak reverse Hölder inequalities

$$\left(⨍_Q |\nabla u|^s\right)^{\frac{1}{s}} \le C \left(⨍_{2Q} |\nabla u|^{\frac{ns}{n+s}}\right)^{\frac{n+s}{ns}} \tag{4.25}$$

for every cube $2Q \subset \Omega$. We could deduce (4.25) directly from (4.13) by applying a local variant of Gehring's Lemma (see Proposition 6.1). However, our approach through global estimates leads to much stronger results; estimate (4.24) is not in any obvious way a corollary of the reverse Hölder inequalities (4.25). The point is that the test function φ in (4.24) is arbitrary (can be supported in a highly irregular set) while inequality (4.25) is a consequence of applying (4.24) to rather special test functions, like in the derivation of (4.18). Estimate (4.24) with s below the natural exponent p also holds and is of crucial importance to establishing removability of singularities of the A-harmonic functions [BIS]. Unfortunately, it would take us a bit afield to present these results here. For a treatment of variational functionals with nearly linear growth we refer the reader to [GIS].

5 Gehring's Lemma Through Maximal Inequalities

In Introduction and Retrospection we have already discussed some ideas related to the maximal functions. In this section we present more details of this alternative approach. This will give us, in particular, another proof of Lemma 3.2, though the points we wish to make here seem to be interesting on their own right.

Let us restate reverse Hölder inequalities of Lemma 3.2 in the language of maximal functions

$$M_p g \leq K M g + M_p h \tag{5.1}$$

The question may arise as to whether this weaker condition still implies (3.15). The following proposition provides an affirmative answer to this question.

Proposition 5.1. *Suppose inequality (5.1) holds for some $g, h \in L^s(\mathbb{R}^n)$ with $p < s \leq p + \frac{p-1}{10^{n+p}K^p}$. Then*

$$\int_{\mathbb{R}^n} g^s \leq 10^{n+p} \int_{\mathbb{R}^n} h^s \tag{5.2}$$

Proof. The main idea of the proof is to use maximal inequalities (2.5). We emphasize the importance of the constants involved in the forthcoming calculation. Repeated application of (2.5) enables us to write

$$\begin{aligned}
\int g^s &= \int (g^p)^{\frac{s}{p}} \leq 2^n \frac{s-p}{s} \int (Mg^p)^{\frac{s}{p}} = 2^n \frac{s-p}{s} \int (M_p g)^s \\
&\leq 2^n \frac{s-p}{s} \int (KMg + M_p h)^s \\
&\leq 2^{n+s-1} \frac{s-p}{s} K^s \int (Mg)^s + 2^{n+s-1} \frac{s-p}{s} \int (Mh^p)^{\frac{s}{p}} \\
&\leq \left(2^{n+s-1} \frac{s-p}{s} K^s\right) \left(3^n 2^s \frac{s}{s-1}\right) \int g^s \\
&\quad + \left(2^{n+s-1} \frac{s-p}{s}\right) \left(\frac{s}{s-p} 3^n 2^{\frac{s}{p}}\right) \int h^s
\end{aligned}$$

Taking into account the restriction on s we may estimate both factors in front of these integrals. An elementary computation shows that

$$\int_{\mathbb{R}^n} g^s \leq \frac{1}{2} \int_{\mathbb{R}^n} g^s + 2^{p+1} 6^n \int_{\mathbb{R}^n} h^s$$

establishing Proposition 5.1.

We leave it to the reader to verify that analogous inequality for maximal functions in a finite cube $\Omega \subset \mathbb{R}^n$ can be handled in much the same way. Indeed, using (2.4) we obtain a variant of Gehring's Lemma in terms of maximal functions.

Proposition 5.2. *Let h be a nonnegative measurable function on a cube $\Omega \subset \mathbb{R}^n$ such that*

$$M_\Omega h^p \leq A_p (M_\Omega h)^p \tag{5.3}$$

at every point of Ω and some $p > 1$. If moreover, $h \in L^s(\Omega)$ with $p < s \leq p + \frac{p-1}{10^{n+p} A_p}$, then

$$⨍_\Omega h^s \leq A_s \left(⨍_\Omega h \right)^s \tag{5.4}$$

Our next goal is to get rid of the unwelcome assumption that the functions g and h belong to $L^s(\mathbb{R}^n)$. This can be achieved by an approximation argument but only if g and h satisfy the nonhomogeneous reverse Hölder inequalities (3.14). Of course these inequalities are is slighly stronger than (5.1).

Let η be an arbitrary nonnegative function of class $C_0^\infty(\mathbb{R}^n)$ whose integral is equal to 1. We consider the following mollifications of g and h

$$G = \eta * g \quad \text{and} \quad H = (\eta * h^p)^{\frac{1}{p}} \tag{5.5}$$

where $*$ means the convolution operator. With the aid of Young's inequality we find that

$$\|G\|_s \leq \|\eta\|_q \|g\|_p \qquad \|H\|_s \leq \|\eta\|_{\frac{s}{p}}^{\frac{1}{p}} \|h\|_p$$

where $\frac{1}{q} = 1 + \frac{1}{s} - \frac{1}{p}$. Thus both G and H belong to $L^s(\mathbb{R}^n)$ for every $s \geq p$. In spite of nonlinearity of conditions (3.14) the above method of mollifying g and h still works.

Proposition 5.3. *The mollified functions $G, H \in L^s(\mathbb{R}^n), s \geq p$, satisfy the same weak reverse Hölder inequalities as g and h; namely,*

$$\left(⨍_Q G^p \right)^{\frac{1}{p}} \leq K ⨍_{2Q} G + \left(⨍_{2Q} H^p \right)^{\frac{1}{p}} \tag{5.6}$$

for all cubes $Q \subset \mathbb{R}^n$.

Proof. Using Minkowski's inequality for integrals we obtain

$$\begin{aligned}
\left(\fint_Q G^p\right)^{\frac{1}{p}} &= \left(\fint_Q \left|\int_{\mathbb{R}^n} \eta(y) g(x-y)dy\right|^p dx\right)^{\frac{1}{p}} \\
&\leq \int_{\mathbb{R}^n} \eta(y) \left(\fint_Q g^p(x-y)dx\right)^{\frac{1}{p}} dy \\
&\leq K \int_{\mathbb{R}^n} \eta(y) \left(\fint_{2Q} g(x-y)dx\right) dy \\
&\qquad + \int_{\mathbb{R}^n} \eta(y) \left(\fint_{2Q} h^p(x-y)dx\right)^{\frac{1}{p}} dy \\
&\leq K \fint_{2Q} G + \left[\int_{\mathbb{R}^n} \eta(y) \fint_{2Q} h^p(x-y)dxdy\right]^{\frac{1}{p}} \\
&= K \fint_{2Q} G + \left(\fint_{2Q} H^p\right)^{\frac{1}{p}}
\end{aligned}$$

as desired.

Having disposed of the reverse Hölder inequalities for G and H, we can now use Proposition 5.1 to obtain

$$\int_{\mathbb{R}^n} G^s \leq 10^{n+p} \int_{\mathbb{R}^n} H^s \tag{5.7}$$

Finally, as η approaches the Dirac distribution we arrive at inequality (3.15) with

$$p < s \leq p + \frac{p-1}{10^{n+p} K^p} \quad \text{and} \quad C = 10^{n+p}. \tag{5.8}$$

This completes the proof of Lemma 3.2.

Remark. Similar method of approximation works when inequalities (3.14) are restricted to the cubes contained in a given domain $\Omega \subset \mathbb{R}^n$. For example, if Ω is bounded and convex we may approximate $g \in L^p(\Omega)$ by the functions

$$g_t(x) = \fint_\Omega g(x - tx + ty)dy, \qquad 0 < t \leq 1.$$

Clearly, $g_t \in C(\overline{\Omega}) \subset L^s(\Omega)$, for all $s \geq p$. As t approaches zero the functions g_t converge to g in L^p-norm. We approximate h by the functions

$$h_t(x) = \left[\fint_\Omega h^p(x - tx + ty)dy\right]^{\frac{1}{p}}$$

In much the same way we show that inequalities (3.14) remain valid for g_t and h_t.

6 Local Inequalities

As illustrated in Section 4, integrability questions for solutions of nonlinear PDEs reduce to weak reverse Hölder inequalities over all cubes in $\mathbb{R}^n$ and, therefore, are handled by Lemma 3.2. There are situations, however, in which it is convenient to work only with cubes contained in an open region $\Omega \neq \mathbb{R}^n$. As an example, see Muckenhoupt weights in [S2]. In this section we show that Lemma 3.2, applies to these local inequalities as well.

Proposition 6.1. *Let Ω be a cube in $\mathbb{R}^n$ and let $g, h \in L^p(\Omega)$, $1 < p < \infty$, be nonnegative functions satisfying:*

$$\left(\fint_Q g^p\right)^{\frac{1}{p}} \leq K \fint_{2Q} g + \left(\fint_{2Q} h^p\right)^{\frac{1}{p}} \tag{6.1}$$

for all cubes $Q \subset 2Q \subset \Omega$. Then for each $0 < \sigma < 1$ and $p < s < p + \frac{p-1}{10^{n+p}4^n K^p}$ we have

$$\left(\fint_{\sigma\Omega} g^s\right)^{\frac{1}{s}} \leq \frac{100^n}{\sigma^{\frac{n}{s}}(1-\sigma)^{\frac{n}{p}}} \left[\left(\fint_\Omega g^p\right)^{\frac{1}{p}} + \left(\fint_\Omega h^s\right)^{\frac{1}{s}}\right] \tag{6.2}$$

This result has enjoyed great popularity with analysts in PDEs.

Proof. Let us introduce two auxiliary functions defined on the entire space $\mathbb{R}^n$

$$G(x) = \rho(x) g(x) \quad \text{and} \quad H(x) = \rho(x) h(x) + \left(\int_\Omega g^p\right)^{\frac{1}{p}} \chi_\Omega(x) \tag{6.3}$$

where $\rho(x) = \operatorname{dist}^{\frac{n}{p}}(x, \mathbb{R}^n - \Omega)$. We first prove the following reverse Hölder inequalities

$$\left(\fint_Q G^p\right)^{\frac{1}{p}} \leq 4^{\frac{n}{p}} K \fint_{2Q} G + 4^{\frac{n}{p}} \left(\fint_{2Q} H^p\right)^{\frac{1}{p}} \tag{6.4}$$

for all cubes $Q \subset \mathbb{R}^n$. Here we may assume without loss of generality that Q meets Ω, otherwise the left hand side is zero. Our derivation of (6.4) falls naturally into two cases.

Case 1. We assume that $3Q \subset \Omega$. By elementary geometric consideration we find that

$$\max\{\rho(x);\ x \in Q\} \leq 4^{\frac{n}{p}} \min\{\rho(x);\ x \in 2Q\}$$

Applying (6.1) yields

$$\begin{aligned}\left(\fint_Q G^p\right)^{\frac{1}{p}} &\leq \max_Q \rho\left(\fint_Q g^p\right)^{\frac{1}{p}}\\ &\leq 4^{\frac{n}{p}} K \min_{2Q} \rho\left(\fint_{2Q} g\right) + 4^{\frac{n}{p}} \min_{2Q} \rho\left(\fint_{2Q} h^p\right)^{\frac{1}{p}}\\ &\leq 4^{\frac{n}{p}} K \fint_{2Q} G + 4^{\frac{n}{p}}\left(\fint_{2Q} H^p\right)^{\frac{1}{p}}\end{aligned}$$

Case 2. We assume that $3Q \not\subset \Omega$. Since Q meets Ω, elementary geometric arguments show that

$$\max\{\rho(x);\ x \in Q\} \leq \max\{\rho(x); x \in 2Q\} \leq |2Q \cap \Omega|^{\frac{1}{p}}$$

Hence we conclude that

$$\begin{aligned}\left(\fint_Q G^p\right)^{\frac{1}{p}} &\leq \max_Q \rho\left(\frac{1}{|Q|}\int_\Omega g^p\right)^{\frac{1}{p}} \leq \left[\frac{|2Q \cap \Omega|}{|Q|}\int_\Omega g^p\right]^{\frac{1}{p}}\\ &\leq 4^{\frac{n}{p}}\left(\fint_{2Q} H^p\right)^{\frac{n}{p}}\end{aligned}$$

Combining these two cases proves inequality (6.4).

We are now in a position to apply Lemma 3.2. Accordingly,

$$\int_{\mathbb{R}^n} G^s \leq 10^{n+s} \cdot 4^{\frac{ns}{p}} \int_{\mathbb{R}^n} H^s$$

for all $p < s \leq p + \frac{p-1}{10^{n+p}4^n K^p}$. These specific numbers have been taken from the proof of Lemma 3.2 as presented in Section 5 and summarized by formulas (5.8). From the definition of the functions G and H it follows that

$$\left(\fint_\Omega \rho^s g^s\right)^{\frac{1}{s}} \leq 50^n \left(\fint_\Omega \rho^s h^s\right)^{\frac{1}{s}} + 50^n \left(\int_\Omega g^p\right)^{\frac{1}{p}} \tag{6.5}$$

This result is slightly more general than (6.2) and, therefore, might be of independent interest. The only point remaining concerns estimates of the distance function ρ in terms of the volume of Ω. For $0 < \sigma < 1$ we find that

$$\inf\{\rho^p(x); x \in \sigma\Omega\} = \left(\frac{1-\sigma}{2}\right)^n |\Omega|$$

Also, it is easily seen that

$$\sup\{\rho^p(x); x \in \Omega\} = \frac{|\Omega|}{2^n}$$

These formulas combined with (6.5) imply (6.2), proving Proposition 6.1.

REFERENCES

[B1] B. Bojarski, *Homeomorphic solutions of Beltrami systems*, Dokl. Akad. Nauk SSSR **102** (1955), 661–664. (Russian)

[B2] B. Bojarski, *Remarks on stability of reverse Hölder inequalities and quasiconformal mappings* Ann. Acad. Sci. Fenn. Ser. A.I. (1985), 291–296.

[BI] B. Bojarski and T. Iwaniec, *Analytical foundations of the theory of quasiconfomal mappings in R^n*, Ann. Acad. Sci. Fenn. Ser. A.I. **8** (1983), 257–324.

[BIS] L. Budney, T. Iwaniec and B. Stroffolini, *Removability of singularities of A-harmonic functions*, Advances in Differential Equations, (to appear).

[BK] S. Buckley and P. Koskela, *Sobolev-Poincaré inequalities for $p < 1$*, Indiana Univ. Math. J. **43** (1994), 221–240.

[BKL] S. Buckley, P. Koskela, and G. Lu, *Subelliptic Poincaré inequalities: the case $p < 1$*, preprint.

[BMR] J. Bastero, M. Milman and F.J. Ruiz, *Reverse Hölder inequalities and interpolation*, preprint, 1996.

[BP] R. Bagby and D. Parson, *Orlicz Spaces and rearranged maximal functions* Math. Nachr. **132** (1987), 15–27.

[BR] J. Bastero and F.J. Ruiz, *Elementary reverse Hölder type inequalities with applications to interpolation of operator theory*, Proc. AMS, (to appear).

[Bu] S. Buckley, *Pointwise multipliers for reverse Hölder spaces*, Studia Mathematica **109** (1994), 23–39.

[CN] D. Cruz-Uribe and C.J. Neugebauer, *The structure of the reverse Hölder classes*, preprint.

[CNO] D. Cruz-Uribe, C.J. Neugebauer and V. Olsen, *The one-sided minimal operator and the one-sided reverse Hölder inequality*, Studia Mathematica.

[Cr] D. Cruz-Uribe, *The class $A^+_\infty(g)$ and the one-sided reverse Hölder inequality*, preprint.

[DS] L. D'Apuzzo and C. Sbordone, *Reverse Hölder inequalities, a sharp result*, Rendiconti di Matematica, Ser. VII **10** (1990), 357–366.

[EM] A. Elcrat and N. Meyers, *Some results on regularity for solutions of nonlinear elliptic systems and quasiregular functions*, Duke Math. J. **42** (1) (1975), 121–136.

[FM] M. Franciosi and G. Moscariello, *Higher integrability results*, Manuscripta Math **52** (1985), 151–170.

[FC] B. Franchi and F. Serra Cassano, *Gehring's Lemma for metrics and higher integrability of the gradient for minimizers of noncoercive variational functionals*, Studia Mathematica, (to appear).

[FS] N. Fusco and C. Sbordone, *Higher integrability of the gradient of minimizers of functionals with nonstandard growth conditions*, Comm. Pure Applied Math. **XLIII** (1990), 673–683.

[G1] F.W. Gehring, *The L^p-integrability of the partial derivatives of a quasiconformal mapping*, Acta Math. **130** (1973), 265–277.

[G2] F.W. Gehring, *Topics in quasiconformal mappings*, Proceedings of the ICM, Berkeley (1986), 62–80.

[GG] M. Giaquinta and E. Giusti, *On the regularity of the minima of variational integrals*, Acta Math. **148** (1982), 31–46.

[Gi] M. Giaquinta, *Multiple integrals in the calculus of variations and nonlinear elliptic systems*, Annals of Math. Stud. No. 105 (1983), Princeton University Press.

[GI] L. Greco and T. Iwaniec, *New inequalities for the Jacobian*, Ann. Inst. H. Poincaré **11** (1994), 17–35.

[GIM] L. Greco, T. Iwaniec, and G. Moscariello, *Limits of the improved integrability of the volume forms*, Indiana Univ. Math. J. **44** (2) (1995), 305–339.

[GIS] L. Greco, T. Iwaniec, and C. Sbordone, *Variational integrals of nearly linear growth*, Differential and Integral Equations, (to appear).

[GM] M. Giaquinta and G. Modica, *Regularity results for some classes of higher order nonlinear elliptic systems*, J. reine angew. Math **311/312** (1979), 145–169.

[HKM] J.T. Heinonen, T. Kilpeläinen, and O. Martio, *Nonlinear potential theory of degenerate elliptic equations* Oxford Univ. Press, 1993.

[I1] T. Iwaniec, *On L^p-integrability in p.d.e and quasiregualr mappings for large exponents* Ann. Acad. Sci. Fenn. Ser. A.I. **7** (1982), 301–322.

[I2] T. Iwaniec, *Projections onto gradient fields and L^p-estimates for degenerated elliptic operators*, Studia Mathematica **75** (1983), 293–312.

[I3] T. Iwaniec, *Some aspects of p.d.e. and quasiregular mappings*, Proceedings of ICM, Warsaw (1983), 1193–1208.

[I4] T. Iwaniec, *p-harmonic tensors and quasiregular mappings*, Annals of Mathematics **136** (1992), 589–624.

[I5] T. Iwaniec, *L^p-theory of quasiregular mappings*, A collection of surveys 1960-1990, Springer Lecture Notes **1508** (1992), 39–64.

[IN] T. Iwaniec and C. Nolder, *The Hardy-Littlewood inequality for quasiregular mappings in certain domains in R^n*, Ann. Acad. Sci. Fenn. Ser. A.I. **10** (1985), 267–282.

[IS] T. Iwaniec and C. Sbordone, *Weak minima of variational integrals*, J. reine angew. Math. **454** (1994), 143–161.

[ISS] T. Iwaniec, C. Scott and B. Stroffolini, *Nonlinear Hodge theory on maniforlds with boundary*, preprint.

[K1] J. Kinnunen, *Higher integrability with weights*, Ann. Acad. Sci. Fenn. Ser. A.I. Math. **19** (1994), 355–366.

[K2] J. Kinnunen, *Sharp results on reverse Hölder inequalities*, Ann. Acad. Sci. Fenn. Ser. A.I. Math., Dissertationes **95**, 1994.

[LMZ] C. Li, A. McIntosh, and K. Zhang, *Higher integrability and reverse Hölder inequalities*, preprint.

[M] B. Muckenhoupt, *Weighted norm inequalities for the Hardy maximal function*, Trans. Amer. Math. Soc. **165** (1972), 207–226.

[Ma] O. Martio, *On the integrability of the derivative of a quasiregular mapping*, Math. Scand. **3** (1974), 43–48.

[Mi1] M. Milman, *A note on Gehring's Lemma*, Ann. Acad. Sci. Fenn. Ser. A.I Math, (to appear).

[Mi2] M. Milman, *A note on interpolation and higher integrability*, Ann. Acad. Sci. Fenn. Ser. A.I Math, (to appear).

[Mi3] M. Milman, *A note on the reverse Hardy inequalities and Gehring's Lemma*, Comm. in Pure and Appl. Math., (to appear).

[Mu] S. Müller, *A surprising higher integrability property of mappings with positive determinant*, Bull. Amer. Math. Soc. **21** (1989), 245–248.

[RR] M.M. Rao and Z.D. Ren, *Theory of Orlicz Spaces*, Marcel Dekker, 1991.

[S1] E.M. Stein, *Note on the class* $L \log L$, Studia Math. **32** (1969), 305–310.

[S2] E.M. Stein, *Harmonic Analysis: real-variable methods, orthogonality, and oscillatory integrals, Chapter V*, Princeton Univ. Press, 1993.

[Sb1] C. Sbordone, *Rearrangement of functions and reverse Hölder inequalities*, Ernio De Giorgi Colloquium Re. Notes in Math., Pitman, **124** (1985), 139–148.

[Sb2] C. Sbordone, *Rearrangement of functions and reverse Jensen inequalities*, Proc. of Symposia in Pure Math. **45** (1986), 325–329.

[Sb3] C. Sbordone, *On some inequalities and their applications to the calculus of variations*, Bolletimo V.M.I. Analisi Funzionale e Applicazionė, Ser. VI **V-C** (1) (1986), 73–94.

[St1] E.W. Stredulinsky, *Higher integrability from reverse Hölder inequalities*, Indiana Univ. Math. J. **29,3** (1980), 408–417.

[St2] E.W. Stredulinsky, *Weighted inequalities and degenerated elliptic partial differential equations*, Lecture Notes in Math. **1074**, Springer-Verlag, Berlin-Heidelberg, New York, Tokyo, 1984.

[W] N. Wiener, *The Fourier Integral and Certain of its Applications*, Cambridge Univ. Press, 1993.

Department of Mathematics
215 Carnegie
Syracuse University
Syracuse, NY 13244
tiwaniec@mailbox.syr.edu

Old and New on the Quasihyperbolic Metric

Pekka Koskela

1 Introduction

Let D be a proper subdomain of $\mathbb{R}^d$. Following Gehring and Palka [GP] we define the quasihyperbolic distance between a pair x_1, x_2 of points in D as the infimum of

$$\int_\gamma \frac{ds}{d(x, \partial D)}$$

over all rectifiable curves γ joining x_1, x_2 in D. We denote the quasihyperbolic distance between x_1, x_2 by $k_D(x_1, x_2)$. As pointed out by Gehring and Osgood [GO], x_1 and x_2 can be joined by a quasihyperbolic geodesic; also see [Mr]. The quasihyperbolic metric is comparable to the usual hyperbolic metric in a simply connected plane domain by the Koebe distortion theorem. For a multiply connected plane domain D these two metrics are comparable if and only if the boundary of D is uniformly perfect as shown by Beardon and Pommerenke [BP]. Gehring and Palka introduced the quasihyperbolic metric as a tool for the study of quasiconformal homogeneity and since then this metric has found a number of applications. Let us simply mention here that this metric plays an important role in the famous extension theorem of Jones [J1] for functions in BMO and that this metric has by now become a standard tool in the study of quasiconformal mappings.

Notice that $k_D(x_1, x_2)$ is comparable to $|x_1 - x_2|/d(x_1, \partial D)$ if $|x_1 - x_2| \leq d(x_1, \partial D)/2$. This case is not very interesting. In the remaining case $k_D(x_1, x_2)$ can be effectively estimated by counting the number $N(x_1, x_2)$ of cubes Q from the Whitney decomposition of D that intersect the quasihyperbolic geodesic joining x_1, x_2. More precisely,

$$N(x_1, x_2)/C \leq k_D(x_1, x_2) \leq CN(x_1, x_2)$$

provided $|x_1 - x_2| \geq d(x_1, \partial D)/2$, where C depends only on d. We remind the reader that the Whitney decomposition of D is a decomposition of D

[1]The author was partially supported by the NSF.

into cubes Q with pairwise disjoint interiors and each of edge length 2^k for some $k \in \mathbb{Z}$ and with $d(Q, \partial D) \operatorname{diam}(Q) \in [1, 4]$ for each Q. For the existence of such a decomposition see [Se].

We shall explain the above applications more carefully and try to give a general overview of the literature dealing with the quasihyperbolic metric. We also include some observations that cannot be found in print. We shall only discuss the theory in the Euclidean space $\mathbb{R}^d$. The quasihyperbolic metric has also been used by Väisälä [Va2] to define and study quasiconformal and quasisymmetric mappings in an abstract Banach space.

The paper is organized as follows. Section 2 deals with the connections between the quasihyperbolic metric and the geometry of domains. In Section 3 we discuss the dimension estimates for the boundaries of domains with a restricted growth of their quasihyperbolic metrics. These estimates have found important applications in complex dynamics. Section 4 contains applications to quasiconformal mappings. In Section 5 we explain a connection between the growth of the quasihyperbolic metric and variational capacity. Section 6 contains various Sobolev type imbeddings that employ the metric. Finally, Section 7 contains a number of further results related to the quasihyperbolic metric.

2 Quasihyperbolic metric and geometry of domains

Martio and Sarvas [MS] introduced the notion of a uniform domain. We call D c–uniform provided each pair of points $x, y \in D$ can be joined by an arc $\gamma \subset D$ satisfying $l(\gamma) \le c|x-y|$ and

$$\min\{l(\gamma'), l(\gamma'')\} \le c \operatorname{dist}(z, \partial D)$$

for all $z \in \gamma$. Here $l(\gamma)$ is the Euclidean arclength of γ and γ', γ'' are the components of $\gamma \setminus \{z\}$. These conditions assert that the length of γ is comparable to the distance between its endpoints and that away from its endpoints γ stays away from ∂D; in particular, points can be joined in D with a curvilinear double cone which is neither too crooked nor too thin.

At the same time Jones [J1] characterized the class of proper subdomains D of $\mathbb{R}^d$ that have the extension property for BMO as the class of domains for which

$$k_D(x, y) \le a j_D(x, y) \qquad \text{for all} \quad x, y \in D;$$

here j_D denotes the distance function defined by

$$j_D(x, y) = \log\left(1 + |x-y| / \min\{\operatorname{dist}(x, \partial D), \operatorname{dist}(y, \partial D)\}\right).$$

By the extension property for BMO we mean that each function that belongs to BMO in the domain D can be extended to a function that belongs to BMO in $\mathbb{R}^d$. Here BMO stands for bounded mean oscillation.

Thereafter Gehring and Osgood [GO] established that these seemingly different classes of domains coincide.

Theorem 2.1. *A proper subdomain $D \subset \mathbb{R}^d$ is $c-$uniform if and only if there is a constant a such that*

$$k_D(x,y) \le a j_D(x,y) \qquad \text{for all} \quad x,y \in D.$$

The constants a, c depend only upon each other.

Notice that we always have $j_D(x,y) \le k_D(x,y)$.

It is known (cf. [GO]) that the quasihyperbolic geodesic joining x and y can be taken as the core of the "double cone" in a uniform domain. Thus it is natural to ask if the quasihyperbolic geodesic is in some sense always the optimal curve for joining a pair of points. This is indeed true in many situations but does not hold always. See the paper [HN] by Heinonen and Näkki for a study on this.

Let us now look at a one-sided version of uniformity. A domain D is called a John domain if there is a fixed point $x_0 \in D$ so that any point $x \in D$ can be joined to x_0 by a curve parametrized by arclength with $\gamma(0) = x_0$ and so that $d(\gamma(t), \partial D) \ge Ct$ for all $0 < t \le l(\gamma)$. Thus any point can be joined to the center point by a twisted cone. Naturally, the center point can be replaced with any other point (one might need to change the constant). For a simply connected planar John domain one can use the quasihyperbolic geodesic as this curve but not in a general infinitely connected plane domain; see the paper [GHM] by Gehring, Hag, and Martio. John domains arise naturally in the iteration of complex polynomials. Carleson, Jones and Yoccoz [CJY] have characterized some of the situations where a Fatou component is a John domain.

Based on Theorem 2.1 one might guess that a one-sided version of the growth condition on the quasihyperbolic metric in Theorem 2.1 could characterize John domains. This condition turns out to be necessary but not sufficient as pointed out by Gehring and Martio [GM1]. In the positive direction, Langmeyer [Ln] has recently characterized John domains by a version of this condition.

There are also many other interesting questions related to the quasihyperbolic geodesics. For various curvature results we refer the reader to the works of Frankel and Tysk [FT] and of Martin and Osgood [Mr] and [MO]. For results related to the Gehring-Hayman inequality, according to which the hyperbolic geodesic between a pair of points in a simply connected plane domain is essentially the shortest curve joining these points in the domain, see the paper by Heinonen and Rohde [HR].

Let us close this section by pointing out that each uniform domain equipped with the quasihyperbolic metric is Gromov hyperbolic [GH] (hence this also holds for any domain that is quasiconformally equivalent to such a domain). This follows by a simple computation. We do not know of any other results in this direction.

Question 2.2. *When is a domain equipped with the quasihyperbolic metric Gromov hyperbolic?*

This natural question would merit some investigation.

3 Growth of the metric and size of the boundary

A recent result of Jones and Makarov [JM] states that the Hausdorff dimension (or the Minkowski dimension) of the boundary of the image of the disk under a uniformly α-Hölder continuous univalent function does not exceed $2 - C\alpha$, where C is a universal constant. Moreover, they show that this statement is sharp in the sense that there exist constants C for which this estimate fails even for small α. The Hölder continuity of a univalent function is equivalent to an appropriate logarithmic growth condition on the hyperbolic metric of the image domain as was shown by Becker and Pommerenke [BcP], and hence the result of Jones and Makarov can be interpreted as a sharp dimension estimate for boundaries of simply connected plane domains that satisfy a logarithmic growth condition on their quasihyperbolic metric; recall that the hyperbolic and the quasihyperbolic metrics are comparable in a simply connected plane domain by the Koebe distortion theorem. Let us now discuss extensions of this result to multiply connected domains and to higher dimensions.

Following Gehring and Martio [GM1] we say that D satisfies a quasihyperbolic boundary condition if

$$k_D(x, x_0) \leq \frac{1}{\alpha} \log \frac{1}{d(x, \partial D)} + C \tag{3.1}$$

for a fixed $x_0 \in D$ and for all $x \in D$. It then follows that D is bounded and that any point in D could act as the point x_0 (with possibly a different constant C). The constant α is the crucial one whereas C plays a minor role. Notice that necessarily $\alpha \leq 1$. We recall that the *Minkowski dimension* of a set $E \subset \mathbb{R}^d$ is defined by

$$\dim_M(E) \equiv \inf\{\delta : \limsup_{r\to 0} M_\delta(E, r) < \infty\},$$

where, for $0 < \delta$ and $0 < r$, $M_\delta(E, r) \equiv \inf\{kr^\delta : E \subset \bigcup_1^k B(x_i, r)\}$. It follows that the Minkowski dimension of E is greater or equal to the usual Hausdorff dimension $\dim_H(E)$ of E, with equality for sufficiently regular sets E.

Smith and Stegenga [SS3] produced the estimate

$$\dim_M(\partial D) \leq d - C\alpha^{d+1}$$

for the dimension of the boundary of a domain satisfying a quasihyperbolic boundary condition by modifying the ideas of Jones and Makarov. However, their estimate is weaker than the result of Jones and Makarov when applied to the case of a simply connected plane domain.

Recently Koskela and Rohde [KR] established the following dimension estimate that for $n = 2$ gives the result of Jones and Makarov.

Theorem 3.1. *Suppose that $D \subset \mathbb{R}^d$ satisfies the quasihyperbolic boundary condition (3.1). Then*

$$\dim_M(\partial D) \leq d - C\alpha^{d-1}$$

with a constant C depending only on d.

Thus a logarithmic growth condition on the quasihyperbolic metric bounds the dimension of the boundary away from d. The problem in establishing an estimate of the above type is in the fact that even though the growth condition implies that each boundary point can be approached from within the domain by a sequence of balls whose radii are comparable to the distance to this boundary point, these balls do not appear in all scales. If we were to have balls in all scales, usual covering arguments would give estimates on the dimension. These balls, however, appear often in a certain averaged sense and one can show that this is already enough for the above estimate.

Cusps show that a bound on the size of the boundary does not guarantee any growth conditions on the quasihyperbolic metric.

Graczyk and Smirnov [GS] have recently shown that the Fatou components of a rational Collet-Eckmann map satisfy a quasihyperbolic boundary condition. Using the dimension estimates of Jones and Makarov [JM] and Smith and Stegenga [SS3] (also see Theorem 3.1) they then concluded that the dimension of the Julia set of a Collet-Eckmann polynomial is bounded away from 2. For related results see the works of Nazarov, Popovici, and Volberg [NPV] and Przytycki and Rohde [PR].

Finally, one could inquire for a growth condition on the quasihyperbolic metric that would force the boundary to have vanishing volume. Very precise results for simply connected plane domains can be found in [JM] but no results in higher dimensions seem to be in print.

4 Boundary behavior of quasiconformal maps

The hyperbolic metric is preserved under conformal maps. An analog of this was established by Gehring and Osgood [GO]: There is a constant $C = C(K, d)$ such that whenever $f : D \to D'$ is K-quasiconformal,

$$k_{D'}(f(x), f(y)) \leq C \max\{k_D(x, y), k_D(x, y)^\alpha\} \tag{4.1}$$

for all $x, y \in D$, where $\alpha = K^{1/(1-d)}$.

The moral of this estimate is that quasiconformal mappings map Whitney cubes to Whitney type objects; that is, if $Q \subset D$ is a cube whose diameter is comparable to the distance from Q to the boundary of D and with center x, then the diameter of $f(Q)$ is comparable to the distance from $f(Q)$ to the boundary of D', and $f(Q)$ contains a cube of comparable diameter and with center $f(x)$. As the quasihyperbolic distance (for large distances) can be essentially computed simply by counting the number of Whitney cubes that touch the quasihyperbolic geodesic, the large scale behavior of the metric under quasiconformal mappings comes as no surprise. The maximum of the quasihyperbolic metric and a power of it appear in (4.1) because quasiconformal maps are not locally Lipschitz but merely locally Hölder continuous.

As observed by Gehring and Martio [GM1] inequality (4.1) and the obvious logarithmic lower bound for the quasihyperbolic metric give the following distortion result.

Theorem 4.1. *Let $f : D \to D'$ be $K-$quasiconformal.*

(1) *Suppose that D' satisfies a quasihyperbolic boundary condition. Then $d(f(x), \partial D') \leq Cd(x, \partial D)^{\alpha}$ for all $x \in D$ for some $0 < \alpha$ that depends only on the given data.*

(2) *Suppose D satisfies a quasihyperbolic boundary condition. Assume that $d(f(x), \partial D') \leq Cd(x, \partial D)^{\alpha}$ for all $x \in D$. Then D' satisfies a quasihyperbolic boundary condition with constants which depend only on the given data.*

Theorem 4.1 has an interpretation in terms of Hölder continuity. Indeed, the given estimation for the distortion of distance to the boundary under f is equivalent to f satisfying a uniform Hölder continuity estimate in all balls contained in D. If the geometry of D is sufficiently nice, this is further equivalent to global Hölder continuity. For a thorough discussion on this topic we refer the reader to [GM1].

Let us next look at the global behavior of the differential of a quasiconformal mapping. By definition [Va1], the differential is locally d-integrable. The celebrated theorems of Bojarski [Bo] and Gehring [G] state that one in fact has local p-integrability with some $p > d$ that depends only on the dilatation K of the map and on d.

The following two results show that somewhat surprisingly the growth of the quasihyperbolic metric is closely related to the behavior at large of the differential.

Theorem 4.2. *Let $D' \subset \mathbb{R}^d$ be a domain satisfying a quasihyperbolic boundary condition. If $D \subset \mathbb{R}^d$ and $f : D \to D'$ is K-quasiconformal, then*

$$\int_D |f'(x)|^p dm < \infty$$

for $d - \epsilon < p < d + \epsilon$. Here the constant $\epsilon > 0$ depends only on d, K, and the data for D'.

Theorem 4.3. *Let $D \subset \mathbb{R}^d$ satisfy a quasihyperbolic boundary condition and let $f : D \to D'$ be quasiconformal. Then $|f'(x)| \in L^p(D)$ for some $p > d$ if and only if D' satisfies a quasihyperbolic boundary condition.*

Theorem 4.3 was proved by Astala and Koskela in [AK]. As Theorem 4.2 extends a corresponding result in [AK] let us indicate how one obtains this extension.

First of all, after a change of variables, we observe that the domain D is irrelevant as it then suffices to study the integrability of the differential of the inverse map g of f over D'; we recall that g is quasiconformal and $|f'|^d \leq KJ_f$ for a quasiconformal map f. The power p then changes to $d - p$. With some work one then replaces the differential of g by the quantity $d(g(x), \partial D)/d(x, \partial D')$; this step is non-trivial and involves the reverse Hölder inequalities for the differential that date back to Gehring's paper [G]. For the details see [AK]. Theorem 4.1 now guarantees that $d(g(x), \partial D) \geq Cd(x, \partial D')^a$ with $a > 0$ depending only on the data. Moreover, using inequality (4.1) and the obvious logarithmic lower bound for the quasihyperbolic metric one computes that $d(g(x), \partial D) \leq Cd(x, \partial D')^{-a}$ with an exponent $a > 0$ as above. Thus $d(g(x), \partial D)/d(x, \partial D')$ can be estimated from above by a negative power and from below by a positive power of $d(x, \partial D')$.

Theorem 4.2 follows when one realizes that the bound on the dimension in Theorem 3.3 implies that $d(x, \partial D')^{-b}$ is integrable for $0 < b < d - \dim_M(\partial D')$.

Theorem 4.3 follows from Theorems 4.1 and 4.2 and the Sobolev imbedding theorem; for the details see [AK].

The quasihyperbolic metric also appears in connection of boundary accessibility. Results from a recent paper of Koskela [K2] imply that except for a small (measured in the unit ball) exceptional set, the boundary points of a domain quasiconformally equivalent to a ball can be reached from within the domain by a subdomain that satisfies a logarithmic growth condition on its quasihyperbolic metric.

Theorem 4.4. *Let $f : B^d \to D$ be quasiconformal. Then there is a set $E \subset S^{n-1}$ of Hausdorff dimension zero so that for each $x \in S^{n-1} \setminus E$, $f(x) \in \partial\Omega_x$ for some domain $\Omega_x \subset D$ that satisfies a quasihyperbolic boundary condition.*

The proof of Theorem 4.4 is essentially the following. One first shows that f is Hölder continuous on all radii that do not terminate in the exceptional set E; the Hölder continuity exponent and the associated constant may depend on the radius. Using quasiconformality one then realizes that this continues to hold if the radius is replaced by a corresponding Stolz cone. The image of this cone under f gives us Ω_x by Theorem 4.1.

For univalent functions of the disk one has better accessibility for a fairly large set on the boundary of the disk. Let $f : B^2 \to D$ be univalent. By a result of Rohde [R], there is a set $E \subset S^1$ of Hausdorff dimension 1 such that f has a radial limit for each $w \in E$ and $f(w) \in \partial\Omega_w$ for some John domain $\Omega_w \subset D$. Rohde obtains this as an application of his work on the boundary behavior of Bloch functions; notice that $\log f'$ is a Bloch function.

Question 4.5. *Let $f : B^d \to D$ be quasiconformal. Is there a set $E \subset S^{n-1}$ of positive Hausdorff dimension (depending only on d and the dilatation of f) so that for each $x \in E$, $f(x) \in \partial\Omega_x$ for some John domain $\Omega_x \subset D$?*

In the plane the answer to Question 4.5 is affirmative as is easily seen using the above Rohde's result and factorization. For interesting related results for univalent functions we refer the reader to the recent paper [Bi] by Bishop.

Let us finally remark that the quasihyperbolic metric has also been used to study the boundary behavior of quasiregular mappings, see e.g. the papers [Vu1-2] by Vuorinen.

5 Conformal capacity and quasihyperbolic metric

If $E, F \subset D$ are disjoint compact sets, we denote the variational or conformal capacity of the the pair E, F relative to D with $cap(E, F; D)$. Thus $cap(E, F; D) = \inf_u \int_D |\nabla u|^n dx$ where the infimum is taken over all continuous Sobolev functions in $D \cup E \cup F$ that satisfy $u = 1$ in E and $u = 0$ in F. If D is a uniform domain, it follows from the extension theorem of Jones [J2] for Sobolev functions that

$$cap(E, F; \mathbb{R}^d) \le M cap(E, F; D) \tag{5.1}$$

for each such pair, where M depends only on D. It is known that for simply connected plane domains (5.1) characterizes uniformity [GM2]. Very recently Herron and Koskela [HK] showed that this statement also holds if the compact sets in (5.1) are required to be disks. Another result from [HK] relates (5.1) to a logarithmic growth condition on the quasihyperbolic metric.

Theorem 5.1. *Suppose $D \subset \mathbb{R}^d$ satisfies a quasihyperbolic boundary condition. Then there is a fixed closed ball $F \subset D$ such that (5.1) holds for all closed balls $E \subset D \setminus F$. Conversely, a bounded domain which is quasiconformally equivalent to a domain satisfying a quasihyperbolic boundary condition and which enjoys this property necessarily satisfies a quasihyperbolic boundary condition.*

Let us quickly give the main ideas of the proof of Theorem 5.1. For simplicity let us consider the case where E, F are far away from from each other and both are of Whitney type; that is their radii are comparable to their distances to the boundary of D. By considering the quasihyperbolic geodesic between the centers x, x_0 of these balls one constructs a chain of k balls with k comparable to $k_D(x, x_0)$ that joins x to x_0. The consecutive balls in this chain have comparable radii and the balls have uniformly bounded overlap. One then estimates the difference of the averages of u over E, F from above by the triangle inequality as the sum of differences of averages over consecutive balls in the chain. The Poincaré inequality can be used to bound these differences from above by the d-norm of $|\nabla u|$ over a slightly larger ball. The estimate (5.1) follows then by using the Hölder inequality for series when one controls the overlap of these larger balls. The second claim is established by reducing the question to Theorem 4.2.

Theorem 5.1 deals with balls $E \subset D \setminus F$. We believe that this geometric restriction on E is superficial.

Conjecture 5.2. *The claim of Theorem 5.1 holds for all continua $E \subset D \setminus F$.*

A positive solution to the above conjecture would imply a rather nice boundary behavior result for quasiconformal mappings, see [HK].

6 Poincaré-type inequalities

We call a domain $D \subset \mathbb{R}^d$ of finite volume a (p, ω)-Poincaré domain provided $\omega > 0$ a.e. in D, $\omega \in L^1(D)$, and

$$||u - u_D||_{L^p(D,\omega)} \leq C||\nabla u||_{L^p(D,\omega)} \tag{6.1}$$

holds for each function $u \in C^1(D)$ with gradient in $L^p(D, \omega)$. Here

$$u_D = \int_D u(x)\omega(x)dx / \int_D \omega(x)dx$$

is the weighted average with respect to the given weight (then (1.1) requires that u_D exists for each u as above). We say that (1.1) holds for each $u \in C^1(D)$ if it holds for each $u \in C^1(D)$ with gradient in $L^p(D, \omega)$. In case of a constant weight function ω we simply say that D is a p-Poincaré domain.

Jerison [Je] proved the following theorem.

Theorem 6.1. *Suppose that D is a plane domain of finite area. Then D is a 2-Poincaré domain provided*

$$\int_D k_D(x, x_0)dx < \infty$$

for some (each) $x_0 \in D$.

Thus the integrability of the quasihyperbolic metric guarantees the validity of a 2-Poincaré inequality.

Let us briefly indicate the main idea behind Theorem 6.1. One fixes a central cube Q_0 and notices that it suffices to estimate the integral of $|u - u_0|$, where u_0 is the average of u over Q_0. Using the Whitney decomposition [Se] one writes D as a union of essentially disjoint Whitney cubes. Let us fix a cube Q different from Q_0 in the decomposition. As the 2-integrals of $|u - u_0|$ over Q_0 and $|u - u_Q|$ over Q are bounded by the 2-integral of the gradient of u by the usual Poincaré inequality for cubes, we conclude that it suffices to estimate the difference of the averages u_0, u_Q (multiplied by the volume of Q). We join the centers of these cubes by a geodesic. Arguing as in the previous section, we cover the geodesic by a sequence of Whitney type balls, estimate the difference of the averages by the triangle inequality and the usual Poincaré inequality, and use Hölder's inequality for series. Thus $|u_0 - u_Q|^2$ is no more than a constant times the quasihyperbolic distance between the centers of the cubes multiplied by the 2-integral of the gradient of u, and the claim follows by summing over Q.

Later on Hurri [H] and Smith and Stegenga [SS2] showed using an idea of Staples [Sa] that the analogous statement holds in any dimension provided we replace the exponent 2 by d and integrate the quasihyperbolic metric to the power $d-1$. This can be proved by the above argument.

It is then natural to inquire if the logarithmic growth condition on the quasihyperbolic metric guarantees a d-Poincaré inequality. This is indeed the case as Smith and Stegenga [SS2] showed that, in this situation, the quasihyperbolic metric is p-integrable for all $p \geq 1$.

The next improvement was the result of Smith and Stegenga [SS3] according to which the logarithmic growth condition implies a p-Poincaré inequality for some $p < d$; notice that this is a stronger conclusion than a d-Poincaré inequality as seen (with some work) using the Hölder inequality. This improved result was reached at with the help of a version of Theorem 6.1 for general p and with the bound for the Minkowski dimension of the boundary (cf. Section 3). It is easy to check that the exponent p necessarily depends on the data. When D is the image of the unit disk under an α-Hölder continuous univalent function, it follows by [JM] (or [KR]) and by [SS3] that any $p > 2 - C\alpha^2$ will work.

Question 6.2. *What is the correct exponent s for α in $p > 2 - C\alpha^s$?*

Not all domains are Poincaré domains but slightly surprisingly each simply connected plane domain D of finite area supports a weighted Poincaré inequality for all $p \geq 1$. This holds with the weight $\omega = J_f^{1/2}$ where f is the Riemann mapping function of D onto the disk. For details see [K1]. We do not know if any domain $D \subset \mathbb{R}^d$ can be made a weighted Poincaré domain by selecting the weight correctly.

Let us next consider a more restrictive Sobolev-Poincaré inequality. Let D, ω, u, u_D be as in (6.1) and assume that $q = kp$, $k > 1$. Then we require that

$$||u - u_D||_{L^q(D,\omega)} \leq C||\nabla u||_{L^p(D,\omega)}. \tag{6.2}$$

If there is a constant $k > 1$ and a weight ω such that (6.2) holds for all $p \geq 1$ we say that D admits a weighted Sobolev-Poincaré inequality. If D is a ball, then one may take $q = pd/(d-p)$ for $1 \leq p < d$ in the constant weight case and thus $k = d/(d-1)$ will work. Recently Buckley and Koskela [BK1] showed that a simply connected plane domain satisfies an unweighted Sobolev-Poincaré with some $1 \leq p < 2$ and $q = 2p/(2-p)$ if and only if D is a John domain (cf. Section 2). The logarithmic growth condition on the hyperbolic metric appears also in connection with the weighted Sobolev-Poincaré inequalities as the following result from [K1] indicates.

Theorem 6.3. *Let D be a simply connected plane domain of finite area. Then D admits a Sobolev-Poincaré inequality with the weight $J_f^{1/2}$, where f is the Riemann mapping function of D, if and only if D satisfies a quasihyperbolic boundary condition.*

Let us now go back to the case $p = 2$ in the plane. This is the borderline case for the usual Sobolev embedding that results in the Trudinger inequality. Smith and Stegenga [SS4] proved that a logarithmic growth condition on the hyperbolic metric guarantees that the domain satisfies a Trudinger inequality, and very recently Buckley and Koskela established a converse [BK2].

We say that D supports a Trudinger inequality if

$$\int_D \exp\left(c|u - u_D|^{d/(d-1)}\right)\, dx \leq 2|D| \tag{6.3}$$

holds whenever u belongs to the Sobolev class $W^{1,d}(D)$, $\int_D |\nabla u|^d\, dx \leq 1$, $u_D = |D|^{-1} \int_D u\, dx$.

Theorem 6.4. *Let D be a simply connected plane domain. Then D supports the Trudinger inequality if and only if D satisfies a quasihyperbolic boundary condition.*

7 Extensions

Recall that the quasihyperbolic metric was defined using the integral

$$\int_\gamma \frac{ds}{d(x, \partial D)^s}$$

with $s = 1$. There have also been some studies on the corresponding metric for $s \neq 1$.

Gehring and Martio [GM1] studied the case $0 < s < 1$. In this case the growth of the metric is tied to the global behavior of functions that satisfy a uniform Hölder condition of order $1 - s$ for all balls B contained in D.

For further studies in this direction see the papers [Lp] by Lappalainen and [AHHL] by Astala, Hag, Hag, and Lappalainen. Buckley and Koskela [BK2] have recently characterized the simply connected planar domains that admit a Sobolev-Poincaré imbedding of the Sobolev class into a Hölder space using a growth condition for such a metric.

Langmeyer [Ln] studies the case $s \geq 1$ in her thesis. She has established results parallel to those of Gehring and Martio [GM1]; there are also some new features for $s > 1$ that do not occur when $s \leq 1$.

These metrics also appear in connection with Poincaré type inequalities with exponents different from d. When one tries to patch up global inequalities from inequalities that hold for cubes, one is again faced with the problem of transporting the averages between different Whitney cubes for a function defined on the domain. See the discussion in the preceding section for the basic idea. This results in integrals of the above type; see e.g. the papers by Hurri [H], Smith, Stegenga, and Stanoyevitch [SS2], [SSS].

Let us now go back to the usual exponent $s = 1$. Let H be a second order divergence form elliptic operator in D with Dirichlet boundary conditions. Banuelos [Ba] shows that the Schrödinger semigroup e^{-tH} is intrinsically ultracontractive if

$$k_D(x, x_0) \leq C_1 \frac{1}{d(x, \partial D)^a} + C_2$$

for some $0 < a < 2$ and if the complement of D is thick at each boundary point of D in the sense of the Newtonian capacity. For the necessary definitions we refer the reader to [Ba]. In the plane his thickness condition is the same as uniform perfectness from Section 1. He gives convincing reasons for the importance of the growth rate of the quasihyperbolic metric in the study of intrinsic ultracontractivity. Related results have also been established by Bass and Burdzy [BB]; see in particular the discussion at page 409.

Another application to partial differential equations we have not discussed is the growth rate of functions that satisfy a Harnack inequality; see for example [HV]. This brings us to behavior at large of positive solutions or supersolutions. By a result of Armitage [A] a positive superharmonic function in a ball in $\mathbb{R}^d$ is p-integrable for any $p < d/(d-1)$. It was recently shown by Lindqvist [Li] that this result has an analog for solutions to a large class of degenerate elliptic equations defined in a domain satisfying a quasihyperbolic boundary condition. Conversely, by the results of Matsumoto [Ms] and Smith and Stegenga [SS3] each simply connected plane domain for which all positive superharmonic functions are p-integrable for some $p > 0$ necessarily satisfies this logarithmic growth condition on the quasihyperbolic metric.

References

[A] D. H. Armitage, *On the extension of superharmonic functions*, J. London Math. Soc. **4** (1971), 365–373.

[AG] K. Astala and F. W. Gehring, *Quasiconformal analogues of theorems of Koebe and Hardy-Littlewood*, Mich. Math. J. **32** (1985), 99-107.

[AHHL] K. Astala, K. Hag, P. Hag and V. Lappalainen, *Lipschitz classes and the Hardy-Littlewood property*, Monats. Math. **115** (1993), 267–279.

[AK] K. Astala and P. Koskela, *Quasiconformal mappings and global integrability of the derivative*, J. Analyse Math. **57** (1991), 203–220.

[Ba] R. Bañuelos, *Intrinsic ultraconductivity and eigenfunction estimates for Schrödinger operators*, J. Funct. Anal. **100** (1991), 181–206.

[BB] R. F. Bass and K. Burdzy, *Lifetimes of conditioned diffusions*, Probab. Theory Relat. Fields **91** (1992), 405–443.

[BP] A. F. Beardon and Ch. Pommerenke, *The Poincare metric of plane domains*, J. London Math. Soc. **18** (1978), 475–483.

[BcP] J. Becker and C. Pommerenke, *Hölder continuity of conformal mappings and non-quasiconformal Jordan curves*, Comment. Math. Helv. **57** (1982), 221–225.

[Bi] C. Bishop, *How geodesics approach the boundary in a simply connected domain*, J. Analyse Math. **64** (1994), 291–335.

[Bo] B. Bojarski, *Homeomorphic solutions of Beltrami systems*, Dokl. Akad. Nauk. SSSR (Russian) **102** (1955), 661–664.

[BK1] S. M. Buckley and P. Koskela, *Sobolev-Poincaré implies John*, Math. Research Letters **2** (1995), 577–593.

[BK2] S. M. Buckley and P. Koskela, *Criteria for imbeddings of Sobolev-Poincaré type*, Internat. Math. Res. Notices **18** (1996), 881–901.

[CJY] L. Carleson, P. Jones and J.-C. Yoccoz, *Julia and John*, Bol. Soc. Bras. Mat. **25** (1994), 1–30.

[FT] S. Frankel and J. Tysk, *Behavior of the Poincaré metric near a fractal boundary*, Complex Variables Theory Appl. **23** (1993), 257–267.

[G] F. W. Gehring, *The L^p-integrability of the partial derivatives of a quasiconformal mapping*, Acta Math. **130** (1973), 265–277.

[GHM] F. W. Gehring, K. Hag and O. Martio, *Quasihyperbolic geodesics in John domains*, Math. Scand. **65** (1989), 75–92.

[GM1] F. W. Gehring and O. Martio, *Lipschitz classes and quasiconformal mappings*, Ann. Acad. Sci. Fenn. Ser. A I Math. **10** (1985), 203–219.

[GM2] F. W. Gehring and O. Martio, *Quasiextremal distance domains and extension of quasiconformal mappings*, J. Analyse Math. **45** (1985), 181–206.

[GO] F. W. Gehring and B. Osgood, *Uniform domains and the quasihyperbolic metric*, J. Analyse Math. **36** (1979), 50–74.

[GP] F. W. Gehring and B. Palka, *Quasiconformally homogeneous domains*, J. Analyse Math. **30** (1976), 172–199.

[GH] E. Ghys and P. de la Harpe, *Sur les Groupes Hyperboliques d'aprés Mikhael Gromov*, Birkhäuser, Progress in Mathematics, Boston-Basel-Berlin, 1990.

[GS] J. Graczyk and S. Smirnov, *Collet, Eckmann, & Hölder*, to appear.

[HN] J. Heinonen and R. Näkki, *Quasiconformal distortion on arcs*, J. Analyse Math. **63** (1994), 19–53.

[HR] J. Heinonen and S. Rohde, *The Gehring-Hayman inequality for quasihyperbolic geodesics*, Math. Proc. Cambridge Philos. Soc. **114** (1993), 393–405.

[HK] D. A. Herron and P. Koskela, *Conformal capacity and the quasihyperbolic metric*, Indiana Univ. Math. J. **45** (1996), 333–359.

[HV] D. A. Herron and M. Vuorinen, *Positive harmonic functions in uniform and admissible domains*, Analysis **8** (1988), 187-206.

[H] R. Hurri, *Poincaré domains in* $\mathbb{R}^n$, Ann. Acad. Sci. Fenn. Ser. A I Math. Diss. **71** (1988), 1–42.

[Je] D. Jerison, *The Poincaré inequality for vector fields satisfying Hörmander's condition*, Duke Math. J. **53** (1986), 503–523.

[J1] P. W. Jones, *Extension domains for BMO*, Indiana Univ. Math. J. **29** (1980), 41–66.

[J2] P. W. Jones, *Quasiconformal mappings and extendability of functions in Sobolev spaces*, Acta Math. **147** (1981), 71–88.

[JM] P. W. Jones and N. G. Makarov, *Density properties of harmonic measure*, Annals of Math. **142** (1995), 427–455.

[K1] P. Koskela, *Natural weights for the Poincaré inequality*, in Proceedings of the Fourth Finnish-Polish Summer School in Complex Analysis at Jyväskylä, Ber. Univ. Jyväskylä Math. Inst. **55** (1993), 55–63.

[K2] P. Koskela, *The radial behavior of a quasiconformal mapping*, Duke Math. J. **74** (1994), 667–679.

[KR] P. Koskela and S. Rohde, *Hausdorff dimension and mean porosity*, Math. Ann., to appear.

[Ln] N. Langmeyer, *The quasihyperbolic metric, growth, and John domains*, University of Michigan, Thesis, 1996

[Lp] V. Lappalainen, Lip_h*-extension domains*, Ann. Acad. Sci. Fenn. Ser. A I Math. Diss. **56** (1985), 1–52.

[Li] P. Lindqvist, *Global integrability and degenerate elliptic equations*, J. Anal. Math. **61** (1993), 283–292.

[Mr] G. Martin, *Quasiconformal and bilipschitz mappings, uniform domains and the quasihyperbolic metric*, Trans. Amer. Math. Soc. **292** (1985), 169–191.

[Ms] M. Masumoto, *Integrability of superharmonic functions on Hölder domains of the plane*, Proc. Amer. Math. Soc. **117** (1993), 1083–1088.

[MO] G. Martin and B. Osgood, *The quasihyperbolic metric and the associated estimates on the hyperbolic metric*, J. Analyse Math. **47** (1986), 37–53.

[MS] O. Martio and J. Sarvas, *Injectivity theorems in plane and space*, Ann. Acad. Sci. Fenn. Ser. A I Math. **4** (1978-79), 383–401.

[MV] O. Martio and J. Väisälä, *Global* L^p*-integrability of the derivative of a quasiconformal mapping*, Complex Variables **9** (1988), 309–319.

[NPV] F. Nazarov, I. Popovici, and A. Volberg, *Domains with the quasihyperbolic boundary condition, domains with the Green's boundary conditions, and estimates of Hausdorff dimension of their boundary*, manuscript.

[PR] F. Przytycki and S. Rohde, *Porosity of Collet-Eckmann Julia sets*, to appear.

[R] S. Rohde, *The boundary behavior of Bloch functions*, J. London Math. Soc. **48** (1993), 488-499.

[SS1] W. Smith and D. A. Stegenga, *A geometric characterization of Hölder domains*, J. London Math. Soc. **35** (1987), 471–480.

[SS2] W. Smith and D. A. Stegenga, *Hölder domains and Poincaré domains*, Trans. Amer. Math. Soc. **319** (1990), 67–100.

[SS3] W. Smith and D. A. Stegenga, *Exponential integrability of the quasihyperbolic metric in Hölder domains*, Ann. Acad. Sci. Fenn. Ser. A I Math. **16** (1991), 345–360.

[SS4] W. Smith and D. A. Stegenga, *Sobolev imbeddings and integrability of harmonic functions in Hölder domains*, in Potential Theory (de Gruyter, Berlin 1992).

[SSS] W. Smith, A. Stanoyevitch and D. A. Stegenga, *Planar Poincaré domains: Geometry and Steiner Symmetrization*, J. Analyse Math. **66** (1995), 137–183.

[Sa] S. G. Staples, *L^p-averaging domains and the Poincaré inequality*, Ann. Acad. Sci. Math. Ser. A I Math. **14** (1989), 103–127.

[Se] E. Stein, *Singular integrals and differentiability properties of functions*, Princeton Univ. Press, 1970.

[Va1] J. Väisälä, *Lectures on n-dimensional quasiconformal mappings*, Lecture Notes in Mathematics 229, Springer-Verlag, Berlin-Heidelberg-New York, 1971.

[Va2] J. Väisälä, *Free quasiconformality in Banach spaces I*, Ann. Acad. Sci. Fenn. Ser. A I Math. **15** (1990), 355–379.

[Vu1] M. Vuorinen, *Conformal invariants and quasiregular mappings*, J. Analyse Math. **45** (1985), 69–115.

[Vu2] M. Vuorinen, *On quasiregular mappings and domains with complete conformal metric*, Math. Z. **194** (1987), 459–304.

Department of Mathematics
University of Jyväskylä
P.O. Box 35, Fin-40351
Jyväskylä, Finland

Kleinian Groups; Eichler Cohomology and the Complex Theory

Irwin Kra

1 Introduction

In this expanded version of an expository lecture delivered at the University of Michigan in honor of Fred Gehring's seventieth birthday, I discuss the complex analytic theory of finitely generated Kleinian groups. The common theme, for all but the last section, is Eichler cohomology as developed in [29] and [30][1]. We follow, however, the notation established in [35]. Our aim is to illustrate the general picture, especially by good examples that can be generalized. Full proofs are found in the articles listed in the bibliography. In order to help the reader,[2] an appendix with some standard definitions and elementary properties has been included.

We fix once and for all Γ, a nonelementary Kleinian group with ordinary set $\Omega = \Omega(\Gamma) \subset \hat{\mathbb{C}} = \mathbb{C} \cup \{\infty\}$. We will usually, but not always, assume that Γ is also finitely generated, since without this assumption there is not much of a theory. Let $\Lambda = \Lambda(\Gamma) = \hat{\mathbb{C}} - \Omega$ be the limit set of Γ, and Ψ a Γ-invariant set of rank one doubly cusped parabolic fixed points of the group. The *extended* ordinary set is

$$\Omega^+ = \Omega \cup \Psi.$$

In many ways the points of Ψ behave as ordinary rather than limit points, and the quotient

$$\Omega^+/\Gamma$$

is naturally viewed as a (perhaps disconnected) Riemann surface with (perhaps) nodes. Let D be a Γ-invariant union of components of Ω and

$$\lambda(z)|dz| = \lambda_D(z)|dz|,$$

---

[0]Research supported in part by NSF Grant DMS 9500557.

[1]See [6] for an alternate approach.

[2]I am grateful to the referee for this and other suggestions that improved and clarified the exposition.

the Poincaré metric of constant negative curvature -1 on (each component of) D. We set D^+ to be the intersection of the "closure" of D in $\hat{\mathbb{C}}$ with Ω^+. We must exercise some care in defining this closure operation. If $x \in \Psi$ and $A \in \Gamma$ is a parabolic generator for the subgroup stabilizing x, then there are two horodiscs in Ω precisely invariant under $< A >$ in Γ. The point x is in D^+ if and only if both of these horodiscs lie in D. A neighborhood of x in D^+ consists of the parabolic fixed point together with the two horodiscs. A neighborhood of its projection is the node (image of the fixed point) along with two discs tangent at the node, if x is stabilized by a cyclic (parabolic) group. If x is stabilized by a $\mathbb{Z}_2$-extension of a cyclic group, then a neighborhood of the node consists of an ordinary disc[3]. These nodes are called *degenerate* in [42], where more relevant information on this subject can be found. A node on D^+/Γ determines one or two punctures and two disjoint horodiscs on D/Γ. We shall say (clearly, an abuse of language) that these punctures and horodiscs are *parts* of the node. In any case, each component of D^+/Γ is a noded Riemann surface with signature – a *noded orbifold.* We note that the genus of this orbifold is, by definition, the genus of the nonsingular surface obtained by replacing a neighborhood of each nondegenerate node by a cylinder.

Finally, we also fix a $q \in \mathbb{Z}$; $q \geq 2$.

Our general aim is to describe the interaction between the group Γ and the surfaces Ω^+/Γ it represents. We are interested in obtaining finiteness statements. The main ingredients are the number of generators of Γ and dimensions of spaces of differentials on Ω/Γ. We will also encounter important intermediaries, vector spaces of cohomology classes for the group that are finite dimensional provided the group is finitely generated. A more far reaching goal is to obtain equations describing the algebraic curves represented by Ω/Γ. A general theory is not available. We discuss special cases in the last two sections of the paper.

2 Spaces of cusp forms

Fix, once and for all, a nonelementary Kleinian group Γ and a Γ-invariant union of components $D \subset \Omega$.

2.1 Automorphic forms

An *automorphic q-form* for Γ on D is a meromorphic function φ on D with the property that

$$\varphi(g(z))g'(z)^q = \varphi(z), \text{ all } g \in \Gamma \text{ and all } z \in D. \tag{2.1}$$

[3]The last two sentences define *nodes* for our purposes.

The automorphic form φ is *integrable* if its *norm*

$$||\varphi||_{D/\Gamma} = \int\int_{D/\Gamma} \lambda^{2-q}(z)|\varphi(z)\frac{dz\overline{dz}}{2}| < +\infty.$$

The Banach space of integrable holomorphic automorphic forms for Γ on D is denoted by $\mathbb{A}_q(D,\Gamma)$. If Γ is the trivial group, we write $\mathbb{A}_q(D)$[4] for $\mathbb{A}_q(D,\Gamma)$. The automorphic form φ is *bounded* provided its *norm*

$$||\varphi||_{\mathbb{B}} = \sup\{\lambda^{-q}(z)|\varphi(z)|;\ z \in D\} < +\infty.$$

The Banach space of bounded automorphic forms for Γ on D is denoted by $\mathbb{B}_q(D,\Gamma)$ ($\mathbb{B}_q(D)$ if Γ is trivial). By a theorem of Bers (see, for example, [32, Chapter III]), the *Petersson scalar product*[5]

$$< \varphi, \psi >_{D/\Gamma} \quad = \quad \int\int_{D/\Gamma} \lambda^{2-2q}(z)\varphi(z)\overline{\psi(z)}\left|\frac{dz\overline{dz}}{2}\right|, \qquad (2.2)$$
$$\varphi \in \mathbb{A}_q(D,\Gamma),\ \psi \in \mathbb{B}_q(D,\Gamma),$$

establishes a topological conjugate $\mathbb{C}$-linear (nonisometric) isomorphism between $\mathbb{A}_q(D,\Gamma)^*$, the topological dual space to $\mathbb{A}_q(D,\Gamma)$, and $\mathbb{B}_q(D,\Gamma)$. The basic ingredients in the proof of Bers' theorem [12] are reproducing kernels for automorphic functions, the Riesz representation theorem, and the Hahn-Banach theorem. If D/Γ is of finite conformal type (this is guaranteed to be the case for groups Γ that satisfy our generic hypothesis (that is, are finitely generated), by Ahlfors' finiteness theorem [4], see also §4), then $\mathbb{A}_q(D,\Gamma)$ and $\mathbb{B}_q(D,\Gamma)$ are the same vector space, the space of *q-cusp* forms, with different norms. In this case, (2.2) endows the underlying vector space with a Hilbert space structure. For most applications, we need the L^1 or L^∞, rather than the L^2, norm on vector spaces of q-forms.

Let us assume that D/Γ is of finite conformal type. In this case, this orbifold can be written as a finite disjoint union

$$D/\Gamma = \cup_{i=1}^N S_i,$$

where each S_i has a hyperbolic signature

$$(p_i, n_i;\ \nu_{1,i}, ..., \nu_{n_i,i}),$$

here $p_i \geq 0$ is the genus of S_i, $n_i \geq 0$ is the number of distinguished points on S_i, $\nu_{j,i}$ is the ramification number of the j-th distinguished point on S_i

[4] Here D is an arbitrary open subset of $\hat{\mathbb{C}}$ with at least three boundary points. We ignore the technical condition that needs to be imposed if $\infty \in D$. We will also usually ignore the modifications of many of our formulae that are required for a parameter or variable assuming the value ∞.

[5] As usual, $< \varphi, \psi >_{D/\Gamma}$ for the trivial group Γ will be abbreviated $< \varphi, \psi >_D$.

(an integer ≥ 2 or the symbol ∞). We may assume that the distinguished points are ordered so that

$$\nu_{1,i} \leq \ldots \leq \nu_{n_i,i}.$$

The hyperbolicity of S_i means that its orbifold Euler characteristic

$$\chi(S_i) = -\left(2p_i - 2 + \sum_{j=1}^{n_i}\left(1 - \frac{1}{\nu_{j,i}}\right)\right)$$

is negative. We record for subsequent use that

$$\text{Area }(S_i) = -2\pi\chi(S_i).$$

2.2 *Basic operators*

Equation 2.1 is a special case of an interesting operation on functions. It is convenient to formalize and generalize the idea involved. Let r and s be two half integers whose sum is an integer. Fix an open set D and a group of Möbius transformations Γ operating discontinuously on it. Let f be a meromorphic function on D. We define an *action* of f on functions φ on $f(D)$ by

$$(f^*_{r,s}\varphi)(z) = \varphi(f(z))f'(z)^r\overline{f'(z)}^s, \ z \in D.$$

We have introduced a contravariant functor, since the identity acts trivially and if g is meromorphic on $f(D)$, then

$$(g \circ f)^*_{r,s} = f^*_{r,s} \circ g^*_{r,s}.$$

We abbreviate $f^*_{r,0}$ by f^*_r and when the subscript r is omitted, it is understood to be the integer q we have fixed.

Many functions exhibit invariance with respect to these operators. Equation 2.1 is, of course, the statement $g^*\varphi = \varphi$. The Poincaré density λ exhibits the following important invariance with respect to (meromorphic) covering maps f:

$$f^*_{\frac{1}{2},\frac{1}{2}}\lambda_{f(D)} = \lambda_D.$$

These operators preserve many structures. For example, if f is univalent, $\Delta = f(D)$, and $G = f\Gamma f^{-1}$ is also a group of Möbius transformations, then

$$f^* : \mathbb{A}_q(\Delta, G) \to \mathbb{A}_q(D,\Gamma) \text{ and } f^* : \mathbb{B}_q(\Delta, G) \to \mathbb{B}_q(D,\Gamma) \qquad (2.3)$$

are isometric isomorphisms with the additional property that (the isomorphisms preserve the Petersson pairing)

$$< \varphi, \psi >_{\Delta/G} = < f^*\varphi, f^*\psi >_{D/\Gamma}, \ \varphi \in \mathbb{A}_q(\Delta, G), \ \psi \in \mathbb{B}_q(\Delta, G). \qquad (2.4)$$

The above facts are simple consequences of the change of variable formulae and the invariance properties of the functions involved (especially of the Poincaré metric). Let $g = f^{-1}$ and apply the last formula to $\varphi \in \mathbb{A}_q(\Delta, G)$ and $g^*\psi$ with $\psi \in \mathbb{B}_q(D,\Gamma)$ to conclude that our family of operators is self adjoint, in the sense that

$$< f^*\varphi, \psi >_{D/\Gamma} = < \varphi, g^*\psi >_{\Delta/G}, \text{ all } \varphi \in \mathbb{A}_q(\Delta, G) \text{ and all } \psi \in \mathbb{B}_q(D,\Gamma).$$

In particular, if f is an involution on D that preserves Γ and D/Γ is of finite conformal type then f^* is a self adjoint unitary operator on the Hilbert space $\mathbb{A}_q(D,\Gamma)$ (with Petersson inner product).

2.3 Reduction to the Fuchsian case

Many, but not the deepest, problems for Kleinian groups reduce to the better understood Fuchsian case. The Banach space theory of automorphic forms falls into this category. We proceed to describe the steps involved in this "reduction." Let G be a nonelementary Kleinian group, and

$$\Delta_1, \ldots, \Delta_j, \ldots$$

a maximal set of G-inequivalent connected components of its region of discontinuity $\Omega(G)$. We let G_j be the stabilizer of Δ_j in G. Then it is routine to verify that

$$\mathbb{A}_q(\Omega, G) \cong \oplus_j \mathbb{A}_q(\Delta_j, G_j) \text{ and } \mathbb{B}_q(\Omega, G) \cong \prod_j \mathbb{B}_q(\Delta_j, G_j).$$

These two isomorphisms reduce many questions about q-forms for a general Kleinian group to the more specialized case of a nonelementary group G with an invariant component Δ, a *function* group. In this case, we let

$$h : \mathbb{H}^2 \to \Delta$$

be a holomorphic universal covering map of Δ by the upper half plane[6] and Γ the *Fuchsian model* of the action of G on Δ; that is, Γ is the set of $f \in \mathrm{PSL}(2,\mathbb{R})$ for which there exits a $g \in G$ with

$$g \circ h = h \circ f. \tag{2.5}$$

Then the orbifolds $\mathbb{H}^2/\Gamma$ and Δ/G are conformally equivalent[7], and the map (even though h is not injective) h^* establishes isometric isomorphisms

[6]The assumption that G be nonelementary guarantees that the boundary of Δ has more than two points.

[7]It should be observed that $\mathbb{H}^2$ need not be a connected component of Γ; this happens when Δ/G has an ideal boundary. In this case, the Fuchsian group Γ is of the second kind, $\Lambda(\Gamma)$ is a proper subset of $\mathbb{R} \cup \{\infty\}$, and $\Omega(\Gamma)$ is connected but not simply connected.

2.3 rewritten here as

$$h^* : \mathbb{A}_q(\Delta, G) \to \mathbb{A}_q(\mathbb{H}^2, \Gamma) \text{ and } h^* : \mathbb{B}_q(\Delta, G) \to \mathbb{B}_q(\mathbb{H}^2, \Gamma)$$

that preserve the Petersson pairing (the translation of 2.4 to the current situation)

$$< \varphi, \psi >_{\Delta/G} = < h^*(\varphi), h^*(\psi) >_{\mathbb{H}^2/\Gamma}, \ \varphi \in \mathbb{A}_q(\Delta, G), \ \psi \in \mathbb{B}_q(\Delta, G).$$

The last equation shows that h^* is Hermitian (it is unitary and its inverse is its adjoint)[8]. The above theory shows that q-differentials for a Kleinian group G (with region of discontinuity Ω) only capture information on the orbifold Ω/G represented by the group, not how the various components of the orbifold interact. In §3 we will discuss an object that also reflects the interaction between the components of Ω/G.

In equation 2.5, $g \in G$ is uniquely determined by $f \in \Gamma$. If we set $g = \chi(f)$, then we obtain a group homomorphism χ of Γ onto G, and the pair (χ, h) is a deformation of Γ. It is a bounded deformation[9] [31] whenever Δ/G is of finite conformal type.

We conclude this subsection with an observation and definition. If S is an orbifold of finite conformal type, then it is a subset of the compact orbifold $\bar{S}$ obtained by filling in the punctures on S (those distinguished points with ramification number ∞)[10].

2.4 Projections to the surface

The last subsection shows that the study of automorphic forms supported on a general union of components D of a general Kleinian group Γ reduces to the case of a function group Γ with an invariant component D. More particularly, we may assume that Γ is Fuchsian and D an invariant disc for Γ. It is hence not surprising that the automorphic q-forms project to objects defined on the quotient orbifold D/Γ. An automorphic q-form φ on D for Γ is projected to the orbifold D/Γ as a q-differential[11] (that is, through the invariance of $\varphi(z)dz^q$). The projection of a form in $\mathbb{A}_q(D, \Gamma)$ may pick up poles at the distinguished points of the orbifold. Assume that D/Γ is of finite conformal type with signature

$$(p, n; \ \nu_1, ..., \nu_n) \tag{2.6}$$

and $\varphi \in \mathbb{A}_q(D, \Gamma)$, then its projection Φ to D/Γ extends to be meromorphic on $\overline{D/\Gamma}$, is regular (holomorphic) at the regular (nondistinguished) points

[8] We may interpret h^* as $(h^{-1})_q^*$ even though h^{-1} is not single valued.

[9] Both terms in *bounded deformation* are defined in [31].

[10] It should noted that $\bar{S}$ may be elementary.

[11] A section of the q-th power of the canonical line bundle on the orbifold.

of D/Γ, and satisfies

$$\mathrm{ord}_{x_i}\Phi \geq -\left[q - \frac{q}{\nu_i}\right],$$ [12]

where x_i is the distinguished point with ramification number ν_i. The Riemann-Hurwitz formula tells us that

$$\deg(\Phi) = \sum_{x \in \overline{D/\Gamma}} \mathrm{ord}_x\Phi = q(2p-2).$$ [13]

The Riemann-Roch theorem then yields (see, for example, [32, Ch. VII] for a proof of this fact and related results) part (b) of

Theorem 1 *Let Γ be a function group with invariant component D.*
(a) *The Banach space $\mathbb{A}_2(D,\Gamma)$ is finite dimensional if and only if D/Γ is of finite conformal type.*
(b) *If D/Γ is of finite conformal type with signature 2.6, then*

$$\dim_{\mathbb{C}} \mathbb{A}_q(D,\Gamma) = (2q-1)(p-1) + \sum_{i=1}^{n}\left[q - \frac{q}{\nu_i}\right]. \tag{2.7}$$

Sufficiency in part (a) follows from the Riemann-Roch theorem as does necessity in the finite genus case; in the infinite genus case, one constructs a square integrable closed 1-form with only one nontrivial period on an arbitrary simple closed curve, and then argues easily that the space of square integrable closed 1-forms is infinite dimensional. Note that if D/Γ is of conformal type (p,n), then

$$\dim_{\mathbb{C}} \mathbb{A}_2(D,\Gamma) = 3p - 3 + n, \tag{2.8}$$

a number that appears prominently in the theory of moduli of Riemann surfaces. It is the dimension of the Teichmüller space $\mathbf{T}(p,n)$ of Riemann surfaces (also orbifolds) of conformal type (p,n). The last equality also holds for surfaces with nodes. We proceed to the definition of regularity of differentials for surfaces with nodes, in order to establish 2.8 for the action of a Kleinian group on the extended ordinary set.

Assume that Γ and D are given with D/Γ of finite conformal type. The vector space $\mathbb{A}_q(D,\Gamma)$ can be characterized as the set of holomorphic q-forms whose projections to D/Γ extend meromorphically to the punctures and have order $\geq -(q-1)$ at each puncture.

---

[12]For $x \in \mathbb{R}$, $[x]$ is the largest integer not exceeding x. We interpret $\left[q - \frac{q}{\infty}\right]$ as $q-1$.

[13]The formula is also valid for $q < 2$. Meromorphicity at the cusps for such a q-form must be properly defined. There is also a version of this formula on the universal cover of the surface which is more general. It applies to multiplicative forms for any character on Γ; not just to the trivial character.

The residue of a meromorphic q-form[14] is not well defined in general. Assume that Φ is a meromorphic q-differential on a Riemann surface S, that

$$\mathrm{ord}_x \Phi \geq -q,$$

and that z is a local coordinate vanishing at x. In this case, if we write locally

$$\Phi = \varphi(z) dz^q,$$

then φ has a power series expansion

$$\varphi(z) = \sum_{i=-q}^{\infty} a_i z^i.$$

We define the *residue* of Φ at x as

$$\mathrm{Res}_x \Phi = a_{-q}.$$

It is routine to check that this is well defined, since we are restricted to coordinates vanishing at x.

The vector space $\mathbb{A}_q(D^+, \Gamma)$ is defined as the set of holomorphic q-forms for Γ on D whose projections Φ to D/Γ extend meromorphically to the punctures, have orders $\geq -(q-1)$ at each puncture on D/Γ that is not part of node (the puncture arises from a parabolic fixed point of Γ not in D^+), have orders $\geq -q$ at each puncture on D/Γ that is part of node, and further if the two punctures that form the node are x_1 and x_2, we require that Φ satisfy the *residue condition*

$$\mathrm{Res}_{x_1} \Phi = (-1)^q \, \mathrm{Res}_{x_2} \Phi$$

at the node[15].

It is also convenient to introduce the vector space $\mathbb{A}_q^+(D, \Gamma)$ consisting of of holomorphic q-forms for Γ on D whose projections to D/Γ extend meromorphically to the punctures and have order $\geq -q$ at each puncture, and its subspace $\mathbb{A}_q^+(D^+, \Gamma)$ consisting of those $\varphi \in \mathbb{A}_q^+(D, \Gamma)$ that satisfy the residue condition at each node. It is obvious that we have the inclusions

$$\mathbb{A}_q(D, \Gamma) \subset \mathbb{A}_q(D^+, \Gamma) \subset \mathbb{A}_q^+(D^+, \Gamma) \subset \mathbb{A}_q^+(D, \Gamma).$$

Only the first of the above four finite dimensional vector spaces has natural metrics on it (making it into a Banach or even Hilbert space); it is of considerable interest to define canonical metrics on the other three spaces. There are no natural candidates. It should be observed that the set of

---

[14] Recall that $q > 1$.

[15] Note that the two punctures are determined by the same parabolic fixed point, but different horodiscs.

nonsingular points in D^+/Γ is, of course, D/Γ which need not be connected even when D^+/Γ is.

Assume that (after normalization) the group Γ contains a parabolic motion $A = \begin{bmatrix} 1 & 1 \\ 0 & 1 \end{bmatrix}$ whose fixed point $\infty \in \Psi$. Assume that A generates the stabilizer of ∞ and that $\infty \in D^+$. In this case the disjoint horodiscs may be chosen as

$$D_1 = \{z \in \mathbb{C};\ \Im z > c_1\} \text{ and } D_2 = \{z \in \mathbb{C};\ \Im z < c_2\},$$

for real constants[16] c_1 and c_2 with $c_2 < 0 < c_1$. If $\varphi \in \mathbb{A}_q(D^+, \Gamma)$, then its Fourier expansions at $\pm\imath\infty$ are of the form

$$\sum_{i=0}^{\infty} a_i \exp(2\pi\imath iz),\ z \in D_1 \text{ and } \sum_{i=0}^{\infty} b_i \exp(-2\pi\imath iz),\ z \in D_2, \tag{2.9}$$

respectively. If the puncture corresponding to the horodisc D_i is denoted by x_i, then

$$\text{Res}_{x_1}\Phi = (2\pi\imath)^{-q} a_0 \text{ and } \text{Res}_{x_2}\Phi = (-2\pi\imath)^{-q} b_0;$$

the residue condition hence reads

$$a_0 = b_0.$$

If x_1 is puncture on D/Γ, then after changing to a conjugate group, we may as above assume that the puncture arises from the action of the cyclic group $< A >$ on D_1. If $\varphi \in \mathbb{A}_q(D, \Gamma)$, then for its Fourier expansion 2.9 on D_1, we have

$$a_0 = 0.$$

If D^+/Γ is connected and this (perhaps[17]) singular orbifold is of finite type with signature 2.6, then formula 2.8 with D replaced by D^+ is valid.

2.5 Poincaré series

We fix a nonelementary Kleinian group Γ and an invariant union of components D of $\Omega(\Gamma)$. The *Poincaré* Θ *operator* is defined on meromorphic functions φ on D by

$$\Theta\varphi = \sum_{g \in \Gamma} (\varphi \circ g)(g')^q = \sum_{g \in \Gamma} g_q^*(\varphi),$$

whenever series converges uniformly and absolutely on compact subsets of D. In this case, $\Theta\varphi$ is a meromorphic automorphic q-form for Γ on D. The

[16]In the last line, $\Im z$ is the imaginary part of z.

[17]In our definitions, Δ^+/Γ is permitted, but not required, to have singularities.

series converges for meromorphic integrable (also more general functions; see, for example, §7.2) functions φ on D, and

$$\Theta : \mathbb{A}_q(D) \to \mathbb{A}_q(D,\Gamma)$$

is $\mathbb{C}$-linear, surjective, of norm

$$||\Theta|| \leq 1.$$

It is quite easy to see that the inclusion

$$\mathbb{B}_q(D,\Gamma) \to \mathbb{B}_q(D)$$

is the adjoint of Θ with respect to the Petersson pairing; that is,

$$<\varphi,\psi>_D = <\Theta\varphi,\psi>_{D/\Gamma}, \text{ all } \varphi \in \mathbb{A}_q(D) \text{ and all } \psi \in \mathbb{B}_q(D,\Gamma).$$

If Γ is a finitely generated Fuchsian group of the first kind and D one of its invariant discs, then (McMullen [45]; see also [8]) for $q = 2$,

$$||\Theta|| < 1.$$

This important result has wide applications, particularly to the theory of extremal quasiconformal mappings, but these applications are too far afield from the topics under discussion.

We have attached a set of Banach spaces to the group Γ. All these spaces are finite dimensional whenever Ω^+/Γ is of finite conformal type. We need to relate this analysis on Ω^+/Γ to the algebraic structure of Γ.

3 Eichler cohomology

We proceed to describe some algebraic formalism, cohomology groups, attached to Γ. Eichler [17] used an action of Möbius transformations on polynomials to attach cohomology groups to certain Fuchsian groups. He then identified spaces of cusp forms with distinguished subspaces of the cohomology groups to study number theoretic questions. Bers [12] used this machinery to obtain finiteness results for Fuchsian groups; Ahlfors [4] made a far reaching generalization of Bers' approach to obtain the definite result on finitely generated Kleinian groups; the Ahlfors finiteness theorem.

3.1 The cohomology group

Let Π_{2q-2} be the $(2q-1)$-dimensional vector space of polynomials (of one complex variable). The group $\mathrm{PSL}(2,\mathbb{C})$ (hence also any of its subgroups;

in particular, a Kleinian group Γ) acts on Π_{2q-2} via the *Eichler action* which is defined as:

$$v\gamma = \gamma_{1-q}^{*}v = (v\circ\gamma)(\gamma')^{1-q},\ v\in\Pi_{2q-2},\ \gamma\in\ \mathrm{PSL}(2,\mathbb{C}).$$

A *cocycle* or *crossed homomorphism* is a mapping

$$\chi:\Gamma\to\Pi_{2q-2}$$

such that

$$\chi(\gamma_1\circ\gamma_2)=\chi(\gamma_1)\gamma_2+\chi(\gamma_2),\ \text{all } \gamma_1 \text{ and } \gamma_2\in\Gamma.$$

Every $v\in\Pi_{2q-2}$ gives rise to a *coboundary* or *principal* crossed homomorphism:

$$\gamma\mapsto v\gamma - v,\ \gamma\in\Gamma.$$

The cohomology group[18], $H^1(\Gamma,\Pi_{2q-2})$, is defined as the vector space of cocycles modulo the subspace of coboundaries. We single out the *parabolic cohomology group*,

$$PH^1(\Gamma,\Pi_{2q-2})\subset H^1(\Gamma,\Pi_{2q-2}),$$

that consists of those cohomology classes that vanish (reduce to coboundaries) on all parabolic cyclic subgroups of Γ. If $D\subset\Omega$ is a union of components, then the *D-parabolic* cohomology classes, denoted by $PH^1_D(\Gamma,\Pi_{2q-2})$, consist of those classes that vanish on all cyclic parabolic subgroups that represent punctures on D/Γ. We have the obvious inclusions

$$H^1(\Gamma,\Pi_{2q-2})\supset PH^1_D(\Gamma,\Pi_{2q-2})\supset PH^1_\Omega(\Gamma,\Pi_{2q-2})\supset PH^1(\Gamma,\Pi_{2q-2}).$$

The basic (and most trivial) result concerning the Eichler cohomology machinery is

Lemma 1 *[13] If Γ is nonelementary and generated by N elements, then*

$$\dim H^1(\Gamma,\Pi_{2q-2})\le(2q-1)(N-1);$$

with equality if Γ is free on N generators.

Thus we have attached a collection of finite dimensional vector spaces to a finitely generated Kleinian group.

---

[18]It is, of course, a vector space.

3.2 Analytic construction of cocycles

We need to connect the function theory of Ω/Γ to the linear algebra for Γ. The key ingredient is the analytic construction of cocycles. Fix a nonelementary Kleinian group Γ. Bers [12] constructed cohomology classes for Fuchsian groups from automorphic forms, using an idea with number theoretic origins [17]. Ahlfors [4] made a conceptual and technical advance over previous approaches. The conceptual advance consisted of constructing cohomology classes from measurable, not just analytic, forms; the technical advance was a delicate approximation argument that has NOT been modified or improved in over 35 years. We proceed to describe Ahlfors' construction.

We define a map [19]

$$\beta^* = \beta_q^* : \mathbb{B}_q(\Omega, \Gamma) \to PH^1(\Gamma, \Pi_{2q-2})$$

as follows. Choose and fix for the rest of the discussion a set $\mathcal{A}$ of $2q-1$ distinct points in Λ:

$$\mathcal{A} = \{a_i;\ i = 1,\ ...,\ 2q-1\}.$$

For $b \in \Lambda - \mathcal{A}$ form the rational function ($z \in \hat{\mathbb{C}}$)

$$f(z,b) = f_{\mathcal{A}}(z,b) = \frac{-1}{\pi(z-b)} \frac{\prod_{i=1}^{2q-1}(b-a_i)}{\prod_{i=1}^{2q-1}(z-a_i)},$$

and observe that for fixed b, $f(\cdot, b) \in \mathbb{A}_q(\Omega)$. Hence we can form its Poincaré series

$$\varphi(z,b) = \sum_{\gamma \in \Gamma} f(\gamma(z), b)\gamma'(z)^q, \ \ z \in \Omega;$$

$\varphi(\cdot, b)$ is an element of $\mathbb{A}_q(\Omega, \Gamma)$. Let $\psi \in \mathbb{B}_q(\Omega, \Gamma)$. Define

$$F(\zeta) = < f(\cdot, \zeta), \psi >_\Omega = < \varphi(\cdot, \zeta), \psi >_{\Omega/\Gamma}, \ \ \zeta \in \mathbb{C}.$$

Using standard potential theoretic estimates, it is easy to verify that the function F

(a) is continuous on $\mathbb{C}$,
(b) vanishes at a_i,
(c) is $O(|z|^{2q-2})$, $z \to \infty$, and
(d) satisfies the generalized inhomogeneous Cauchy-Riemann equation

$$\frac{\partial F}{\partial \bar{z}} = \lambda_\Omega^{2-2q} \bar{\psi}.$$

[19] The letter β is chosen in honor of Bers; the reasoning behind the name and notation (use of the star *) will become clear later.

Further, Weyl's lemma and invariance of λ and ψ show that
(e)
$$\chi(\gamma)(\zeta) = F(\gamma(\zeta))\gamma'(\zeta)^{1-q} - F(\zeta),\ \zeta \in \mathbb{C},\ \gamma \in \Gamma,$$
is an entire function.
Hence from (c) that $\chi(\gamma)$ is an element of Π_{2q-2}. Thus
(f)
$$\gamma \mapsto \chi(\gamma),\ \gamma \in \Gamma,$$
defines a cocycle whose cohomology class (is independent of the choices made and) is denoted by $\beta^*(\psi)$. It is not hard to conclude that the cohomology class $\beta^*(\psi)$ vanishes on all parabolic elements of Γ. The delicate mollifier of Ahlfors [4] is needed to show that
(g) β_2^* is injective. From this it is easy to conclude that
(h) β_q^*, $q > 2$, is also injective if each component of Ω/Γ is either of finite conformal type or simply connected.

Although quasiconformal mappings have not appeared in the above discussion, the original studies were motivated in great part by infinitesimal quasiconformal theory. In this connection see section §6. Before discussing the surjectivity of β^* (in the next subsection) we turn to

3.3 The structure theorems for $H^1(\Gamma, \Pi_{2q-2})$ and $PH^1(\Gamma, \Pi_{2q-2})$

An *Eichler integral (of order)* $1-q$ is a holomorphic function F on D with the property that for all $\gamma \in \Gamma$, the function $v = \gamma_{1-q}^* F - F$ is the restriction to D of an element of Π_{2q-2}. Note that when D is not connected, we are requiring that v be the restriction of the <u>same</u> polynomial to each component of D. The map
$$\gamma \mapsto v = \gamma_{1-q}^* F - F,\ \gamma \in \Gamma,$$
defines a cocycle, the *period* of F, whose cohomology class we denote by $\alpha(F)$. It is easily checked that as a consequence of the definition of Eichler integrals
$$\gamma_q^* \circ \frac{d^{2q-1}}{dz^{2q-1}} = \frac{d^{2q-1}}{dz^{2q-1}} \circ \gamma_{1-q}^*,\ \text{all } \gamma \in \text{ PSL}(2, \mathbb{C}).$$

It follows that the differential operator $\frac{d^{2q-1}}{dz^{2q-1}}$ maps Eichler integrals to automorphic forms. The Eichler integral F is *bounded* (*quasibounded*) if
$$\frac{d^{2q-1}F}{dz^{2q-1}} \in \mathbb{A}_q(D,\Gamma)\ \ (\mathbb{A}_q^+(D,\Gamma))\,.$$

The last definition makes sense only if D/Γ is of finite conformal type, which we assume for the remainder of this subsection. The space of bounded (quasibounded) Eichler integrals modulo Π_{2q-2} is denoted by $\mathbb{E}^b_{1-q}(D,\Gamma)$ ($\mathbb{E}^c_{1-q}(D,\Gamma)$). Because we are dealing with Eichler integrals, the quasiboundedness assumption

$$\frac{d^{2q-1}F}{dz^{2q-1}} \in \mathbb{A}^+_q(D,\Gamma)$$

has the stronger implication

$$\frac{d^{2q-1}F}{dz^{2q-1}} \in \mathbb{A}_q(D^+,\Gamma).$$

One last map

$$\beta : H^1(\Gamma,\Pi_{2q-2}) \to \mathbb{A}_q(D,\Gamma)^*$$

will complete the list of ingredients needed for a description of $H^1(\Gamma,\Pi_{2q-2})$ and $PH^1_D(\Gamma,\Pi_{2q-2})$. Let χ be a cocycle representing a cohomology class in $H^1(\Gamma,\Pi_{2q-2})$. Using a partition of unity, we can construct a C^∞-function F on D that represents the cocycle χ in the sense that

$$\chi(\gamma) = \gamma^*_{1-q}F - F, \text{ all } \gamma \in \Gamma.$$

Set

$$\mu = \frac{\partial F}{\partial \bar{z}}.$$

The partition of unity can be so chosen that

$$l(\varphi) = \int\int_{D/\Gamma} \mu(z)\varphi(z)dz\overline{dz}, \ \varphi \in \mathbb{A}_q(D,\Gamma),$$

defines a bounded linear functional $\beta(\chi)$ on $\mathbb{A}_q(D,\Gamma)$. We have the following

Theorem 2 *[29], [30] If D/Γ is of finite conformal type, then*

$$O \to \mathbb{E}^b_{1-q}(D,\Gamma) \xrightarrow{\alpha} PH^1_D(\Gamma,\Pi_{2q-2}) \xrightarrow{\beta} \mathbb{A}_q(D,\Gamma)^* \to 0$$

and

$$O \to \mathbb{E}^c_{1-q}(D,\Gamma) \xrightarrow{\alpha} H^1(\Gamma,\Pi_{2q-2}) \xrightarrow{\beta} \mathbb{A}_q(D,\Gamma)^* \to 0$$

are exact sequences. If we identify $\mathbb{B}_q(D,\Gamma)$ with the dual space $\mathbb{A}_q(D,\Gamma)^*$ to $\mathbb{A}_q(D,\Gamma)$ via the Petersson scalar product, then β^* becomes the splitting map of β.

Alternatively, the *period map* pd $= (\alpha \oplus \beta^*)$ establishes (surjective) isomorphisms

$$\text{pd} : \mathbb{E}^b_{1-q}(D,\Gamma) \oplus \mathbb{B}_q(D,\Gamma) \to PH^1_D(\Gamma,\Pi_{2q-2})$$

and

$$\text{pd} : \mathbb{E}^c_{1-q}(D,\Gamma) \oplus \mathbb{B}_q(D,\Gamma) \to H^1(\Gamma,\Pi_{2q-2}).$$

The fact that for a fixed Γ we have many candidates for D is extremely useful for applications.

It is of interest to obtain alternate characterizations of the various spaces of Eichler integrals as well as good necessary and sufficient conditions for the operator β^* to be injective and surjective. Only partial results (in addition to those described here) are available [46], [47] and [49]. The last theorem provides a satisfactory and useful description of the cohomology group $H^1(\Pi_{2q-2}, \Gamma)$ and the partially parabolic subgroups $PH^1_D(\Pi_{2q-2}, \Gamma)$. We do not have a comparable description of the totally parabolic subgroup $PH^1(\Pi_{2q-2}, \Gamma)$.

4 Ahlfors' finiteness theorem (and its quantitative versions – the Bers area theorems)

Ahlfors' finiteness theorem *[4] If Γ is finitely generated, then Ω^+/Γ is of finite conformal type.*

The fact that β^* is injective for $q = 2$ tells us that each connected component of Ω/Γ is of finite conformal type, and that this space has finitely many nonrigid connected components. There remains the possibility of infinitely many thrice punctured and/or marked spheres appearing in Ω/Γ. This possibility was overlooked in [4]. The situation was remedied by [14], [26] and [5]. The presence of infinitely many thrice marked spheres can also be easily eliminated by the use of the structure theorem for the Eichler cohomology groups with D chosen appropriately.

Bers' area theorem 1 *[13] If Γ is nonelementary and generated by N elements, then*

$$\textit{Area}\ (\Omega/\Gamma) \leq 4\pi(N-1).$$

The proof follows immediately from the observation that

$$\lim_{q\to\infty} \frac{\dim \mathbb{B}_q(\Omega, \Gamma)}{q} = \frac{\text{Area } (\Omega/\Gamma)}{2\pi}.$$

Similar methods yield

Bers' area theorem 2 *[13] If Γ is a nonelementary finitely generated function group with invariant component Δ, then*

$$\textit{Area}\ (\Omega/\Gamma) \leq 2\ \textit{Area}\ (\Delta/\Gamma).$$

It is of interest to bound K, the number of connected components of Ω/Γ, in terms of the number of generators N of Γ. Since, as Bers observed, the minimal area of a connected hyperbolic orbifold is $\frac{\pi}{21}$, it follows immediately that

$$K \leq 84(N-1).$$

Ahlfors made the observation that for all function groups Γ with invariant component Δ, the dimension inequality

$$\dim \mathbb{A}_4(\Delta,\Gamma) + \dim \mathbb{A}_6(\Delta,\Gamma) \geq 1, \tag{4.1}$$

leads to the improved estimate on the number of components

$$K \leq 18(N-1).$$

Maskit has conjectured that

$$K \leq 2(N-1).$$

Despite some important work towards this conjecture in [2], the problem of finding sharp estimates and equalities for K remains open. Another interesting open problem is to obtain an alternate characterization of K^+, the number of connected components of Ω^+/Γ.

5 Sullivan's finiteness theorem

It is difficult to construct Eichler integrals, elements of $\mathbb{E}^c_{1-q}(\Omega,\Gamma)$. Following an idea due to Ahlfors [7], Sullivan [50] (see also [34]) constructed a quasibounded Eichler integral for each parabolic fixed point of the group. Let $\alpha \in \Lambda$ be a parabolic fixed point with parabolic stabilizer P_α.[20] Form

$$\Psi(\zeta,\alpha) = \sum_{\gamma\in\Gamma/P_\alpha} \frac{\gamma'(\alpha)^q}{\gamma(\alpha)-\zeta}, \quad \zeta \in \Omega. \tag{5.1}$$

The series converges for $q \geq 3$ and also for $q = 2$ if α is cusped or P_α has rank two, and defines a quasibounded Eichler integral. The Eichler integral $\Psi(\cdot,\alpha)$ may, however, vanish identically. This happens only if Γ_α/P_α is nontrivial. It is nonzero for either $q = 4$ or 6. (It is tempting to try to explain the appearance of these two integers in connection here with Eichler integrals and in connection with the dimensions of spaces of cusp forms in 4.1.) As a consequence we obtain that the number of inequivalent parabolic fixed points of Γ is $\leq 18(N-1)$. It is an open problem to obtain a sharp bound.

6 Stability criteria

Assume that Γ is finitely generated. The set of homomorphisms of Γ into $\mathrm{PSL}(2,\mathbb{C})$ modulo conjugation by Möbius transformations forms a complex variety. The tangent space at the identity of this variety is roughly

[20] The stabilizer of α in Γ, Γ_α is, in general, a finite cyclic extension of P_α.

$H^1(\Gamma, \Pi_2)$, because the tangent space to the Lie group $\mathrm{PSL}(2, \mathbb{C})$ is the Lie algebra $\mathrm{sl}(2, \mathbb{C})$, of 2×2 matrices of trace 0, which can be canonically identified with Π_2. If β_2 is surjective, then every parabolic homomorphism of Γ into $\mathrm{PSL}(2, \mathbb{C})$ (a homomorphism χ is *parabolic* if parabolic elements of Γ are mapped by χ to parabolic elements in $\mathrm{PSL}(2, \mathbb{C})$) sufficiently close to the identity is a conjugation by a quasiconformal self map of $\hat{\mathbb{C}}$ [25].

7 Vanishing and spanning theorems for Poincaré and relative Poincaré series

For a finitely generated Kleinian group Γ, the Poincaré series operator

$$\Theta : \mathbb{A}_q(\Omega) \to \mathbb{A}_q(\Omega, \Gamma)$$

has a finite dimensional range and infinite dimensional domain. It is hence of interest to determine necessary and sufficient conditions for the vanishing of the Poincaré series of an integrable holomorphic function. We can make considerable progress, if we assume that the function is rational. We need to develop some additional machinery.

7.1 The dual space to $H^1(\Gamma, \Pi_{2q-2})$

Let F be a function on a Γ-invariant set S. We say F *represents* a cocycle χ on S if

$$\chi(\gamma)(z) = F(\gamma(z))\gamma'(z)^{1-q} - F(z), \text{ for all } \gamma \in \Gamma, \text{ all } z \in S.$$

Choose $2q$ distinct loxodromic[21] (for simplicity[22]) fixed points of elements of Γ:

$$b, \ \mathcal{A} = \{a_1, \ ..., \ a_{2q-1}\};$$

where b is a fixed point of (the loxodromic motion) $B \in \Gamma$, and a_j is fixed by $A_j \in \Gamma$. Let χ be a cocycle representing a class in $H^1(\Gamma, \Pi_{2q-2})$. By adding a coboundary to χ, we may and do assume that

$$\chi(A_j)(a_j) = 0.$$

We define a function F on set of loxodromic fixed points of Γ that represents χ on this set. We let

$$F(a_j) = 0,$$

---

[21] A Möbius transformation conjugate to $\begin{bmatrix} \kappa & 0 \\ 0 & \kappa^{-1} \end{bmatrix}$, with $|\kappa| \neq 1$.

[22] One can also choose certain parabolic and elliptic fixed points; but not points with trivial stabilizers.

and

$$l_{b:A}(\chi) = F(b) = \frac{\chi(B)(b)}{B'(b)^{1-q} - 1}.$$

The linear functionals

$$\{l_{b:A}; b \text{ is a loxodromic fixed point } b \neq a_j\},$$

span $H^1(\Gamma, \Pi_{2q-2})^*$. See [33] for details.

We proceed to define another family of linear functionals that span the dual space of $H^1(\Gamma, \Pi_{2q-2})$. Let $A \in \Gamma$ be loxodromic with fixed points α and β. Define $l_A \in H^1(\Gamma, \Pi_{2q-2})^*$ as follows. Let χ be a cocycle. Write

$$\chi(A)(\zeta) = \sum_{j=0}^{2q-2} a_j \frac{(\zeta - \alpha)^j (\zeta - \beta)^{2q-2-j}}{(\alpha - \beta)^j}.$$

Set

$$l_A(\chi) = a_{q-1}.$$

The linear functionals

$$\{l_A;\ A \in \Gamma \text{ is loxodromic}\},$$

are well defined and span ([36] and [48]) $H^1(\Gamma, \Pi_{2q-2})^*$.

For the last two families of linear functionals ([34] and [41]), we consider a parabolic motion $A \in \Gamma$, with fixed point α. We define

$$l_A(\chi) = \chi(A)(\alpha),\ \chi \in H^1(\Gamma, \Pi_{2q-2}).$$

The linear functional l_A is well defined, and for fixed $\chi \in H^1(\Gamma, \Pi_{2q-2})$,

$$l_A(\chi) = 0 \text{ for all parabolic } A \in \Gamma \text{ if and only if } \chi \in PH^1(\Gamma, \Pi_{2q-2}).$$

Similarly, if A and B generate a rank two parabolic subgroup of Γ, we can define an element $l_{A,B} \in PH^1(\Gamma, \Pi_{2q-2})^*$ as follows. Assume that $\alpha \in \mathbb{C}$ is the common fixed point of the elementary group $< A, B >$, and that A is given in normal form as $\begin{bmatrix} 1+\alpha\tau & -\alpha^2\tau \\ \tau & 1-\alpha\tau \end{bmatrix}$ ($\alpha = \alpha(A)$ is the fixed point of A and $\tau = \tau(A)$ is its *translation length*). Let χ be a cocycle representing a cohomology class in $PH^1(\Gamma, \Pi_{2q-2})$. Adding a coboundary, if needed, we may assume that

$$\chi(A) = 0.$$

Then

$$\chi(B)(z) = l_{A,B}(\chi) \left(\frac{-1}{\tau(z-\alpha)^2} \right)^{1-q},\ z \in \mathbb{C},$$

for a unique $l_{A,B}(\chi) \in \mathbb{C}$.

7.2 Poincaré and relative Poincaré series

(a) Poincaré series of rational functions. For $\psi \in \mathbb{B}_q(\Omega, \Gamma)$, we have the important relation [33]

$$l_{b:\mathcal{A}}(\beta^*(\psi)) = 2 < \varphi(\cdot, b), \psi >_{\Omega/\Gamma} . \tag{7.1}$$

We conclude:

$$\text{if } l_{b:\mathcal{A}} = 0, \text{ then } < \varphi(\cdot, b), \psi >_{\Omega/\Gamma} = 0,$$

all $\psi \in \mathbb{B}_q(\Omega, \Gamma)$, and thus

$$\varphi(\cdot, b) = 0.$$

The left hand side of 7.1 is algebraic data; the right hand side is analytic data. It says

$$l_{b:\mathcal{A}}|_{\beta^*(\mathbf{A}_q(\Omega,\Gamma))} = 0 \text{ if and only if } \varphi(\cdot, b) = 0.$$

In particular, if β^* is surjective, then

$$l_{b:\mathcal{A}}|_{PH^1(\Gamma,\Pi_{2q-2})} = 0 \text{ if and only if } \varphi(\cdot, b) = 0.$$

This last equation is an algebraic test for determining when the infinite sum $\varphi(\cdot, b)$ is identically zero. The last formula is valid as soon as the a_j and b are limit points of Γ. It becomes a (finite) algorithm for deciding when a Poincaré series vanishes only when the limit points are algebraic in nature; for example, fixed points of elements of Γ. By linearity, we obtain formulae that can be used to get bases for spaces of cusp forms.

(b) Relative Poincaré series associated to a cyclic loxodromic subgroup. Let $A \in \Gamma$ be a loxodromic motion with (finite) fixed points α and β. Set

$$g(z) = \frac{(\alpha - \beta)^q}{(z - \alpha)^q (z - \beta)^q}, \; z \in \hat{\mathbb{C}},$$

and define the *relative Poincaré series associated to A*:

$$\varphi_A(z) = \sum_{<A>\backslash\Gamma} g(\gamma(z))(\gamma'(z))^q, \; z \in \Omega.$$

Then for all $\psi \in \mathbb{B}_q(\Gamma)$

$$l_A(\beta^*(\psi)) = \frac{-1}{\pi} < \varphi_A, \psi >_{\Omega/\Gamma} .$$

The above equation is the counterpart for relative Poincaré series of 7.1 for Poincaré series. It can be used to obtain vanishing theorems for relative Poincaré series and bases for spaces of cusp forms [36].

(c) Relative Poincaré series associated to a maximal parabolic subgroup. For parabolic $A \in \Gamma$ with fixed point $\alpha \in \mathbb{C}$, define

$$\psi_A(\zeta) = \sum \frac{\gamma'(\zeta)^q}{(\gamma(\zeta) - \alpha)^{2q}}, \ \zeta \in \Omega, \tag{7.2}$$

where the sum is over representatives γ for the right cosets $P_\alpha \setminus \Gamma$ and P_α is the subgroup of Γ consisting of the identity and those parabolic elements that fix α. The form ψ_A is a holomorphic (modular) form, but not a cusp form, if α represents a puncture on Ω/Γ. The infinite sum 7.2 defining the relative Poincaré series converges precisely when the sum 5.1 defining the corresponding Eichler integral converges. There is a simple and easily verifiable relation between these two series:

$$(2q-1)! \ \psi_A(\zeta) = \frac{d^{2q-1}}{d\zeta^{2q-1}} \ \Psi(\zeta, \alpha).$$

Consider a sequence of finitely generated Kleinian groups, $\{\Gamma_n\}$, that converges algebraically to Γ. Assume that $A_n \in \Gamma_n$ is loxodromic with

$$A = \lim_{n\to\infty} A_n.$$

Then the two fixed points of A_n coalesce to the single fixed point of A. If properly normalized, ψ_A should, under proper hypothesis, be the limiting case φ_{A_n}. This program has, however, not been carried out. So, it should be regarded as a conjecture.

(d) Relative Poincaré series associated to a rank two parabolic subgroup. The above construction applies, in particular, when P_α has rank two; hence we have produced a relative Poincaré series, a cusp form, associated to each rank two parabolic subgroup $< A, B >$ of Γ. It leads to the useful formula

$$l_A(\beta^*(\psi)) = -\imath \left(\frac{-1}{\tau(A)}\right)^q < \psi_A, \psi >_{\Omega/\Gamma}, \text{ all } \psi \in \mathbb{B}_q(\Omega, \Gamma).$$

The vanishing of the form ψ_A is a rather delicate matter. It depends on the order of Γ_α/P_α and the accessibility of the limit point α. It can happen that Eichler integral $\Psi(\cdot, \alpha)$ is not an element of Π_{2q-2}, but reduces to such a polynomial on each component of Ω. See [34] and [41].

8 Good coordinates on moduli spaces

The first constructions of complex analytic coordinates for Teichmüller spaces where either not global [3] or not intrinsic [11] (they depended on the choice of a base point). All intrinsic global coordinates for these spaces are based on producing a correct family of Kleinian groups to represent Riemann surfaces. Maskit's idea of using terminal *b*-groups [43] has lead to the most usable set of coordinates; the same groups, in disguised form at times, appear in most work on intrinsic coordinates for moduli spaces [16], [15], [39] and [40]. We illustrate these ideas in the simplest (hyperbolic) case; that of once punctured tori. Let $s \in \mathbb{C}$. The motions

$$A = \begin{bmatrix} -1 & -2 \\ 0 & -1 \end{bmatrix} \text{ and } C_s = \begin{bmatrix} s & 1 \\ 1 & 0 \end{bmatrix}$$

generate a group Γ_s. If $\Im s$ is sufficiently large (> 2 will do), then Γ_s is a terminal *b*-group representing a once punctured torus on its (simply connected) invariant component $\Delta = \Delta_s$ and a thrice punctured sphere elsewhere. Hence the same holds for s in a simply connected domain $\mathbb{T}$ which is a model for the Teichmüller space $\mathbf{T}(1,1)$ of Riemann surfaces of finite conformal type $(1,1)$ (once punctured tori). For $s \in \mathbb{T}$, Γ_s is a free group on the two generators A and C_s, the motion A is *accidental* parabolic (it does NOT represent a puncture on Δ/Γ_s), and

$$P = C_s^{-1} A^{-1} C_s A = \begin{bmatrix} 1 & -2 \\ 2 & -3 \end{bmatrix}$$

represents the puncture on Δ/Γ_s.

Although the construction of the group Γ_s is purely algebraic (for many s, so is the construction of a fundamental domain for Γ_s; but not the construction of Δ_s itself); the determination of the parameters s for which Γ_s is a good group must involve analysis (transcendental operations). The easiest way to determine the region $\mathbb{T}$ starts with the observation that

$$s = \text{ tr } C_s.$$

We select a value of s_o that produces a good group; for example, $s_o = 4\imath$. Consider the set of Beltrami coefficients μ for the group Γ_{s_o}; these are measurable functions μ on $\hat{\mathbb{C}}$ with supremum norm at most one and satisfying the equivariance relation

$$\gamma^*_{-1,1}\mu = \mu, \;\; \gamma \in \Gamma_{s_o}.$$

The normalized μ-conformal automorphism w^μ conjugates Γ_{s_o} onto a Γ_s[23].

The above program can be generalized to obtain a holomorphic injective map [43], [39]

$$\Phi : \mathbf{T}(p,n) \to \mathbb{T}^{3p-3+n}$$

that defines, what I have called, in [40], the *horocyclic* coordinates on the finite dimensional Teichmüller spaces.

---

[23]We choose the trace of the SL$(2,\mathbb{C})$ representatives of $C_s = w^\mu C_{s_o}(w^\mu)^{-1}$ with positive imaginary part.

It is an open and difficult problem to describe the "shape" of the Teichmüller space (that is, of $\Phi(\mathbf{T}(p,n))$) in horocyclic coordinates. This question should be interpreted quite broadly. We describe some particular aspects of this question. The construnction of $\Phi(\mathbf{T}(p,n))$ depends on a graph $\mathcal{G}$. It leads naturally to a deformation space (intermediate moduli space) $\mathbb{D}(\mathcal{G})$ constructed from the graph $\mathcal{G}$. The points of $\mathbb{D}(\mathcal{G})$ represent nonsingular surfaces of type (p,n) along with singular surfaces obtained by squeezing a set of curves on such surfaces (which curves are permitted to be squeezed is controlled by the graph). The manifold $\mathbb{D}(\mathcal{G})$ is a partial complition of $\Phi(\mathbf{T}(p,n))/\mathbb{Z}^{3p-3+n}$, where the lattice $\mathbb{Z}^{3p-3+n}$ acts on $\Phi(\mathbf{T}(p,n))$ by Dehn twists. See [40] for details. For example, we would like to compute the Kobayashi metric of $\mathbb{D}(\mathcal{G})$ as well as its group of biholomorphic automorphisms. Except for the obvious answers in the one dimensional situation, nothing is known in this area. The moduli space $\mathbb{D}(\mathcal{G})$ has a distinguished point. It always contains the origin 0 of $\mathbb{C}^{3p-3+n}$. We do not know if the origin must be kept fixed by every biholomorphic automorphism of $\mathbb{D}(\mathcal{G})$, when the dimension of this moduli space is at least two.

We return to the study of once punctured tori. One good tool is Eichler cohomology. The one dimensional cohomology group $PH^1(\Gamma_s, \Pi_2)$ can be represented by the cocycles χ with values

$$\chi(A) = 0, \ \ \chi(C_s)(z) = az \text{ for } z \in \mathbb{C},$$

with $a \in \mathbb{C}$ arbitrary. Hence

$$l_{C_s} \neq 0,$$

and the relative Poincaré series

$$\varphi_{C_s}(z) = s^2 \sum_{<C_s>\backslash\Gamma_s} \frac{\gamma'(z)^2}{[\gamma(z)^2 - s\gamma(z) + 1]^2}, \ \ z \in \Omega,$$

is a basis for the one dimensional space of cusp forms

$$\mathbb{A}_2(\Delta, \Gamma_s) = \mathbb{A}_2(\Omega, \Gamma_s).$$

We can now compute the period of the compactification $\overline{\Delta/\Gamma_s}$ of Δ/Γ_s. The function (defined on Δ)

$$\psi = \sqrt{\varphi_{C_s}}$$

projects to a nontrivial holomorphic one form on the torus $\overline{\Delta/\Gamma_s}$. The axes of A and C_s in Δ project to a canonical homology basis on $\overline{\Delta/\Gamma_s}$. Hence the period τ of this torus with homology basis can be computed. Choose f such that $f' = \psi$, then

$$\tau = \frac{f(C_s(z_o)) - f(z_o)}{f(A(z_o)) - f(z_o)},$$

with $z_o \in \Delta$ arbitrary.

(The function f is an abelian integral (an Eichler intregral of order 0[24] for the group Γ_{s_o}. It is of interest to obtain alternate constructions of this function.) The above formula can be used to determine the shape of $\mathbb{T}$ (in this one dimensionsal situation). We know that the range of the injective holomorphic map $s \mapsto \tau$ is $\mathbb{H}^2$. Hence $\mathbb{T}$ is the range of the inverse map (defined) on $\mathbb{H}^2$. The determination of the period τ from the horocyclic coordinate s involves two transcendental operations: the infinite sum in computing the relative Poincaré series φ_{C_s} and the integration involved in obtaing f from ψ. We shall see in the next section that one of these transcendental calculations (integration) can be eliminated.

The horocyclic coordinates for $\mathbf{T}(p,n)$ discussed briefly above should be regarded as the complex analytic version of the real analytic Fenchel-Nielsen coordinates (see, for example, [1]). It is of considerable interest to understand the relation between these sets of coordinates as well as their relation to the pleating coordinates of [27].

9 Algebraic curves represented by Kleinian groups

To describe the curves represented by a Kleinian group Γ, we must obtain Γ-invariant meromorphic functions on $\Omega(\Gamma)$. One way of obtaining such functions is as ratios of automorphic forms. We illustrate the method for the case of once punctured tori. We continue with the notation of the last section. To obtain meromorphic functions on $\overline{\Delta/\Gamma_s}$, we need to produce, in view of the existence of the cusp form φ_{C_s}, a holomorphic quadratic differential for Γ_s whose projection to Δ/Γ_s has a double pole at the puncture. The relative Poincaré series

$$\Phi(z) = \sum_{\gamma \in <P> \backslash \Gamma_s} \frac{\gamma'(z)^2}{(\gamma(z)-1)^4},\ z \in \Delta,$$

will do. Now

$$\mathcal{P} = \frac{\Phi}{\varphi_{C_s}}$$

is a function of degree two[25] on $\overline{\Delta/\Gamma_s}$. Identify the puncture on Δ/Γ_s with the origin 0 on the torus $\overline{\Delta/\Gamma_s}$. Since by Riemann-Roch or more elementary elliptic function theory

$$\mathcal{P} = a\wp + b,\ a \in \mathbb{C} - \{0\},\ b \in \mathbb{C},$$

[24]Not covered by the general theory described above for the structure of the cohomology groups.

[25]By abuse of language, we identify Γ_s-invariant functions on Δ with their projections to Δ/Γ_s.

where $\wp$ is the Weierstrass $\wp$-function, and the three points in Δ

$$x_1 = 1 + \frac{s}{2},\ x_2 = -1 + \sqrt{1+s^2} \text{ and } x_3 = \imath$$

project to the three proper half periods on $\overline{\Delta/\Gamma_s}$, the λ-invariant (its calculation, in contrast to the calculation of τ of the previous section, does not involve integration) for this marked torus can be computed as

$$\lambda = \frac{\mathcal{P}(x_3) - \mathcal{P}(x_2)}{\mathcal{P}(x_1) - \mathcal{P}(x_2)},$$

and the rougher $\jmath$-invariant for the torus by the classical formula

$$\jmath = \frac{4}{27}\frac{(1-\lambda+\lambda^2)^3}{\lambda^2(1-\lambda)^2}.$$

It follows that the torus $\overline{\Delta/\Gamma_s}$ is also described as the algebraic curve

$$y^2 = (x-e_1)(x-e_2)(x-e_3),$$

where $e_j = \mathcal{P}(x_j)$. The method described above can be used only for a very limited class of groups; mostly groups that represent punctured (or marked) spheres, punctured tori, and more generally punctured and/or marked hyperelliptic surfaces; see for example, [37], [38] and [28]. In trying to expand this method, we run into a formidable problem. The divisor of a Poincaré series is difficult (impossible, perhaps) to compute. In special cases other methods are available. We turn now to onc of these special situations.

10 Uniformizations of surfaces represented by subgroups of SL(2, z) via θ-constants

For subgroups Γ' of the modular $\Gamma = \mathrm{PSL}(2,\mathbb{Z})$, we can use θ-functions to produce Γ'-invariant automorphic functions on the upper half plane. This section is a report on work in progress. The foundations for this study are [23] and [21]. The material for this section is mostly from [20]; additional results are described in [18], [19] and [22].

We need to introduce some more notation. As before z represents a complex variable, a point in $\mathbb{C}$. The symbol τ is reserved for a parameter[26], a point in the upper half plane $\mathbb{H}^2$. The semi-direct product $\mathbb{Z}^2 \propto \mathrm{SL}(2,\mathbb{Z})$

[26] Another name for a variable; especially in this situation.

acts on $\mathbb{C} \times \mathbb{H}^2$ as follows. Let $(z,\tau) \in \mathbb{C} \times \mathbb{H}^2$, $(n,m) \in \mathbb{Z}^2$ and $\gamma = \left[\begin{smallmatrix} a & b \\ c & d \end{smallmatrix}\right] \in \mathrm{SL}(2,\mathbb{Z})$, then

$$(n,m)(z,\tau) = (z+n+m\tau, \tau),$$

and

$$\gamma(z,\tau) = \left(\frac{z}{c\tau+d}, \frac{a\tau+b}{c\tau+d}\right).$$

In this setting, points of $\mathbb{R}^2$ are called *characteristics*; they will be denoted symbolically as $\chi = \left[\begin{smallmatrix} \epsilon \\ \epsilon' \end{smallmatrix}\right] \in \mathbb{R}^2$. We also consider an action of $\mathrm{SL}(2,\mathbb{Z})$ on characteristics; it is given by

$$\chi\gamma = \begin{bmatrix} a\epsilon + c\epsilon' - ac \\ b\epsilon + d\epsilon' + bd \end{bmatrix}.$$

Riemann's theta theta function with characteristic χ is defined by the convergent series

$$\theta[\chi](z,\tau) = \sum_{n\in\mathbb{Z}} \exp 2\pi\imath \left[\frac{1}{2}\left(n+\frac{\epsilon}{2}\right)^2 \tau + \left(n+\frac{\epsilon}{2}\right)\left(z+\frac{\epsilon'}{2}\right)\right].$$

The semi-direct product of $\mathbb{Z}\oplus\mathbb{Z}$ with $\mathrm{SL}(2,\mathbb{Z})$ acts on the set of θ-functions. The function of one variable $\theta[\chi](\cdot,\tau)$ is almost invariant under the group of translations:

$$\theta\begin{bmatrix} \epsilon \\ \epsilon' \end{bmatrix}(z+n+m\tau,\tau) = \exp\left\{\frac{n\epsilon - m\epsilon'}{2} - mz - \frac{m^2}{2}\tau\right\}\theta\begin{bmatrix} \epsilon \\ \epsilon' \end{bmatrix}(z,\tau). \tag{10.1}$$

The value of the theta function at $z=0$ is called a *theta constant*. It is, of course, a holomorphic function on $\mathbb{H}^2$. We abbreviate

$$\theta[\chi](0,\cdot) = \theta[\chi].$$

The transformation of theta functions under the modular group is more complicated:

$$\frac{\exp\left\{\pi\imath\left(\frac{-cz^2}{c\tau+d}\right)\right\}\theta\begin{bmatrix} \epsilon \\ \epsilon' \end{bmatrix}\left(\frac{z}{c\tau+d}, \frac{a\tau+b}{c\tau+d}\right)}{\theta\begin{bmatrix} a\epsilon + c\epsilon' - ac \\ b\epsilon + d\epsilon' + bd \end{bmatrix}(z,\tau)} = \kappa(\chi,\gamma)(c\tau+d)^{\frac{1}{2}}, \tag{10.2}$$

where $\kappa(\chi,\gamma)$ is a constant of absolute value one. It is a root of unity if χ is rational. The action of $\mathrm{SL}(2,\mathbb{Z})$ on theta constants is easier to understand:

$$\theta[\chi](0,\gamma(\tau))\gamma'(\tau)^{\frac{1}{4}} = \kappa(\chi,\gamma)\theta[\chi\gamma](0,\tau).$$

Whereas 10.1 is an almost immediate consequence of the definitions, 10.2 is a deeper and perhaps more fundamental identity. Its proof depends on a partial differential equation satisfied by theta functions (the *heat* equation), the fact that the sum of the residues of an elliptic function is zero, and lots of computations. Computations also yield useful formulae for κ.

As a sample of the possibilities, we uniformize surfaces represented by a family of subgroups of PSL(2, $\mathbb{Z}$). We fix a positive odd prime k. The *level k principal congruence subgroup* $\Gamma(k)$ of Γ is defined as the set of matrices $\begin{bmatrix} a & b \\ c & d \end{bmatrix} \in \mathrm{PSL}(2,\mathbb{Z})$ with

$$a \equiv 1 \bmod k \equiv d, \;\; b \equiv 0 \bmod k \equiv c;$$

we view it as acting on $\mathbb{H}^2$. We introduce for $l = 0, \ldots, \frac{k-3}{2}$ the function

$$\varphi_l(\tau) = \theta \begin{bmatrix} \frac{2l+1}{k} \\ 1 \end{bmatrix} (0, k\tau), \;\; \tau \in \mathbb{H}^2.$$

The collection of functions $\{\varphi_l\}$ exhibit some remarkable properties.

(a) They span a $\frac{k-1}{2}$-dimensional vector space $V(k)$ of forms for the <u>same</u> factor of automorphy of weight $\frac{1}{4}$; that is,

$$\varphi_l(\gamma(\tau))\; c_\gamma \;(\gamma')^{\frac{1}{4}}(\tau) = \varphi_l(\tau), \;\; \gamma \in \Gamma(k), \;\; \tau \in \mathbb{H}^2;$$

where c_γ is a constant of absolute value one that is independent of τ and l. The proof of this fact uses, of course, the transformation formula for theta constants, a "cheap trick" involving the change of variable from τ to $k\tau$, and explicit formulae for κ.

(b) A theta constant does not vanish on the upper half plane unless it is identically zero. For most of the useful cases, the theta constant extends by continuity to the cusps, $\mathbb{Q} \cup \{\infty\}$. The definition of the theta constants leads, under suitable hypothesis, more or less immediately to the evaluation of the order of the theta constant at $\imath\infty$. The action of the modular group on theta constants can then be used to compute the order of vanishing of the theta constant at an arbitrary cusp[27]. The Riemann surface $\mathbb{H}^2/\Gamma(k)$ has $\frac{k^2-1}{2}$ punctures; we can distinguish $\frac{k-1}{2}$ of these from the others. The divisors of the φ_l are "supported" at the distinguished punctures of $\mathbb{H}^2/\Gamma(k)$ in the sense that the order of vanishing is constant ($\frac{1}{8}$) at all the nondistinguished punctures. Although the orders of these forms at the cusps are not integral but only rational, the orders of ratios are integers and the ratios are hence well defined functions on the compact surface $\overline{\mathbb{H}^2/\Gamma(k)}$.

(c) Actually, $V(k)$ is Hilbert space with the Petersson inner product.

[27]In the case under consideration, we have the additional complication as a result of modifying the function by replacing the argument τ by $k\tau$.

(d) The functions φ_l form an orthogonal basis for $V(k)$.
(e) The group $\Gamma/\Gamma(k)$ of conformal automorphisms of $\mathbb{H}^2/\Gamma(k)$ acts as a group of projective isometries of $\mathbf{P}V(k)$, projectivized $V(k)$.
(f) The field of meromorphic functions on $\overline{\mathbb{H}^2/\Gamma(k)}$, $\mathcal{K}(\overline{\mathbb{H}^2/\Gamma(k)})$, is generated by (the projections to $\mathbb{H}^2/\Gamma(k)$ of) the $\Gamma(k)$-invariant meromorphic functions

$$\frac{\varphi_1}{\varphi_0} \text{ and } \frac{\varphi_2}{\varphi_1}.$$

For special cases involving small values of k, we can produce more accurate descriptions of the surfaces; for example, we can show by these methods that $\overline{\mathbb{H}^2/\Gamma(7)}$ is the Klein surface

$$X^3Z + Y^3X + Z^3Y = 0.$$

The last equation is equivalent to the θ-identity

$$\left(\omega^{-3}\theta\begin{bmatrix}\frac{1}{7}\\1\end{bmatrix}\right)^3\left(\omega^9\theta\begin{bmatrix}\frac{5}{7}\\1\end{bmatrix}\right)+\left(\omega^{-3}\theta\begin{bmatrix}\frac{1}{7}\\1\end{bmatrix}\right)\left(\omega^9\theta\begin{bmatrix}\frac{3}{7}\\1\end{bmatrix}\right)^3$$

$$+\left(\omega^9\theta\begin{bmatrix}\frac{3}{7}\\1\end{bmatrix}\right)\left(\omega^9\theta\begin{bmatrix}\frac{5}{7}\\1\end{bmatrix}\right)^3 = 0,\ \omega = \exp\left(\frac{\pi\imath}{28}\right).$$

In general we can map the surface into projective space. The assignment

$$\tau \mapsto (\varphi_0(\tau),\ \varphi_1(\tau),\ \ldots,\ \varphi_{\frac{k-5}{2}}(\tau),\ \varphi_{\frac{k-3}{2}}(\tau)),\ \tau \in \mathbb{H}^2,$$

defines a holomorphic submersion[28]

$$\Phi : \overline{\mathbb{H}^2/\Gamma(k)} \to \mathbf{P}\mathbb{C}^{\frac{k-3}{2}} \cong \mathbf{P}V(k)$$

(the same map is also defined by a subset of holomorphic one forms on the surface $\overline{\mathbb{H}^2/\Gamma(k)}$); the resulting homomorphism

$$\Theta : \Gamma/\Gamma(k) \to \text{ Aut } \mathbf{P}V(k)$$

is injective.

The results discussed extend to the case of composite k; also to other families of subgroups of $\mathrm{PSL}(2,\mathbb{Z})$. Automorphic forms are also constructed from the partial derivative with respect to z (the first variable) of θ-functions evaluated at $z = 0$. The theories for even and odd k have different flavors.

---

[28]We conjecture it is an embedding.

11 Appendix: A short glossary of definitions and a few brief remarks

We are dealing with groups Γ of Möbius transformations acting on the extended complex plane $\hat{\mathbb{C}} = \mathbb{C} \cup \{\infty\}$. Good general references for the subject are the classic work [24] and the modern treatment [44]; some of the geometric aspects are discussed in [9] and many of the analytic aspects in [32]. The group Γ acts *discontinuosly* at $z_o \in \hat{\mathbb{C}}$ if there is a neighborhood U of z_o such that $\gamma(U)$ is disjoint from U for all but finely many $\gamma \in \Gamma$. The *ordinary set* or *region of disconinuity* Ω of Γ consists of the set points at which the group acts discontinuosly. The group Γ is *Kleinian*[29] if Ω is not empty. The *limit set* Λ of Γ is the complement of Ω in $\hat{\mathbb{C}}$. It is always the closure of the fixed points of elements of infinite order of the group. It is either empty, contains one, two or infinitely many points. If Λ is finite, then Γ is called *elementary*. For nonelementary groups, the limit set is closed and perfect (hence uncountable). It is useful to introduce the following terminology. Let X be a subset of $\hat{\mathbb{C}}$ and $H \subset G$, groups of Möbius transformations. We say that X is *precisely invariant* under H in G if $g(X)$ equals X for all $g \in H$ and $g(X)$ is disjoint from X for all $g \in G - H$.

The fixed points of parabolic[30] elements of a Kleinian group play an important role in the theory. Let $x \in \Lambda$ be a parabolic fixed point. Then Γ_x, the stabilizer of x in Γ, is a finite cyclic extension of a purely parabolic group[31] P_x. A *horodisc* for x is a closed disc[32] $D \subset \hat{\mathbb{C}}$ with x on its boundary, contained, except for x, in Ω, and precisely invariant under P_x in Γ. There are several distinct possibilities for the structure of Γ near x:

(A) The subgroup P_x has rank two. In this case $\Gamma_x/P_x \cong \mathbb{Z}_n$ with $n = 1$, 2, 3, 4 or 6.

(B) The subgroup P_x has rank one. In this case $\Gamma_x/P_x \cong \mathbb{Z}_n$ with $n = 1$ or 2. There are now four mutually exclusive possibilities:

(B1) The fixed point x represents two punctures on Ω/Γ. In this case, there are two disjoint horodiscs for x.

(B2) The fixed point x represents one puncture on Ω/Γ, and there are two disjoint horodiscs for x. In this case $n = 2$, and one can choose the

[29] This is classical terminology, and such a group is discrete. But not every discrete group of Möbius trasnsformations is Kleinian in the classical definition. In more modern terminology a *Kleinian* group (in dimension three) is a discrete group of Möbius transformations. It automatically acts discontinuosly on all $\mathbb{H}^3$, the upper half space in $\mathbb{R}^3$. In this terminology a group is called *of the first (second) kind* if Ω is empty (non empty).

[30] Motions with one fixed point.

[31] Consisting of the identity and parabolic elements only.

[32] A half plane is a considered a (projective) disc.

disjoint horodiscs D_1 and D_2 so that

$$\gamma(D_1) = D_2, \text{ for all } \gamma \in \Gamma_x - P_x.$$

(B3) The fixed point x represents one puncture on Ω/Γ, and there is precisely one maximal horodisc for x.

(B4) The fixed point x does not represent any puncture on Ω/Γ. In this case x has no horodiscs.

Possibilities (B3) and (B4) indicate some degeneracy in the group, [10]. In cases (B1) and (B2), we say that the fixed point x is *cusped*.

The *rank* of the parabolic fixed point x is the rank of P_x. The parabolic fixed point is *accessible* if it lies on the boundary of some component of Ω. By abuse of language a connected component of Ω is called a *component* of Γ, and Γ is a *function group* if it has an invariant component; that is, a component which is fixed (as a set) by all elements of the group. A Kleinian group Γ is *Fuchsian* if Ω contains a disc that is invariant under the entire group. In particular, a function group whose invariant component is a disc is Fuchsian. The converse is not true. A nonelementary function group whose invariant component is simply connected is called a *b*-group; in honor of Lipman Bers and the expectation that every such group is on the boundary of a finite dimensional Teichmüller space.

References

[1] W. Abikoff, *The Real Analytic Theory of Teichmüller Space*, Lecture Notes in Math., vol. 820, Springer-Verlag, 1980.

[2] ———, *The Euler characteristic and inequalities for Kleinian groups*, Proc. Amer. Math. Soc. **97** (1986), 593–601.

[3] L.V. Ahlfors, *The complex analytic stucture of the space of closed Riemann surfaces*, Analytic Functions, Princeton University Press, 1960, pp. 45–66.

[4] ———, *Finitely generated Kleinian groups*, Amer. J. Math. **86** (1964), 413–429 and **87** (1965), 759.

[5] ———, *Eichler integrals and Bers' area theorem*, Mich. Math. J. **15** (1968), 257 263.

[6] ———, *The structure of a finitely generated Kleinian group*, Acta Math. **122** (1969), 1–17.

[7] ———, *Some remarks on Kleinian groups*, Lars Valerian Ahlfors: Collected Papers Volume 2 1954-1979, Birkhäuser, 1982, pp. 316–319.

[8] D.E. Barrett and J. Diller, *Poincaré series and holomorphic averaging*, Invent. Math. **110** (1992), 23–27.

[9] A.F. Beardon, *The Geometry of Discrete Groups*, Graduate Texts in Mathematics, vol. 91, Springer-Verlag, 1983.

[10] A.F. Beardon and B. Maskit, *Limit points of Kleinian groups and finite sided fundamental polyhedra*, Acta Math. **132** (1974), 1–12.

[11] L. Bers, *Spaces of Riemann surfaces*, Proc. Internat. Congress Math. (Edinborough, 1958), Cambridge University Press, 1960, pp. 349–361.

[12] ———, *Automorphic forms and Poincaré series for infinitely generated Fuchsian groups*, Amer. J. Math. **87** (1965), 196–214.

[13] ———, *Inequalities for finitely generated Kleinian groups*, J. d'Analyse Math. **18** (1967), 23–41.
[14] ———, *On Ahlfors' finiteness theorem*, Amer. J. Math. **89** (1967), 1078–1082.
[15] ———, *Finite dimensional Teichmüller spaces and generalizations*, Bull. Amer. Math. Soc. **(N.S.)5** (1981), 131–172.
[16] C.J. Earle and A. Marden, *Geometric complex coordinates for Teichmüller space*, to appear.
[17] M. Eichler, *Eine Verallgemeinerung der Abelschen Integrale*, Math. Z. **67** (1957), 267–298.
[18] H.M. Farkas and Y. Kopeliovich, *New theta constant identities*, Israel J. Math. **82** (1993), 133–140.
[19] ———, *New theta constant identities II*, Proc. Amer. Math. Soc. **123** (1995), 1009–1020.
[20] H.M. Farkas, Y. Kopeliovich, and I. Kra, *Uniformization of modular curves*, Comm. Anal. Geom. **4** (1996), 207–259.
[21] H.M. Farkas and I. Kra, *Automorphic forms for subgroups of the modular group. II: Composite level*, to appear in J. d'Analyse Math.
[22] ———, *Theta constant identities with applications to combinatorial number theory*, submitted for publication.
[23] ———, *Automorphic forms for subgroups of the modular group*, Israel J. Math. **82** (1993), 87–131.
[24] L.R. Ford, *Automorphic Functions*, Chelsea, 1951.
[25] F. Gardiner and I. Kra, *Quasiconformal stability of Kleinian groups*, Indiana Univ. Math. J. **21** (1972), 1037–1059.
[26] L. Greenberg, *On a theorem of Ahlfors and conjugate subgroups of Kleinian groups*, Amer. J. Math. **89** (1967), 56–68.
[27] L. Keen and C. Series, *Pleating coordinates for the Maskit embedding of the Teichmüller space of punctured tori*, Topology **32** (1993), 719–749.
[28] Y. Komori, *On the automorphic functions for Fuchsian groups of genus two*, RIMS (Kyoto Univ.) preprint 963 (January 1994).
[29] I. Kra, *On cohomology of Kleinian groups*, Ann. of Math. **89** (1969), 533–556.
[30] ———, *On cohomology of Kleinian groups: II*, Ann. of Math **90** (1969), 575–589.
[31] ———, *Deformations of Fuchsian groups. II*, Duke Math. J. **38** (1971), 499–508.
[32] ———, *Automorphic Forms and Kleinian Groups*, Benjamin, 1972.
[33] ———, *On the vanishing of Poincaré series of rational functions*, Bull. Amer. Math. Soc. **8** (1983), 63–66.
[34] ———, *On the cohomology of Kleinian groups IV. The Ahlfors-Sullivan construction of holomorphic Eichler integrals*, J. d'Analyse Math. **43** (1983/84), 51–87.
[35] ———, *On the vanishing of and spanning sets for Poincaré series for cusp forms*, Acta Math. **153** (1984), 47–116.
[36] ———, *Cusp forms associated to loxodromic elements of Kleinian groups*, Duke Math. J. **52** (1985), 587–625.
[37] ———, *On algebraic curves (of low genus) defined by Kleinian groups*, Ann. Polonici Math. **46** (1985), 147–156.

[38] ———, *Uniformization, automorphic forms and accessory parameters*, RIMS (Kyoto Univ.) Kokyuroku **571** (1985), 54–84.
[39] ———, *Non-variational global coordinates for Teichmüller space*, Holomorphic Functions and Moduli II, Springer-Verlag, 1988, pp. 221–249.
[40] ———, *Horocyclic coordinates for Riemann surfaces and moduli spaces. I: Teichmüller and Riemann spaces of Kleinian groups*, J. Amer. Math. Soc. **3** (1990), 499–578.
[41] ———, *Cusp forms associated to rank 2 parabolic subgroups of Kleinian groups*, Proc. Amer. Math. Soc. **111** (1991), 803–814.
[42] I. Kra and B. Maskit, *Pinched two component groups*, to appear in Analysis and Topology (a volume dedicated to S. Stoilow).
[43] B. Maskit, *Moduli of marked Riemann surfaces*, Bull. Amer. Math. Soc. **80** (1974), 773–777.
[44] ———, *Kleinian Groups*, Grundlehren der mathematischen Wissenschaften, vol. 287, Springer-Verlag, 1988.
[45] C. McMullen, *Amenability, Poincaré series and quasiconformal maps*, Invent. Math. **97** (1989), 95–127.
[46] M. Nakada, *Cohomology of finitely generated Kleinian groups with an invariant component*, J. Math. Soc. Japan **28** (1976), 699–711.
[47] ———, *Surjectivity of the Bers map*, Tôhoku Math. J. **28** (1976), 257–266.
[48] W. Parry, *H^1 and symmetric tensors*, Duke Math. J. **52** (1985), 577–586.
[49] D.C. Sengupta, *On cohomology of finitely generated function groups*, J. d'Analyse Math. **63** (1994), 1–17.
[50] D. Sullivan, *A finiteness theorem for cusps*, Acta Math. **147** (1981), 289–299.

State University of New York at Stony Brook
Stony Brook, NY 11794

On the Geometry of Kleinian Groups

Gaven J. Martin

1 Introduction

This paper is intended as a survey of the recent joint work of F.W. Gehring with the author and is an expanded version of the lecture given at the conference "Quasiconformal Mappings and Analysis."

Although quasiconformal mappings are an important tool in the study of Kleinian groups, they do not play a great role in what follows. However we were led to problems in the geometry of Kleinian groups through our studies of quasiconformal groups and their more general topological counterparts, convergence groups [18].

The theory of discrete groups and their orbit spaces, Riemann surfaces and hyperbolic 3–folds, is a deep and beautiful subject with many important applications in analysis, geometry and low dimensional topology. Hyperbolic geometry is the natural geometry in two dimensions and, due to the work of Thurston and others, we are coming to realize that this is also the case in three dimensions.

We have made no attempt here to present a general survey of the theory of Kleinian groups, preferring instead to concentrate on our own work on geometric properties of these groups and our partial description of certain analytic spaces of two generator discrete groups. We recommend the books [1, 4, 40, 46, 48] as good places to begin for readers who want to learn more about Kleinian groups.

Here is an outline of the topics we intend to cover.

- **Kleinian groups**
- **2–Generator Discrete Groups**
- **Discreteness and Commutators**

[1]Supported in part by grants from the Australian Research Council and the New Zealand Marsden Fund.

- **Polynomial Trace Identities**
- **Disk Coverings**
- **Discreteness Criteria and Arithmeticity**
- **Applications**
  - Volume
  - Geodesics
  - Degeneration
  - Margulis Constant
  - Triangle subgroups

2 Kleinian Groups

Let us begin by recalling a few basic definitions.

- Hyperbolic 3–space is the space

$$\mathbb{H}^3 = \{(x_1, x_2, x_3) \in \mathbb{R}^3 : x_3 > 0\} \tag{2.1}$$

  endowed with the complete Riemannian metric $ds = |dx|/x_3$ of constant negative curvature equal to -1.

- A *Kleinian group* Γ is a discrete subgroup of the orientation preserving isometry group $Isom^+(\mathbb{H}^3)$ of hyperbolic space which does not contain an abelian subgroup of finite index.

- *Discrete* means that the identity element of Γ is isolated in the topology of local uniform convergence on $\mathbb{H}^3$ in the induced Euclidean metric.

- A hyperbolic 3–*orbifold* (or simply 3–fold) is the orbit space of a Kleinian group,

$$\mathcal{Q} = \mathbb{H}^3/\Gamma. \tag{2.2}$$

  This orbit space is a hyperbolic 3–*manifold* if Γ is torsion free (i.e. does not contain any elements of finite order)

Thurston's geometrization conjecture asserts that any 3–manifold can be canonically cut up into geometric pieces. Each piece is a manifold locally modeled on one of eight possible geometries. The most prevalent geometry is hyperbolic. Each hyperbolic piece is the orbit space of a Kleinian group. Therefore we expect that Kleinian groups play a similar role in 3–dimensions to that played by Fuchsian groups in two dimensions, where every surface, with a few simple exceptions, is the orbit space of a Fuchsian group.

A fundamental invariant of a hyperbolic 3–fold is its volume. Yet, in general, relatively little is known about the set of volumes of hyperbolic 3–folds, at least in comparison with the situation in two dimensions where the signature formula [4] gives the complete spectrum of possible volumes. From the work of Jørgensen and Thurston we know that the set of volumes is well ordered of type ω^ω. That is, there is a smallest volume and a smallest volume that is the limit of other volumes, then a next smallest limit volume and so forth. Every limit volume is only a limit from below. Some extremal volumes have been identified, for instance the smallest volume arithmetic orbifold [9] and only very recently the smallest volume arithmetic manifold [10], the smallest volume noncompact orbifold and (non–oriented) non-compact manifold are also known [42, 2], as is the smallest volume limit orbifold [3]. There are a few other extremals identified as well, see [22] for a discussion of these.

The celebrated Mostow rigidity theorem [45] tells us that for finite volume hyperbolic 3–folds, isometric invariants are homotopy invariants. Thus the volume of a (finite volume) hyperbolic 3–fold is a topological invariant. Because there is an absolute lower bound for the volume of a hyperbolic 3–fold, together with the rigidity theorem we see that the order of the automorphism group of a hyperbolic 3–fold is bounded above by the ratios of the volume of the orbifold and the minimal possible 3–fold volume. In two dimensions this is the well known $84g-84$ theorem of Hurwitz which arises because the orbit space of the $(2,3,7)$ Fuchsian triangle group is uniquely of minimal hyperbolic co–area. The identification of the minimal volume hyperbolic 3–fold is an important problem in the geometry of discrete groups with significant consequences.

There are other interesting absolute constants associated with Kleinian groups. For instance the Margulis constant (defined later), the minimal distance between components of the singular set (the points at which local branching occurs in the orbit space), constants associated with the length spectrum of geodesics, distances between embedded totally geodesic surface and so forth. The identification of these constants poses very interesting problems in hyperbolic geometry. The techniques and ideas used in their solution broadens our understanding of the geometry of discrete groups and the structure of hyperbolic manifolds in general.

3 2–Generator Discrete Groups

Let $\mathbb{M}$ denote the group of all Möbius transformations of the extended complex plane $\overline{\mathbb{C}} = \mathbb{C}\cup\{\infty\}$. We associate with each Möbius transformation

$$f = \frac{az+b}{cz+d} \in \mathbb{M},\ ad-bc=1, \tag{3.3}$$

the matrix

$$A = \begin{pmatrix} a & b \\ c & d \end{pmatrix} \in \mathrm{SL}(2,\mathbb{C}) \tag{3.4}$$

and set $\mathrm{tr}(f) = \mathrm{tr}(A)$ where $\mathrm{tr}(A)$ denotes the trace of the matrix A. Next for each f and g in $\mathbb{M}$ we we let $[f,g]$ denote the multiplicative commutator $fgf^{-1}g^{-1}$. We call the three complex numbers

$$\beta(f) = \mathrm{tr}^2(f) - 4,\ \beta(g) = \mathrm{tr}^2(g) - 4,\ \gamma(f,g) = \mathrm{tr}([f,g]) - 2 \tag{3.5}$$

the parameters of the two generator group $\langle f,g\rangle$ and write

$$\mathrm{par}(\langle f,g\rangle) = (\gamma(f,g), \beta(f), \beta(g)). \tag{3.6}$$

These parameters are independent of the choice of matrix representations for f and g in $\mathrm{SL}(2,\mathbb{C})$ and they determine $\langle f,g\rangle$ uniquely up to conjugacy whenever $\gamma(f,g) \neq 0$.

The elements of a discrete group, other than the identity, fall into three types

- *elliptic*: $\beta(f) = -4\sin^2(q\pi/p)$ and f is conjugate to $z \mapsto \zeta z$, ζ a root of unity.
- *parabolic*: $\beta(f) = 0$ and f is conjugate to $z \mapsto z+1$
- *loxodromic*: $\beta(f) \notin [-4,0]$ and f is conjugate to $z \mapsto \lambda z$, $|\lambda| \neq 1$.

Each Möbius transformation of $\overline{\mathbb{C}}$ extends uniquely via the Poincaré extension to an orientation preserving isometry of hyperbolic 3–space when we make the usual identification $\overline{\mathbb{C}} = \partial\mathbb{H}^3$. In this way discrete groups are identified with Kleinian groups.

We view the space of all two generator Kleinian groups as a subset of the three complex dimensional space $\mathbb{C}^3$ via the map

$$\langle f,g\rangle \to \mathrm{par}(\langle f,g\rangle) = (\gamma,\beta,\beta'). \tag{3.7}$$

A fundamental problem is to describe the space of all points in $\mathbb{C}^3$ corresponding to discrete 2–generator groups.

There is a projection from the three complex dimensional space of discrete groups to the two complex dimensional slice $\beta' = -4$ which preserves discreteness and which arises roughly because we can assume one generator has order two. This is only a property of two generator groups [6]. Hence

$$(\gamma,\beta,\beta')\ \text{discrete} \Rightarrow (\gamma,\beta,-4)\ \text{discrete} \tag{3.8}$$

and away from a small finite set of exceptional parameters this projection preserves the property of being nonelementary as well.

We define

$$\mathcal{K} = \{(\gamma,\beta) \in \mathbb{C}^2 : (\gamma,\beta,-4)\ \text{are parameters of a Kleinian group.}\} \tag{3.9}$$

Next set

$$\mathcal{K}_\beta = \{\gamma \in \mathbb{C} : (\gamma, \beta) \in \mathcal{K}.\} \tag{3.10}$$

A theorem of Jørgensen [37] shows that $\mathcal{K}$ and $\mathcal{K}_\beta$ are closed and that components of either space consist entirely of isomorphic groups. Our plan is to describe, for each fixed β, that subset of the complex plane consisting of the commutator parameters for Kleinian groups, $\mathcal{K}_\beta$, and subsequently to describe the fiber in $\mathbb{C}^3$ over each point of $\mathcal{K}$.

We have made substantial progress along these lines when one of the generators is elliptic of order $p \geq 3$, that is when

$$\beta = -4\sin^2(\pi/p), \quad p \geq 3. \tag{3.11}$$

Computer pictures show that the spaces $\mathcal{K}$ and $\mathcal{K}_\beta$ have highly fractal boundary and thus it would seem impossible to obtain a complete description of these spaces. However, using the classification of the two generator Fuchsian groups, see [47] and the references therein, the totally real part of $\mathcal{K}$ is described in [31]. This description is possible because, although not every real point of $\mathcal{K}$ represents a Fuchsian group, the groups corresponding to real points usually have low index Fuchsian subgroups.

The spaces $\mathcal{K}$ and $\mathcal{K}_\beta$ admit an interesting linear symmetry group. In particular in [32] we establish the following linear S_3 action on $\mathcal{K}$

$$\begin{array}{lllllll} (\gamma,\beta) & \mapsto & (\gamma,\beta) & (\gamma,\beta) & \mapsto & (\beta-\gamma,\beta) \\ (\gamma,\beta) & \mapsto & (-\beta-4,\gamma-\beta-4) & (\gamma,\beta) & \mapsto & (\beta-\gamma,-\gamma-4) \\ (\gamma,\beta) & \mapsto & (\gamma,\gamma-\beta-4) & (\gamma,\beta) & \mapsto & (-\beta-4,-\gamma-4) \end{array} \tag{3.12}$$

The description of the totally real part of $\mathcal{K}$ shows this to be the largest real linear action on $\mathcal{K}$. These symmetries can be very useful, especially since they are invertible. However there are a number of nonlinear mappings from $\mathcal{K}$ into itself. The simplest of these are certain shifted Chebyshev polynomials which act by scalar multiplication: Set

$$P_n(z) = 2T_n(1 + z/2) - 2 \tag{3.13}$$

where T_n is the nth Chebyshev polynomial of the first kind. Then away from the roots of P_n we have

$$(\gamma,\beta) \in \mathcal{K} \quad \Rightarrow \quad P_n(\beta)(\gamma,\beta) \in \mathcal{K}. \tag{3.14}$$

This action is induced by the mapping $\langle f, g\rangle \mapsto \langle f^n, g\rangle$. There are other natural ways to produce mappings of $\mathcal{K}$ to itself. One might, for instance, consider the group generated by the Nielsen transformations

$$\begin{array}{lll} \langle f,g\rangle & \mapsto & \langle g,f\rangle \\ \langle f,g\rangle & \mapsto & \langle f^{-1},g\rangle \\ \langle f,g\rangle & \mapsto & \langle fg,g\rangle \end{array} \tag{3.15}$$

generating the automorphism group of the free group on two symbols. However in order that these moves act on $\mathcal{K}$ the second generator needs to be of order two. This rules out the first move. The second is easily seen to act as the identity on $\mathcal{K}$ while the Fricke identity and the fact that Nielsen transformations preserve the commutator parameter implies that the third move is the symmetry of order two $(\gamma, \beta) \mapsto (\gamma, \gamma - \beta - 4)$ that we have already listed. However Nielsen transformations are useful when studying the full space of two generator discrete groups in $\mathbb{C}^3$. We shall describe other important polynomial mappings of $\mathcal{K}$ into itself later.

Every loxodromic or elliptic element f in a discrete group fixes two points of $\overline{\mathbb{C}}$ and the closed hyperbolic line joining these two fixed points is called the *axis* of f, denoted $\mathrm{ax}(f)$. Such an element f translates along this axis by an amount $\tau(f)$, the *translation length*, and rotates by an angle $\theta(f)$, the *holonomy*, about this axis. The translation length of an elliptic element in a discrete group is 0 while its holonomy is a rational multiple of π. The following formulas relate these quantities and the parameters.

$$\begin{aligned} \beta(f) &= 4\sinh^2\left(\frac{\tau(f)}{2} + \frac{i\theta(f)}{2}\right) \\ \cosh(\tau(f)) &= \frac{|\beta(f)+4| + |\beta(f)|}{4} \\ \cos(\theta(f)) &= \frac{|\beta(f)+4| - |\beta(f)|}{4} \end{aligned} \tag{3.16}$$

Also the parameters conveniently encode various other geometric quantities.

Lemma 3.1 *Let f and g be Möbius transformations with distinct pairs of fixed points. Then*

$$\begin{aligned} \frac{4\gamma(f,g)}{\beta(f)\beta(g)} &= \sinh^2(\delta \pm i\theta) \\ \cosh(2\delta) &= \left|\frac{4\gamma(f,g)}{\beta(f)\beta(g)} + 1\right| + \left|\frac{4\gamma(f,g)}{\beta(f)\beta(g)}\right| \\ \cos(2\theta) &= \left|\frac{4\gamma(f,g)}{\beta(f)\beta(g)} + 1\right| - \left|\frac{4\gamma(f,g)}{\beta(f)\beta(g)}\right| \end{aligned} \tag{3.17}$$

where $\delta = \delta(f,g)$ is the hyperbolic distance between the axes of f and g and $\theta = \theta(f,g)$ is the angle between the spheres or hyperplanes which contain $\mathrm{ax}(f)$ *or* $\mathrm{ax}(g)$ *and the common perpendicular of* $\mathrm{ax}(f)$ *and* $\mathrm{ax}(g)$.

The number $\delta(f,g)$ is referred to as the *axial distance* between f and g.

4 Discreteness and Commutators

It has long been known in the theory of discrete subgroups of Lie groups that studying the behaviour of commutators, and in particular iterated commutators, provides good geometric information. Jordan used the iterated commutator in his theorem concerning the order of the finite non-cyclic subgroups of the orthogonal group, as did Bieberbach in his theorem on crystallographic groups. Siegel, Shimizu, and Leutbecher used the same ideas to establish constraints for Fuchsian groups and groups with a parabolic generator. More recently Jørgensen studied the same sequence of commutators to obtain his inequality [37]. This was extended slightly by Brooks and Matelski [5] who studied the higher order cycles associated to the same sequence. See also [25].

If Γ is a Kleinian group and $f \in \Gamma$, then (generically) the set

$$\{\gamma(f,g) : g \in \Gamma\} \tag{4.18}$$

is discrete in $\mathbb{C}$. Here generically means whenever f is loxodromic, or Γ is geometrically finite and so forth. There are however cases where the statement is not true. The discreteness of this set in $\mathbb{C}$ can be seen roughly because $\gamma(f,g)$ represents a loop in the orbit space "based" by f. Thus the terms are comparable with those of the Poincaré series [39]. It is certainly true that if Γ is not discrete, then there is a sequence $\{g_j\} \in \Gamma$ such that

$$0 \neq \gamma(f,g_j) \to 0 \qquad j \to \infty. \tag{4.19}$$

Conversely if there is a sequence of commutators, based by a fixed f which accumulates at 0, then the group Γ cannot be discrete. It is on these simple ideas that most necessary discreteness criteria are based. A question then arises of how to generate sequences of commutators in a group of Möbius transformations.

To answer this question we shall now discuss a very interesting family of polynomial trace identities which can be used to generalize the approaches mentioned above in the case of Kleinian groups.

Let $\langle a,b\rangle$ be the free group on the two letters a and b. We say that a word $w \in \langle a,b\rangle$ is a *good word* if w can be written as

$$w = a^{s_1}b^{r_1}a^{s_2}b^{r_2}\ldots a^{s_{m-1}}b^{r_{m-1}}a^{s_m} \tag{4.20}$$

where $s_1 \in \{\pm 1\}$, $s_j = (-1)^{j+1}s_1$ and $r_j \neq 0$ but are otherwise unconstrained. The good words start and end in a and the exponents of a oscillate in sign. The following theorem is a key tool used in the study of the parameter spaces of discrete groups.

Theorem 4.1 *Let $w \in \langle a, b\rangle$ be a good word. Then there is a monic polynomial p_w of two complex variables having integer coefficients and with the following properties: if*

$$a \mapsto g \quad \text{and} \quad b \mapsto f \tag{4.21}$$

generates a homomorphism $\langle a, b\rangle \to \mathbb{M}$, then

$$\gamma(f, w) = p_w(\gamma, \beta). \tag{4.22}$$

Except for a few simple cases, corresponding to representations of elementary discrete groups, the theorem implies that

$$\gamma \in \mathcal{K}_\beta \quad \Rightarrow \quad p_w(\gamma) \in \mathcal{K}_\beta. \tag{4.23}$$

There are two things to note. The first is that if we assume that $a^2 = 1$, then the alternating sign condition is redundant and every word is good. The second thing is that there is a natural semigroup operation on the good words. If $w_1 = w_1(a, b)$ and $w_2 = w_2(a, b)$ are good words, then so is

$$w_1 * w_2 = w_1(w_2(a, b), b). \tag{4.24}$$

It is not too difficult to see that

$$p_{w_1 * w_2}(\gamma, \beta) = p_{w_1}(p_{w_2}(\gamma, \beta), \beta) \tag{4.25}$$

which corresponds to polynomial composition in the first slot. If we fix a homomorphism $a \mapsto g$ and $b \mapsto f$, then we can suppress β and the semigroup operation is polynomial composition. Evidently

$$w^{*n} = w * w * \cdots * w \tag{4.26}$$

has polynomial

$$p_{w^{*n}}(z) = p_w^n(z), \tag{4.27}$$

the n–fold iterate. In this way we find that we can use results from complex dynamics to describe the space $\mathcal{K}$, see [25]. We have shown that for a given polynomial p_w associated to a good word w that (generically) the subset $\{p_w^n(\gamma(f, g)) : n \in \mathbb{Z}_+\}$ of $\mathbb{C}$ should be discrete. In dynamical terms this means that $\gamma(f, g)$ should not lie in the filled in Julia set of the polynomial p_w unless it is an eventually periodic point. The eventually periodic points are sometimes discrete groups, for instance there is a two generator representation of the $(2, 3, 7)$ triangle group in which the commutator parameter lies on a period two attracting cycle [25] and a representation of a $\mathbb{Z}_2$ extension of the figure eight knot complement group with parabolic generator has commutator $(1+\sqrt{-3})/2$ which is preperiodic for the polynomial $p(z) = z^2$ which arises from the good word aba^{-1} when b is parabolic.

Let us give two simple examples of word polynomials and how they generate inequalities. First is the classical example:

$$w = aba^{-1} \qquad p_w(z) = z(z-\beta). \tag{4.28}$$

Here $z = \gamma(f, g)$ and $\beta = \beta(f)$ are parameters of the representation $a \mapsto g$ and $b \mapsto f$. We treat z as a variable. Then, as noted above, if f is parabolic we have $\beta(f) = 0$, $p_w(z) = z^2$ and the sequence $p_w^n(z)$ is discrete only if $|z| \geq 1$ or $z = 0$. Since $z = 0$ implies that $\langle f, g\rangle$ is elementary, this is the well known Shimizu–Leutbecher inequality:

Lemma 4.2 *If $\langle f, g\rangle$ is a Kleinian group and if f is parabolic, then*

$$|\gamma(f,g)| \geq 1. \tag{4.29}$$

The result is sharp.

Upon noting the easy fact that

$$p_w(\{|z| \leq 1-|\beta|\}) \subset \{|z| \leq 1-|\beta|\} \tag{4.30}$$

so that the disk $\{|z| \leq 1-|\beta|\}$ lies in the filled in Julia set, we can obtain Jørgensen's inequality.

Lemma 4.3 *If $\langle f, g\rangle$ is a Kleinian group, then*

$$|\gamma(f,g)| + |\beta(f)| \geq 1. \tag{4.31}$$

However Jørgensen's inequality describes only a part of the filled in Julia set of p_w about the attracting (when $|\beta(f)| < 1$) fixed point 0. Substantial improvements can be made, see [25].

The next example comes from the good word

$$w = aba^{-1}b^{-1}a \qquad p_w(z) = z(1+\beta-z)^2. \tag{4.32}$$

Following the method described above we find firstly in the case that f has order 6, so that $\beta = -1$,

Lemma 4.4 *If $\langle f, g\rangle$ is a Kleinian group and f is elliptic of order* 6, *then*

$$|\gamma(f,g)| \geq 1. \tag{4.33}$$

And then more generally we have

Lemma 4.5 *If $\langle f, g\rangle$ is a Kleinian group, then*

$$|\gamma(f,g)| + |1+\beta(f)| \geq 1 \tag{4.34}$$

unless $\gamma(f,g) = 1+\beta(f)$.

The exceptional case $\gamma = 1+\beta$, corresponding to the preimage of 0 under p_w, represents certain groups generated by elliptics of order two and three [21]. Notice too that $z = 1+\beta$ is the fixed point of the good word polynomial $z \mapsto z(z-\beta)$ discussed above.

In general one can use these ideas to find inequalities describing portions of the complement of the space of two generator discrete groups by either producing inequalities or by studying such things as the filled in Julia sets of the polynomial semigroup generated by the good words. The theory of holomorphic motions indicates how to put together information as β varies. In practice we do both. Initially some omitted region about the origin is described using dynamics. This region is then propagated via the polynomials corresponding to the good words.

As another example of how the good word polynomials can be used in describing the space $\mathcal{K}$ we note the following. Let $\ell_{\mathbb{C}}$ denote the complex line

$$\ell_{\mathbb{C}} = \{(z,z) \in \mathbb{C}^2 : z \in \mathbb{C}\}. \tag{4.35}$$

Then

$$\ell_{\mathbb{C}} \cap \mathcal{K} = \{-1,-2,-3\}. \tag{4.36}$$

To see this note that by (28) if $(\gamma,\gamma) \in \mathcal{K}$ is represented by $\langle f,g\rangle$, then $\langle f, gfg^{-1}\rangle$ has parameters $(0,\beta,\beta)$ and is a discrete elementary group fixing a point of $\mathbb{C}$. Since $\langle f,g\rangle$ is nonelementary, it easily follows that f is not loxodromic or parabolic. Then considering the noncyclic elementary groups generated by two elliptics of the same order and with a common fixed point we find the only the possibilities are $\gamma = \beta \in \{-1,-2,-3\}$ and Maskit showed in [41] that these all occurred.

5 Disk Coverings

It is not to hard in general to find some round disk $\mathbf{D}$ about the origin which lies in the complement of $\mathcal{K}_\beta$. Indeed in [20] and [7] it is shown that if $\gamma \in \mathcal{K}_\beta$, then

$$|\gamma(\gamma-\beta)| \geq 0.19806\ldots \tag{5.37}$$

unless $\gamma = \beta \in \{-1,-2,-3\}$. Hence the disk of radius r_0 about the origin, where

$$r_0 = \frac{1}{2}(\sqrt{|\beta|^2+0.7922} - |\beta|) > 0, \tag{5.38}$$

always lies in the complement of $\mathcal{K}_\beta$. Although this estimate is sharp, most often there is a substantially larger such disk. For instance we saw above that if $\beta = 0$ or $\beta = -1$ then the unit disk has this property.

Now suppose that p_w is the polynomial associated to a good word w. Then the deleted neighbourhood

$$N_w = \{z \in \mathbb{C} : p_w(z) \in \mathbf{D} \setminus \{0\}\} \tag{5.39}$$

of the roots of p_w cannot contain the commutator parameter γ for any discrete group because if it did so, then $p_w(\gamma) \in \mathbf{D} \setminus \{0\}$ would also be the parameter of a discrete nonelementary group, a contradiction (there are a few elementary exceptions which we do not discuss here). A root of p_w may be the commutator parameter of a Kleinian group $\langle f, g \rangle$ for which $\langle f, w \rangle$ is elementary with a fixed point in $\overline{\mathbb{C}}$. Note however that the roots of p_w need not correspond to discrete groups. They are only *candidates* for discrete groups. We can now extend the region in the complement of $\mathcal{K}_\beta$ by looking at the preimage of N_w under either p_w or any other polynomial associated to a good word. In this way we build up a region in the complement of $\mathcal{K}_\beta$.

Let us now study a special, but very interesting case. Suppose that f is elliptic of order $p \geq 3$. Thus we investigate the space $\mathcal{K}_{-3}$. We find an omitted neighbourhood about the origin. For example, the disk of radius $0.2469\ldots$ is the largest such disk. On its boundary is a representation of the $(2,3,7)$–triangle group. We then use the good word polynomials to propagate this disk and its preimages in the complement of $\mathcal{K}_{-3}$. In this way we obtain the following picture of a portion of the complement of $\mathcal{K}_{-3}$.

Possible values for commutator parameter when n=3

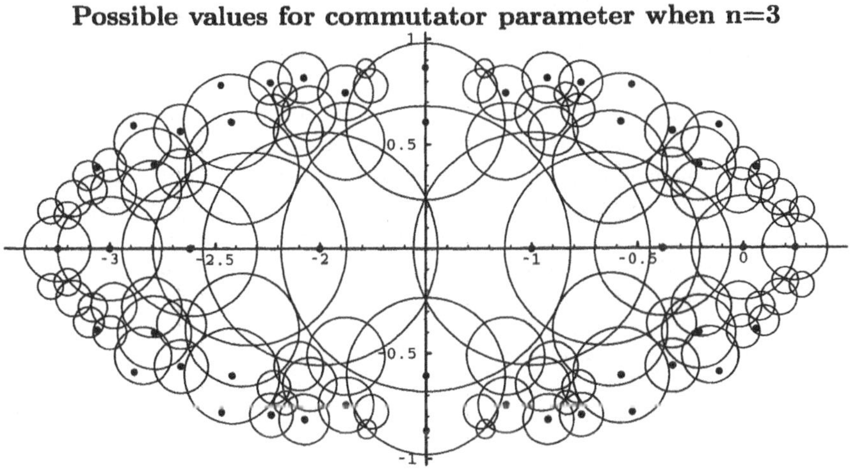

In the picture the dots mark *all* the commutator parameters for two generator Kleinian groups with one generator elliptic of order three. Each such group represents an isolated point not only in $\mathcal{K}_\beta$ but also in $\mathcal{K}$. Each disk is mapped into the initial disk about the origin by a word polynomial. Away from this region is the space of Schottky groups generated by elliptics of order two and three. (If γ is sufficiently large, e.g. $|\gamma| \geq 4$, then γ is the commutator of a Kleinian group). Thus the complement of $\mathcal{K}_{-3}$ is bounded. The examples found as isolated points from this procedure provide lots of different interesting groups. For instance the isolated points are rigid groups, groups which are isolated in the algebraic topology. They can be of finite co–volume, web groups with triangle stabilizers or (improbably) geometrically infinite groups.

Theorem 5.1 *Let f be elliptic of order $p \geq 3$, $\beta = \beta(f)$ and $\gamma(f,g)$ an isolated point of $\mathcal{K}_\beta$. Then $\langle f, g\rangle$ is a rigid group.*

There are natural orthogonal coordinates which one can use to study $\mathcal{K}_\beta$. These are

- $\delta(f,g)$: the axial distance between f and g
- $\theta(f,g)$: the angle between the axes of f and g.

The points of constant axial distance form ellipses with foci 0 and β while the points of constant angle form hyperbolas, orthogonal to the ellipses. In this way we can begin to list the groups in order of axial distance. However, in order to proceed we need a way of deciding if the candidates for commutator parameters of discrete groups (the roots of polynomials etc) actually do correspond to discrete groups.

6 Discreteness criteria and arithmeticity

We discover isolated points of $\mathcal{K}_\beta$ as roots of a polynomial with integer coefficients. In order to assert that these points correspond to discrete groups, we must establish their discreteness or otherwise. We in a good situation to apply an arithmeticity criterion for discreteness. In [21] we gave a simple criteria when the polynomial was quadratic. More recently in joint work with C. Maclachlan and A. Reid the following very useful discreteness criteria was developed and used.

Theorem 6.1 *Let f be elliptic of order $p \geq 3$ and*

$$\mathrm{par}(\langle f,g\rangle) = (\gamma, \beta, -4). \tag{6.40}$$

Suppose

1. *that γ is a root of $p_\beta(\gamma) = p(\gamma,\beta)$ with integer coefficients and monic as a polynomial in γ,*

2. *that all roots of*

$$q(z) = \Pi_{(k,n)=1, 2\leq k\leq n/2}\, p(\gamma, \beta(f^k)) \tag{6.41}$$

are real, and

3. *that if $\gamma, \overline{\gamma}$ are complex, then all other roots of p_β lie in the interval $(\beta, 0)$*

Then $\langle f, g\rangle$ is a discrete subgroup of an arithmetic Kleinian group.

The description we have so far obtained of the space $\mathcal{K}_\beta$ suggests a few natural conjectures.

CONJECTURES

1. Every isolated commutator parameter of a Kleinian group in $\mathcal{K}_\beta$ is the root of some good word polynomial
2. There are no components other than the isolated points and the unbounded component consisting of Schottky groups.
3. There are only finitely many two generator arithmetic Kleinian groups with elliptic generators [1].

The first two conjectures suggest that the method of describing the complement of the space $\mathcal{K}_\beta$ is complete. The first conjecture is valid if the commutator parameter is that of a group which is not free on the two generators. It is therefore equivalent to the conjecture that there are no rigid, discrete and faithful representations of $\mathbb{Z}_2 * \mathbb{Z}_n$, $n \geq 3$. Any such representation would have to be an isolated geometrically infinite group and it is conjectured that there are no such Kleinian groups. The third conjecture is suggested by observations. It has been verified whenever the degree of the trace field over $\mathbb{Q}$ is at least 4 and in other special cases.

Below we have tabulated some of the results obtained by the disk covering method in the case of an elliptic generator of order 3 or 4. More details can be found in our work [27, 36] where the cases $p = 5, 6, 7$ are also dealt with, see also [19, 21]. The tables give all possible commutator parameters for Kleinian groups $G_{p,i}$ generated by an elliptic of order 2 and an elliptic of order $p = 3$ or $p = 4$ for which the distance between the axes of the two elliptics is less than the given value. (They actually contain more groups as one can see from looking at them). For a more detailed account of the calculations on which these diagrams are based, see [23]. In each table the number $\gamma_{p,i}$ denotes the approximate value of the commutator parameter for the group $G_{p,i}$ and $\delta_{p,i}$ denotes the distance between the two elliptic generators of order 2 and p. With further applications in mind we also indicate whether the axis of the elliptic of order p is simple in $G_{p,i}$, whether the group is arithmetic or not, and the minimal co–volume arithmetic group in which the group $G_{p,i}$ embeds. Note that if $G_{p,i}$ is arithmetic, then any group in which it embeds is arithmetic and so the volume bound given is an absolute lower bound for the co–volume of any Kleinian group containing elliptics of order 2 and p with this axial distance and angle.

[1]This has recently been verified in joint work with C. Maclachlan

Table 1 - Groups $G_{3,i}$ $\delta \leq 0.295$

i	$\gamma_{3,i}$	$\delta_{3,i}$	Simple	Arithmetic	Min. Co–Vol.
1	-1	0	No	Yes	- -
2	$-.3819$	0	No	Yes	- -
3	$-1.5+.6066i$	.1970	No	Yes	0.0390...
4	$-.2118+.4013i$	.2108	No	Yes	0.0408...
5	$-.5803+.6062i$	.2337	No	Yes	0.0661...
6	$-1.1225+.7448i$	.2448	Yes	Yes	0.0785...
7	$-.3376+.5622i$	.2480	Yes	Yes	0.0785...
8	$-.9236+.8147i$	.2740	No	Yes	0.1274...
9	$-1.5+.8660i$	.2746	No	Yes	0.0845...
10	$-.7672+.7925i$	.2770	Yes	Yes	0.0659...
11	$.0611+.3882i$	.2788	No	Yes	0.1028...
12	$.2469$	.2831	Yes	Yes	$(2,3,7)$-Δ
13	$-.5284+.7812i$	.2944	No	?	?
14	$-.1153+.5897i$	.2970	No	Yes	0.1654...

Table 2 - Groups $G_{4,i}$ $\delta \leq 0.535$

i	$\gamma_{4,i}$	$\delta_{4,i}$	Simple Axis	Arithmetic	Min. Co–Vol
1	-1	0	No	Yes	- -
2	$-.5+.8660i$	.4157	Yes	Yes	0.1268...
3	$-.1225+.7448i$	.4269	No	Yes	0.1374...
4	$-1+i$	.4406	No	Yes	0.2289...
5	$-.6588+1.1615i$	.5049	Yes	Yes	0.2968...
6	$.2327+.7925i$	.5225	Yes	Yes	0.2968...
7	$-.2281+1.1151i$	.5297	Yes	Yes	0.0661...
8	$.4196+.6062i$	.5297	No	Yes	0.0661...
9	i	.5306	No	Yes	0.1526...
10	$.6180$	.5306	Yes	Yes	Fuchsian
11	$-1+1.2720i$	.5306	Yes	No	0.0717...
12	$-.4063+1.1961i$	.5345	No	Yes	0.4475...
13	$.7881+.4013i$	.6130	No	Yes	0.3475...

7 Applications

In this last section we present various applications of our description of the space $\mathcal{K}_\beta$. First we discuss our progress in determining the Kleinian group of minimal co–volume, or equivalently determining the minimal volume hyperbolic 3–fold.

7.1 Volumes

Let Γ be a Kleinian group. A subset F of $\mathbb{H}^3$ is said to be *precisely invariant* under Γ if for each $g \in \Gamma$ either

$$g(F) = F \quad \text{or} \quad g(F) \cap F = \emptyset. \tag{7.42}$$

Given a set $F \subset \mathbb{H}^3$ we denote by Γ_F the *stabilizer* of F in Γ. That is

$$\Gamma_F = \{g \in \Gamma : g(F) = F\}. \tag{7.43}$$

A *collar* of radius $t > 0$ about a hyperbolic line ℓ is the set

$$\mathcal{C} = \mathcal{C}(t) = \mathcal{C}(\ell, t) = \{x \in \mathbb{H}^3 : \rho(x, \ell) < t\}. \tag{7.44}$$

An elliptic element $f \in \Gamma$ in a Kleinian group is called *simple* if $\mathrm{ax}(f)$ is precisely invariant under Γ. If an elliptic element $f \in \Gamma$ is not simple, then G contains either a tetrahedral, octahedral, icosahedral or a parabolic subgroup. Here the *tetrahedral* subgroup is isomorphic to the alternating group A_4, the *octahedral* subgroup is isomorphic to the symmetric group S_4, the *icosahedral* subgroup is isomorphic to the alternating group A_5 and a *parabolic* subgroup is a group conjugate to an infinite group of euclidean isometries (a wallpaper group in the case of finite co–volume).

Simple elliptic elements have precisely invariant collars about their axes, while, in the co–compact case, nonsimple elliptic elements lie in a maximal finite spherical triangle group A_4, A_5 or S_4. Each such group stabilizes a finite point of $\mathbb{H}^3$ and there is a precisely invariant ball about such a fixed point [15] [29]. The tables of §4 are used to bound from below the size of these precisely invariant sets. We then obtain volume estimates based on the following elementary observation

Lemma 7.1 *Let G be a Kleinian group and $F \subset \mathbb{H}^3$ a precisely invariant subset. Then*

$$Vol(\mathbb{H}^3/\Gamma) \geq Vol(F/\Gamma_F). \tag{7.45}$$

The point to this observation is of course that most often Γ_F is a far simpler group to understand than Γ, and with a judicious choice of F the estimate need not be too bad, for instance if F were a fundamental domain, $\Gamma_F = Id$ and we would have equality. In our case we shall, of course, not be so lucky. The set F will most often be a solid hyperbolic cylinder and we will have to estimate the co–volume of F/Γ_F. This estimate will depend on the radius of F and should be bounded below by a number independent of Γ. Note that it is not immediate that any such bound should exist!

Define the function

$$\mathcal{V}(t) = \frac{\sqrt{3}}{2} \tanh(t) \cosh(2t) \operatorname{arcsinh}^2(\sinh(t)/\cosh(2t)). \tag{7.46}$$

Our first result is a volume estimate in terms of the radius of a precisely invariant collar, [30].

Theorem 7.2 *If G is a Kleinian group and if $C(t)$ is a precisely invariant collar with respect to G, then*

$$Vol(\mathbb{H}^3/G) \geq \frac{1}{2}\mathcal{V}(t). \tag{7.47}$$

Moreover if G is torsion free or if G contains no primitive 2–torsion, then

$$Vol(\mathbb{H}^3/G) \geq 2\mathcal{V}(t). \tag{7.48}$$

There is currently substantial interest in hyperbolic 3–manifolds which contain an embeded tubular neighborhood of radius $(\log 3)/2$. This is largely due to Gabai's version of the rigidity theorem [16]. As a consequence of the previous volume estimate we have

Corollary 7.3 *If M^3 is a hyperbolic 3–manifold containing an embeded tubular neighborhood of radius $(\log 3)/2$ about a closed geodesic, then*

$$Vol(M^3) \geq \frac{5}{2\sqrt{3}} \operatorname{arcsinh}^2\left(\frac{\sqrt{3}}{5}\right) = 0.16668\ldots. \tag{7.49}$$

Recently Gabai, Meyerhoff and N. Thurston announced a classification (via computer) of all closed finite volume hyperbolic 3–manifolds which do not contain a $(\log 3)/2$ embeded collar [17] and all those found have larger volume than the Weeks manifold. Then this classification has the implication that if M^3 is a hyperbolic 3–manifold, then $\mathcal{V}(M^3) \geq 0.16668\ldots$ and as another consequence that if G is a Kleinian group of minimal co–volume, then G contains no torsion free subgroup of index n, $n \leq 4$. Important volume estimates for hyperbolic 3–manifolds satisfying certain topological conditions (e.g. positivity of the first Betti number) are given in [13, 14].

For higher order elliptics we have quite reasonable estimates [2], [28].

Theorem 7.4 *Let Γ be a Kleinian group with an elliptic element of order $p \geq 7$. Then*

$$Vol(\mathbb{H}^3/\Gamma) \geq V_p = \frac{\sqrt{3}}{8}\frac{c_p - 1}{c_p^2}\left(\sqrt{1+c_p^2} - 1\right) \tag{7.50}$$

where

$$c_p = \cosh(2r_p) = \frac{1}{2\sin^2(\pi/p)} - 1. \tag{7.51}$$

This formula gives a volume bound which depends only on p and it is easy to see that it is increasing in p. We tabulate a few of the values below (we have also included improved estimates for the low order cases).

[2] These estimates have recently been superseded

Volume estimates for group with p–torsion, $p \geq 6$

Table 3

p	Lower Bound	Conjectured Extremal
6	0.0846...	0.0846...
7	0.0934...	0.1771...
8	0.1146...	0.2144...
9	0.1315...	0.2365...
10	0.1448...	0.2510...
20	0.1950...	0.2927...
100	0.2156...	0.3048...
1000	0.2164...	0.3053...
∞	0.2165...	0.3053...

Theorem 7.5 *Let Γ be a Kleinian group containing an elliptic of order $p \geq 6$. Then*

$$Vol(\mathbb{H}^3/\Gamma) \geq Vol(\mathbb{H}^3/PGL(2, O(\sqrt{-3})) \approx 0.0846\ldots. \tag{7.52}$$

Summarizing the current state of our knowledge we have

Theorem 7.6 *Let Γ be a minimal co–volume Kleinian group. Then Γ is either the arithmetic $\mathbb{Z}_2$–extension of the orientation preserving subgroup of the reflection group with Coxeter diagram* $3-5-3$ *with co–volume approximately* $0.03905\ldots$, *or*

- Γ *is not torsion free and contains no torsion free subgroup of index* ≤ 4.
- Γ *contains only torsion of order* 2 *and* 3
- Γ *contains only a single conjugacy class of* 3*–torsion and this lies inside a maximal dihedral subgroup of order* 6

Here maximal means that the dihedral subgroup lies in no larger finite subgroup of Γ.

7.2 Geodesics

In a hyperbolic 3–fold two closed geodesics that meet cannot both be short. If two closed geodesic meet at an angle θ, then the relative lengths of the geodesics should depend on θ basically because each geodesics represents a different element of the topology of the 3–fold. If the angle between the geodesics is small, then from standard ODE theory, the geodesics stay close for a while. There is a quantification of this [35]

Theorem 7.7 *There is an absolute constant* μ,

$$0.121 < \mu, \tag{7.53}$$

with the following property: If α_i, $i = 1, 2$ *are two geodesics in a hyperbolic 3-fold meeting at an angle* θ, $0 \leq \theta \leq \pi/2$, *then*

$$\sqrt{\sinh(\ell_1)\sinh(\ell_2)}\sin^{2/3}(\theta) \geq \mu \tag{7.54}$$

where ℓ_i *is the length of* α_i. *The exponent* $2/3$ *is best possible. If* $\alpha_1 = \alpha_2$ *(geodesics with self intersection) then*

$$\sinh(\ell)\sin(\theta) \geq \mu. \tag{7.55}$$

Again the exponent 1 *of* $\sin(\theta)$ *is best possible.*

Corollary 7.8 *A nonsimple closed geodesic in a hyperbolic 3-fold has length at least* .122.

7.3 Degeneration

In a sequence of discrete groups converging to a limiting discrete group, it is not possible that loxodromic elements degenerate into elliptic elements (say by having their translation length tend to 0). It is possible that loxodromic elements degenerate into parabolic elements, so both the holonomy and the translation length must simultaneously tend to zero. The question is how does holonomy die with translation length? An answer can be obtained from the following theorem [24]

Theorem 7.9 *Let* $\langle f, g\rangle$ *be a Kleinian group,* g *not of order* 2. *Then*

$$\|f\|^2\|g\|^2 \geq \frac{\sqrt{3}\sin^2(\theta(f)/2)}{4\pi\tau(f)} \approx \frac{\sqrt{3}}{8\pi}\frac{\theta^2}{\tau} \tag{7.56}$$

this is asymptotically sharp.

Thus if $f_i \to f$ and $g_i \to g$ and if $\langle f_i, g_i\rangle$ is Kleinian, then $\langle f, g\rangle$ is Kleinian [37] and we must have $\theta^2(f_i)/\tau(f_i)$ and $\theta^2(g_i)/\tau(g_i)$ bounded above. Further refinements of this result can be found in [8].

7.4 Margulis Constant

In their proof of Selberg's hypothesis Kazdan and Margulis found an absolute constant with important properties. This existence of this constant has found to have many interesting applications and it is known as the Margulis constant.

Theorem 7.10 *There is an absolute constant c_M such that if*

$$\Gamma = \langle g_1, g_2, \ldots, g_n \rangle \tag{7.57}$$

is discrete and $x \in \mathbb{H}^3$ then

$$\max\{\rho(g_i(x), x) : i = 1, 2, \ldots, n\} < c_M \tag{7.58}$$

implies that Γ is elementary (virtually abelian).

Thus at least one generator of a discrete nonelementary group (Kleinian group) must move any given point of hyperbolic space a fixed distance. The constant c_M has important geometric implications and is the key tool in dividing a hyperbolic 3–manifold into its "thick" and "thin" pieces. The thick pieces are where there is nontrivial topology, while the thin pieces are where the manifold is topologically the product $S^1 \times \mathbb{R}^2$. Important progress in determining the Margulis constant for certain torsion free 2–generator Kleinian groups was given in by Culler and Shalen [12]. They prove that $\log 3$ is a Margulis number for torsion free Kleinian groups with first Betti number at least three. We have from [33] and [34]

Theorem 7.11

- *The Margulis constant is achieved in a two or three generator group.*
- *Among the three generator groups the constant is achieved in a group generated by three elliptics of order 2 and is less than .132.*
- *For two generator groups the following holds: If $\langle f, g \rangle$ is discrete nonelementary and f is elliptic or parabolic, then*

$$\max\{\rho(f(x), x), \rho(g(x), x)\} \geq c_2 \tag{7.59}$$

where

$$c_2 = \operatorname{arccosh}\left(\frac{2\sqrt{8+2\sqrt{5}}-1-\sqrt{5}}{6-\sqrt{5}}\right) \approx 0.1829 \tag{7.60}$$

the result is sharp.

We have results for two generator Kleinian groups in general. Unfortunately as yet they are not quite good enough to determine this constant completely.

7.5 Triangle Subgroups

Many extremal examples of hyperbolic 3–manifolds and of 3–orbifolds are geometrically realized by (subgroups of) Kleinian groups with triangle subgroups. This occurs most often when studying problems of minimal volume manifolds with totally geodesic surfaces of a given genus. It is an interesting fact that like elliptics elements of order 7 or more in a Kleinian group whose axes are always simple, the hyperbolic plane stabilized by certain triangle groups has a precisely invariant neighbourhood.

Theorem 7.12 *Let Δ_p and Δ_q (possibly $p = q$) be $(2,3,p)$ and $(2,3,q)$–triangle subgroups of a Kleinian group Γ stabilizing hyperbolic planes Π_p and Π_q respectively. Then*

- *If $p \neq q$, then $\rho(\Pi_p, \Pi_q) \geq \operatorname{arccosh}(\sqrt{3}) \approx 1.14$*
- *If $p = q$, then $\rho(\Pi_p, \Pi_q) \geq \operatorname{arccosh}\left(1 + \frac{2\sin^2(\pi/p)}{1-4\sin^2(\pi/p)}\right)$*

Each estimate is best possible.

Among them the $(2,3,p)$ triangle groups contain every surface group. From this theorem and the examples of [11] one finds totally geodesic surface which are presumably as close together as is possible, without actually intersecting (which they may do). Notice that the result implies that algebraically two triangle subgroups of the same Kleinian group can intersect in only a finite cyclic subgroup.

REFERENCES

[1] B. Apanasov, S.L. Krushkal and N.A. Gusevskii, *Kleinian Groups and uniformization in examples and problems* Translations of Mathematical Monographs **62**, Amer. Math. Soc. 1987.

[2] C. Adams, *The noncompact hyperbolic 3–manifold of minimal volume* Proc. Amer. Math. Soc. **100** (1987), 601–606.

[3] C. Adams, *Limit volumes of hyperbolic 3–orbifolds* J. Diff. Geom. **34** (1991), 115–141.

[4] A. F. Beardon, *The geometry of discrete groups*, Springer–Verlag, 1983.

[5] R. Brooks and J.P. Matelski, *The dynamics of 2–generator subgroups of $SL(2,\mathbb{C})$* Discontinuous groups and Riemann surfaces, Princeton Univ. Press, 1981.

[6] C. Cao, *On three generator Möbius groups* NZ J. Math. **23** (1994), 111–120.

[7] C. Cao, *Some trace inequalities for discrete groups of Möbius transformations* Proc. Amer. Math. Soc. (1995) (to appear).

[8] C. Cao, F.W. Gehring and G.J. Martin, *Lattice constants and a lemma of Zagier*, Contemporary Mathematics (to appear).

[9] T. Chinburg and E. Friedman, *The smallest arithmetic hyperbolic 3-orbifold*, Invent. Math. **86** (1986), 507–527.

[10] T. Chinburg, E. Friedman, K. N. Jones and A. W. Reid, *The smallest volume arithmetic hyperbolic 3-manifold* (in preparation).

[11] M. D. E. Conder and G. J. Martin, *Cusps, triangle groups and hyperbolic 3–folds*, J. Austral. Math. Soc. **55** (1993), 149–182.

[12] M. Culler and P. Shalen, *Paradoxical decompositions, 2–generator Kleinian groups, and volumes of hyperbolic 3–manifolds*, J. Amer. Math. Soc. **5** (1992), 231–288.

[13] M. Culler and P. Shalen, *The volume of a Hyperbolic 3–manifold with Betti number 2*, preprint.

[14] M. Culler and P. Shalen, *Volumes of hyperbolic Haken Manifolds* **I**, preprint.

[15] D.A. Derevnin and A.D. Mednykh, *Geometric properties of discrete groups acting with fixed points in Lobachevsky space*, Soviet Math. Dokl. **37** (1988), 614–617.

[16] D. Gabai, *On the geometric and topological rigidity of hyperbolic 3–manifolds*, Annals of Math., (to appear).

[17] D. Gabai, R. Meyerhoff and N. Thurston, Personal Communication.

[18] F.W. Gehring and G.J. Martin, *Discrete quasiconformal groups, I*, Proc. London. Math. Soc. (3) **55** (1987), 331–358.

[19] F.W. Gehring and G.J. Martin, *Axial distances in discrete Möbius groups*, Proc. Natl. Acad. Sci. USA **89** (1992), 1999–2000.

[20] F.W. Gehring and G.J. Martin, *Some universal constraints for discrete Möbius groups*, Paul Halmos, Celebrating 50 years of mathematics, Ed. F.W. Gehring and J.H. Ewing, Springer-Verlag, 1991, 205–220.

[21] F.W. Gehring and G.J. Martin, *Commutators, collars and the geometry of Möbius groups*, J. d'Analyse Math. **63** (1994), 175–219.

[22] F. W. Gehring and G. J. Martin, *On the minimal volume hyperbolic 3-orbifold*, Math. Res. Letters **1** (1994), 107–114.

[23] F.W. Gehring and G.J. Martin, *Inequalities for discrete groups and Möbius transformations*, J. Reine Angew. Math. **418** (1991), 31–76.

[24] F.W. Gehring and G.J. Martin, *Chebyshev polynomials and discrete groups*, Proc. Intl. Conf. in Complex Analysis, International Press (1992), 114–125.

[25] F.W. Gehring and G.J. Martin, *Iteration theory and Discreteness in Kleinian groups*, Bull. Amer. Math. Soc. **2** (1989), 57–63.

[26] F.W. Gehring and G.J. Martin, 6–*Torsion and hyperbolic volume*, Proc. Amer. Math. Soc. **117** (1993), 727–735.

[27] F.W. Gehring and G.J. Martin, *Commutator spectra for discrete groups with an elliptic generator*, in preparation.

[28] F.W. Gehring and G.J. Martin, *The volume of hyperbolic 3–folds with p–torsion, $p \geq 6$*, to be submitted.

[29] F.W. Gehring and G.J. Martin, *Tetrahedral, octahedral and icosahedral points in a discrete group*, in preparation.

[30] F.W. Gehring and G.J. Martin, *Precisely invariant collars and the volume of hyperbolic 3–folds*, to be submitted.

[31] F.W. Gehring and G.J. Martin, *Discrete groups with real parameters*, to be submitted.

[32] F.W. Gehring and G.J. Martin, *Discreteness in Kleinian groups*, in preparation.

[33] F.W. Gehring and G.J. Martin, *The Margulis constant for Kleinian groups, I*, Ann. Acad. Sci. Fenn. Math. **21** (1996), 439–462.

[34] F.W. Gehring and G.J. Martin, *The Margulis constant for Kleinian groups, II*, in preparation.

[35] F.W. Gehring and G.J. Martin, *Geodesics in hyperbolic 3–folds*, Michigan Math. J. (to appear).

[36] F.W. Gehring, C. Maclachlan, G.J. Martin and A.W. Reid, *Arithmeticity, discreteness and volume*, Trans. Amer. Math. Soc. (to appear).

[37] T. Jørgensen, *On discrete groups of Möbius transformations*, Amer. J. Math. **78** (1976), 739–749.
[38] G.J. Martin, *Triangle subgroups of Kleinian groups*, Comment. Math. Helv. **71** (1996), 339–361.
[39] G.J. Martin, *The commutator spectrum and conjugacy classes of a discrete subgroup of* $SL(2,\mathbf{C})$. Indian J. Math. **35** (1993), 155–170.
[40] B. Maskit, *Kleinian groups*, Springer–Verlag, 1987.
[41] B. Maskit, *Some special 2–generator Kleinian groups*, Proc. Amer. Math. Soc. **106** (1989), 175–186.
[42] R. Meyerhoff, *The cusped hyperbolic 3-orbifold of minimum volume*, Bull. Amer. Math. Soc. **13** (1985), 154–156.
[43] R. Meyerhoff, *A lower bound for the volume of hyperbolic 3–manifolds*, Can. J. Math. **39** (1987), 1038–1056.
[44] J.W. Morgan and H. Bass, *The Smith Conjecture*, Academic Press, 1984.
[45] G.D. Mostow, *Quasiconformal mappings in space and the rigidity of hyperbolic space forms*, Inst. Hautes Études Sci. Publ. Math. **34** (1968), 53–104.
[46] J.G. Ratcliffe, *Foundations of hyperbolic manifolds*, Springer–Verlag, 1994.
[47] G. Rosenberger, *All generating pairs of all two generator Fuchsian groups*, Arch. Math. **46** (1986), 198–204.
[48] W. P. Thurston, *The geometry and topology of three-manifolds*, Princeton University Lecture Notes, 1976.

Australian National University
Canberra
Australia

University of Auckland
Auckland
New Zealand

Old and New on the Schwarzian Derivative

Brad Osgood

1 Introduction

Let f be an analytic function. We define its *Schwarzian derivative* to be

$$Sf = \left(\frac{f''}{f'}\right)' - \frac{1}{2}\left(\frac{f''}{f'}\right)^2 = \frac{f'''}{f'} - \frac{3}{2}\left(\frac{f''}{f'}\right)^2. \qquad (1.1)$$

The classical notation is $\{f, z\}$, or $\{w, z\}$ if we write $w = f(z)$, and is due to Cayley in an 1880 paper [19]. Cayley was also the one to honor Schwarz with the appellation.

With its ungainly appearance the Schwarzian has always been an acquired taste, but through its varied applications it has been one acquired by many complex analysts. The 'Old' of the story is pretty stable by now, but I am pleased to report that the 'New' keeps changing. Some of the newest considered here is part of ongoing joint work with M. Chuaqui, Ch. Pommerenke and D. Stowe. I am happy to acknowledge them.

The Schwarzian is associated foremost with Möbius transformations and with differential equations. To recall some of the virtues of Möbius transformations, they map circles to circles and they are the only conformal maps of the sphere to itself. These are each global characterizations of Möbius transformations, while an infinitesimal characterization is the vanishing of the Schwarzian.To see this, if

$$f(z) = \frac{az+b}{cz+d}, \quad ad - bc = 1,$$

then

$$f'(z) = \frac{1}{(cz+d)^2},$$

[1]Parts of this paper are based on lectures given at Washington University, St. Louis; the University of Paris, Orsay; Berkeley; and the Technion, Israel. The author is grateful to friends and colleagues at those institutions for their hospitality and interest.

so

$$u(z) = f'(z)^{-1/2} = cz + d.$$

Hence

$$\begin{aligned} 0 = u'' &= \frac{3}{4}(f')^{-5/2}(f'')^2 - \frac{1}{2}(f')^{-3/2}f''' \\ &= -\frac{1}{2}(f')^{-1/2}\left(-\frac{3}{2}\left(\frac{f''}{f'}\right)^2 + \frac{f'''}{f'}\right) \\ &= -\frac{1}{2}f'(z)^{-1/2}Sf(z), \end{aligned} \tag{1.2}$$

which delivers the Schwarzian to us and proves that it vanishes identically exactly for the Möbius transformations.

The calculation in (1.2) shows that if f is any analytic function and

$$u = (f')^{-1/2}. \tag{1.3}$$

then we have

$$u'' + \frac{1}{2}(Sf)u = 0. \tag{1.4}$$

Conversely, if u is a solution of

$$u'' + qu = 0,$$

where q is analytic, and if

$$f(z) = \int_{z_0}^{z} u^{-2}(\zeta)\, d\zeta \tag{1.5}$$

then $Sf = 2q$.

The Schwarzian has important invariance properties under composition with Möbius transformations. If T is Möbius then

$$S(T \circ f) = Sf, \tag{1.6}$$

that is, Sf is invariant under composition with Möbius transformations on the left. On the right, if T is Möbius then

$$S(f \circ T) = (S(f) \circ T)(T')^2. \tag{1.7}$$

These are included in a more general formula holding for the composition of any two functions:

$$S(f \circ h) = (S(f) \circ h)(h')^2 + Sh. \tag{1.8}$$

We note the consequence that if $Sf = Sh$ then f is a Möbius transformation of h, and vice versa. The chain rule for the Schwarzian (1.8) is also due to Cayley in the same paper mentioned earlier. It is possible to explain this formula in a very general and conceptual way, as we shall do in Section 4.

All this has been by way of preliminaries, and we shall return to each of these points. The rest of this paper is divided into three parts:

- The Classical Period;
- The Modern Period;
- The Post-Modern, Abstract Expressionist, Neo-Classical Period.

For our purposes, the Classical Period begins roughly in 1836 and ends around 1880. Since we are only celebrating Fred Gehring's 70'th birthday, his work does not appear there. But during the Modern Period, which picks up about 1949, he has made several fundamental contributions, and it will be a pleasure to present them.

All of our discussion in the Modern Period concerns univalent functions and quasiconformal mappings. The book 'Univalent functions and Teichmüller spaces' by O. Lehto, [60], is so thorough and so clear that any friend of the Schwarzian should constantly turn to it in good times and bad. Though there is a considerable overlap, especially in highlighting some of Fred's contributions, in this paper we have tried to cover different aspects and new results. Our choice of topics is not the only one possible, and the author apologizes for omissions. In particular, we do not touch at all the uses of the Schwarzian in one-dimensional dynamics, see for instance [28], [85], [88].

The final, Post-Modern section has to do with generalizations of the Schwarzian to conformal mappings of Riemannian manifolds.

2 The Classical Period

The sources I have followed lead to a paper of Kummer in 1836 [59] as the first published appearance of what was later called the Schwarzian. Kummer observed the following: If y_1 and y_2 are independent solutions of a general second order differential equation

$$y'' + py' + qy = 0 \tag{2.1}$$

then their quotient $f = y_1/y_2$ satisfies

$$Sf = 2q - \frac{1}{2}p^2 - p'. \tag{2.2}$$

This identity has the honor of surviving as the last exercise in Ahlfors' complex analysis book.

Of course, *the* second order equation of interest is the hypergeometric equation

$$z(1-z)y'' + (c - (a+b+1)z)y' - aby = 0, \tag{2.3}$$

and this was really the subject of Kummer's study. The corresponding Schwarzian equation is

$$Sf(z) = \frac{1-\lambda^2}{2z^2} + \frac{1-\mu^2}{2(1-z)^2} + \frac{1-\lambda^2-\mu^2+\nu^2}{2z(1-z)}, \tag{2.4}$$

where

$$\lambda = 1 - c, \quad \mu = c - a - b, \quad \nu = a - b. \tag{2.5}$$

Kummer wanted to know how the solutions to the hypergeometric equation transform under changes of variable, in particular under Möbius transformations. He knew that the Schwarzian vanishes for Möbius transformations and has the invariance property (1.6).

Why look at the quotient of two solutions? The hypergeometric equation is interesting because of its singularities at 0, 1 and ∞. The fact that one can locate the singularities at these points already uses properties of Möbius transformations, and the now standard form of the hypergeometric equation (2.3) is due to Gauss. The singularities are interesting because of monodromy; what happens to solutions under analytic continuation? If y_1 and y_2 are independent solutions of the hypergeometric equation then analytic continuation will lead to linear combinations of y_1, y_2, say $y_1 \mapsto \alpha y_1 + \beta y_2$ and $y_2 \mapsto \gamma y_1 + \delta y_2$. Thus for the quotient

$$f = \frac{y_1}{y_2} \mapsto \frac{\alpha y_1 + \beta y_2}{\gamma y_1 + \delta y_2} = \frac{\alpha f + \beta}{\gamma f + \delta}.$$

Continuing the solution to the associated Schwarzian equation leads to a Möbius transformation of the original solution, and, moreover, this is the most general form of a solution to the Schwarzian equation.

The study of the monodromy group of the hypergeometric equation becomes a study of a group of Möbius transformations. Among other things, this brings new elements of geometry into the picture, a point of view which proved to be extremely helpful to Schwarz. Just as important for the later theory, especially in the work of Poincaré, is that the 'inverse function' $z = h(w)$ of a global solution $w = f(z)$ to the original differential equation will then be an automorphic function with respect to this group: $h(Tw) = h(w)$, where T is an element of the group.

In several papers in the 1860's, L. Fuchs raised a question that attracted a considerable amount of interest: For what parameters a, b and c are all solutions of the hypergeometric equation (2.3) algebraic functions? He also raised more general questions about algebraic solutions of differential equations having algebraic coefficients, much in the tradition of questions on algebraic functions and their integrals. See the article by Gray [48] for an appreciation of Fuchs' work and for complete references.

Fuchs' problem was solved for the hypergeometric equation by H. A. Schwarz in an 1873 paper [83]. He used pretty much all he knew about conformal mapping and analytic continuation, much of which he had developed in earlier papers, see especially his 1868 and 1869 papers in his collected works [84]. In particular, his reflection principle allowed him to study analytic continuation geometrically. He also singled out the Schwarzian for special attention, calling it $\Psi(w, z)$ and leading Cayley to associate it with Schwarz's name.

Here is Schwarz's approach to the problem, highly abbreviated. Assuming that any local solution has at worst algebraic singularities, a condition which is easy to check, all solutions will be *algebraic functions* if and only if analytic continuation around the singularities produces only a finite number of branches. Thus the first key observation is that all solutions will be algebraic functions if and only if the monodromy group is *finite*. Passing to the associated Schwarzian equation, one is then asking first to determine the finite groups of Möbius transformations, and second to determine which parameters λ, μ and ν lead to these groups.

To address the first point, via stereographic projection one wants to know the finite groups of motions of the sphere. These are:

1. the identity;
2. finite cyclic groups;
3. dihedral groups;
4. the tetrahedral group;
5. the octahedral group;
6. the icosahedral group.

See [47] for a geometric treatment of this result, and [11] for a purely algebraic proof.

How does one determine when the parameters of the equation will lead to one of these groups? Schwarz proved that a solution to the (yet to be so named) Schwarzian equation (2.4) maps the upper half-plane onto the interior of a triangle bounded by circular arcs. The circles meet at angles $\pi\lambda$, $\pi\mu$, and $\pi\nu$, which are simple differences of the parameters a, b, c in the original hypergeometric equation, (2.5). Analytic continuation of the solution is accomplished by reflecting in the real axis and in the sides of the triangles, and elements of the monodromy group are formed by a composition of an even number of reflections. Aside from the cyclic groups, the monodromy group will be finite when the angles are those of a spherical triangle whose vertices are fixed points of three rotations of the sphere, any pair of which generate a finite group of rotations.

Schwarz found exactly fifteen cases when the monodromy groups are finite. Here is his list of angles λ, μ, and ν:

λ	μ	ν	Group
1/2	1/2	$1/n$	Dihedral
1/2	1/3	1/3	Tetrahedral
2/3	1/3	1/3	
1/2	1/3	1/4	Octahedral
2/3	1/4	1/4	
1/2	1/3	1/5	
2/5	1/3	1/3	
2/3	1/5	1/5	
1/2	2/5	1/5	
3/5	1/3	1/5	Icosahedral
2/5	2/5	2/5	
2/3	1/3	1/5	
4/5	1/5	1/5	
1/2	2/5	1/3	
3/5	2/5	1/3	

This is actually a list of equivalence classes of angles. Two triples (λ, μ, ν) and (λ', μ', ν') are equivalent if $\lambda' = \pm\lambda + p$, $\mu' = \pm\mu + q$ and $\nu' = \pm\nu + r$, where p, q and r are integers with $p + q + r \equiv 0 \bmod 2$. If the angles are equivalent then the monodromy groups are isomorphic. The parameters in the list are the smallest in a given equivalence class.

Schwarz also studied the case when the solutions have single-valued inverses. This occurs when the reflected triangles tessellate the plane or the disk. Recall that this is important because these functions are then automorphic for the monodromy group.

λ	μ	ν	Group	Order
1	$1/n$	$1/n$	Cyclic	n
1/2	1/2	$1/n$	Dihedral	$2n$
1/2	1/3	1/3	Tetrahedron	12
1/2	1/3	1/4	Octahedron	24
1/2	1/3	1/5	Icosahedron	60

Schwarz's work was followed up by several people. Fuchs, in 1876, considered the question of algebraic solutions to higher order differential equations. In his famous memoir on the icosahedron in 1884, Klein studied the problem of algebraic solutions of second order equations with more that three singular points; see, however, [11] for a critique of Klein's claims. In 1880 Cayley's paper [19] appeared, christening the Schwarzian, introducing the notation $\{f, z\}$ and proving the chain rule (1.8). He began by stating: "I propose in this Memoir to consider the whole theory: and in particular give some additional developments in regard to the Polyhedral Functions."

This is as far as I want to pursue this line except to say that the questions raised so long ago are far from being completely answered. I've traced forward (no doubt with gaps) to the 1979 paper of Baldassari and Dwork [11]. The most direct predecessor to their work is a long paper of N. Katz in 1972, [55]. Katz considers a very general conjecture of Grothendieck on algebraic solutions of differential equations coming from algebraic geometry. According to my informal scheme for dividing up the history of the Schwarzian, I would say that this bypasses the Modern Period and heads straight into the Abstract Expressionist phase. I am grateful to Professor Katz for providing me with recent references to work on this problem.

Notes: Kummer's identity (2.2) when $p = 0$ says that the ratio $f = y_1/y_2$ of two independent solutions of $y'' + qy = 0$ satisfies $Sf = 2q$. This fact is what one most often sees in treatments of the Schwarzian. This can be reconciled with the approach in the Introduction by taking a pair of solutions to $y'' + qy = 0$ whose Wronskian is identically 1.

Another way of linking the Schwarzian to second order differential equations is to treat $Sf = 2q$ as a Riccati equation

$$w' - \frac{1}{2}w^2 = 2q$$

for

$$w = \frac{f''}{f'}.$$

If we now let v be an antiderivative of w (in this case $v = \log f'$) and

$$u = e^{-v/2} = (f')^{-1/2}$$

then u satisfies $u'' + qu = 0$ as before. Later we will see a higher dimensional version of this.

The projective geometry of a second order equation is discussed in Arnold's book [7]. The solutions to a Riccati equation also have a 'projective structure' (rather than forming a vector space) since any solution is a Möbius transformation of three given solutions; see, for instance, [53].

Along with Gray's article [48] for a general history of the problem of differential equations with algebraic solutions, we also refer to a very clear exposition of Schwarz's work by Goursat in [47], and to Hille's book [53] for many other uses of the Schwarzian in the Classical Period.

3 The Modern Period

A highlight of the Classical Period was the discovery of Schwarz that ratios of solutions to the hypergeometric equation give conformal mappings of the upper half plane onto curvilinear triangles. Thus via differential equations there is a beautiful relationship between the Schwarzian and conformal mapping. I would like to define 'The Modern Period' by emphasizing more general connections to conformal mapping.

Whatever the earliest reference to the Schwarzian might be, the central result of the Modern Period is a theorem of Nehari proved in 1949, [66]. In that paper Nehari established a connection between the growth of the Schwarzian and univalence.

Theorem 3.1 (Nehari) *If f is analytic and univalent in the unit disk then*

$$|Sf(z)| \leq \frac{6}{(1-|z|^2)^2}. \tag{3.1}$$

If f is analytic and locally univalent in the disk with

$$|Sf(z)| \leq \frac{2}{(1-|z|^2)^2} \tag{3.2}$$

then f is univalent in the disk.

The Koebe function

$$k(z) = \frac{z}{(1-z)^2} = z + 2z^2 + 3z^3 + \cdots$$

has

$$Sk(z) = \frac{-6}{(1-z^2)^2},$$

which shows that the necessary condition for univalence is sharp. The constant 2 in the sufficient condition is also best possible, for the function

$$f(z) = \left(\frac{1-z}{1+z}\right)^{i\gamma}, \quad \gamma > 0$$

is not univalent in the disk, and

$$Sf(z) = \frac{2(1+\gamma^2)}{(1-z^2)^2}.$$

This example was given by Hille [52] as an addendum to Nehari's paper.

The necessary and sufficient conditions for univalence are different in character and we shall treat them separately. We let N denote the class of univalent functions satisfying the sufficient condition (3.2). We let $\mathbf{D}$ denote the unit disk.

Given the history up to the time of Nehari's paper, as much as I have followed it, I would say that his results were quite unanticipated and surprising. (Though the necessary condition for univalence (3.1) was actually first proved by Krauss [58].) The subject then went in two directions. One line of research aimed at establishing other necessary and sufficient conditions for univalence depending on the Schwarzian or related quantities. Another was opened up by an important result of Ahlfors and Weill [5] in 1962 linking the growth of the Schwarzian to quasiconformal extension.

Theorem 3.2 (Ahlfors-Weill) *If*

$$|Sf(z)| \leq \frac{2t}{(1-|z|^2)^2} \tag{3.3}$$

for some $0 \leq t < 1$ *then* f *has a* $\frac{1+t}{1-t}$*-quasiconformal extension to the sphere.*

Since then, quasiconformal mappings have been front and center. A simply connected domain Ω is a *quasidisk* if it is the image of $\mathbf{D}$ under a quasiconformal mapping of the sphere. If f is a K-quasiconformal mapping then Ω is a K-quasidisk and $\partial\Omega$ is a K-quasicircle. Thus the Ahlfors-Weill theorem says that a function satisfying (3.3) maps the disk onto a $\frac{1+t}{1-t}$-quasidisk.

3.1 Necessary Conditions for Univalence

The bound (3.1) is equivalent to a coefficient inequality for univalent functions. Specifically, if

$$h(z) = z + \frac{b_1}{z} + \ldots \tag{3.4}$$

is univalent in the exterior of the disk then

$$|b_1| \leq 1. \tag{3.5}$$

The source of (3.5) is Gronwall's Area Theorem, see [30]. From this, if

$$f(z) = z + a_2 z^2 + a_3 z^3 + \cdots$$

is univalent in the disk then

$$\frac{1}{f(\frac{1}{z})} = z - a_2 + (a_2^2 - a_3)\frac{1}{z} + \cdots.$$

and hence

$$|a_2^2 - a_3| \leq 1.$$

This is

$$|Sf(0)| \leq 6.$$

Renormalizing with the Koebe transform,

$$F(z) = \frac{f\left(\dfrac{z+\zeta}{1+\bar{\zeta}z}\right) - f(\zeta)}{(1-|\zeta|^2)f'(\zeta)} = z + \cdots,$$

then gives

$$(1-|\zeta|^2)^2 Sf(\zeta) = SF(0),$$

and thus

$$(1-|\zeta|^2)^2|Sf(\zeta)| \leq 6.$$

More grandly, (3.1) can also be obtained as a simple special case of the Grunsky-Golusin inequalities, as in [79], p. 67, These inequalities can themselves be regarded as a generalization of the area theorem, so perhaps this route is not so different. The same destination can be reached from a different starting point via the relation between the Schwarzian and the Bergman kernel function.

Let Ω be a bounded, simply connected domain, with Green's function $g_\Omega(z,\zeta)$. Following Schiffer, and the presentation in [15], we define the Bergman kernel to be

$$K_\Omega(z,\bar{\zeta}) = -\frac{2}{\pi}\frac{\partial^2 g(z,\zeta)}{\partial z \partial \bar{\zeta}}, \tag{3.6}$$

and the L-kernel to be

$$L_\Omega(z,\zeta) = -\frac{2}{\pi}\frac{\partial^2 g(z,\zeta)}{\partial z \partial \zeta}. \tag{3.7}$$

The kernel K_Ω is an analytic function of z while L_Ω has a double pole at $z=\zeta$ and may be written

$$L_\Omega(z,\zeta) = \frac{1}{\pi}\frac{1}{(z-\zeta)^2} - l_\Omega(z,\zeta),$$

where $l_\Omega(z,\zeta)$ is analytic.

For the unit disk,

$$g_{\mathbf{D}}(z,\zeta) = \log\left|\frac{1-\bar{\zeta}z}{z-\zeta}\right|, \qquad K_{\mathbf{D}}(z,\bar{\zeta}) = \frac{1}{\pi}\frac{1}{(1-\bar{\zeta}z)^2}, \tag{3.8}$$

$$L_{\mathbf{D}}(z,\zeta) = \frac{1}{\pi}\frac{1}{(z-\zeta)^2}, \qquad l_{\mathbf{D}}(z,\zeta) = 0. \tag{3.9}$$

Let f be a conformal mapping of Ω onto a domain Ω'. From $g_{\Omega'}(f(z),f(\zeta))$
$= g_\Omega(z,\zeta)$ one finds that

$$K_{\Omega'}(f(z),\overline{f(\zeta)})f'(z)\overline{f'(\zeta)} = K_\Omega(z,\bar{\zeta}), \tag{3.10}$$

and that

$$\begin{aligned} l_{\Omega'}(f(z),f\zeta))f'(z)f'(\zeta) &= l_\Omega(z,\zeta) + \frac{1}{\pi}\left\{\frac{f'(z)f'(\zeta)}{(f(z)-f(\zeta))^2} - \frac{1}{(z-\zeta)^2}\right\} \\ &= l_\Omega(z,\zeta) + U(z,\zeta), \end{aligned} \tag{3.11}$$

where

$$\Phi(z,\zeta) = \frac{1}{\pi}\log\frac{f(z)-f(\zeta)}{z-\zeta}, \qquad U(z,\zeta) = \frac{\partial^2\Phi(z,\zeta)}{\partial z\partial\zeta}.$$

The Schwarzian enters the picture through the identity

$$U(z,z) = \frac{1}{6\pi} Sf(z). \tag{3.12}$$

Inequalities involving the various kernel functions arise typically by forming the L^2-norm of a quadratic form and using reproducing properties of K_Ω and orthogonality properties of L_Ω. The simplest such inequality is

$$|l_\Omega(z,z)| \le K_\Omega(z,\bar{z}). \tag{3.13}$$

Using the transformation properties of the kernel functions together with (3.12), this spreads out to

$$|l_\Omega(z,z) + \frac{1}{6\pi} Sf(z)| \le K_\Omega(z,\bar{z}), \tag{3.14}$$

where f is a conformal mapping of Ω. When Ω is the unit disk we are back to

$$|Sf(z)| \le \frac{6}{(1-|z|^2)^2}.$$

From (3.13),

$$|Sf(z)| \le 12\pi K_\Omega(z,\bar{z}) \tag{3.15}$$

when f is a conformal mapping of Ω. We want to write this differently.

Recall that the Poincaré metric $\lambda_\Omega |dw|$ of a simply connected domain Ω is defined by

$$\lambda_\Omega(f(z))|f'(z)| = \lambda_{\mathbf{D}}(z) = \frac{1}{1-|z|^2}, \tag{3.16}$$

where $f\colon \mathbf{D} \to \Omega$ is a conformal mapping of the unit disk onto Ω. It follows from (3.8) and (3.10) that

$$K_\Omega(z,\bar{z}) = \frac{1}{\pi}\lambda_\Omega(z)^2. \tag{3.17}$$

Thus

$$|Sf(z)| \le 12\lambda_\Omega(z)^2. \tag{3.18}$$

The inequality (3.18) was given by Lehto, [60]. He deduced it directly from the result for the disk using the chain rule (1.8) and the conformal invariance of the Poincaré metric. It is sharp.

But much more is true. Beardon and Gehring proved in [12] that the inequality (3.18) holds on *any* domain.

Theorem 3.3 (Beardon-Gehring) *Let f be analytic and univalent in a domain Ω. Then*

$$|Sf(z)| \le 12\lambda_\Omega(z)^2 \tag{3.19}$$

for all $z \in \Omega$. The constant 12 *is best possible for any given connectivity n.*

(The Poincaré metric can be defined just as in (3.16) by using an analytic covering map of the disk onto the domain Ω.)

To prove the Beardon-Gehring theorem one first needs (3.15) for finitely connected domains. This was the original setting of the Bergman-Schiffer paper [15] and the result is contained there. However, for finitely connected domains one must work with the *reduced* Bergman kernel (and associated kernels) on the space of L^2-integrable *derivatives* of analytic functions (so the functions have single valued integrals). The kernel functions on this space are not defined simply in terms of the Green's function as in (3.6) and (3.7), and it is somewhat remarkable that the relevant identities and inequalities continue to hold. But it works, and (3.15) obtains. Using an approximation and limiting argument it then follows that $|Sf(z)| \leq 12\pi K_\Omega(z,\bar{z})$ when Ω is not of class O_{AD}, *i.e.* Ω supports nonconstant analytic functions with finite Dirichlet norm.

Next, while equality holds in (3.17) for a simply connected domain, for multiply connected domains Beardon and Gehring proved that

$$K_\Omega(z,\bar{z}) \leq \frac{1}{\pi}\lambda_\Omega(z)^2, \quad \Omega \notin O_{AD}.$$

This follows from a chain of inequalities on solutions of extremal problems, starting with the kernel function and ending with the square of the Poincaré metric. These serve to establish (3.19) for domains not in O_{AD}, and the passage to arbitrary multiply connected domains is again via a limiting argument. As Beardon and Gehring point out, the framework for the main argument, from K_Ω to λ_Ω^2, was in the paper by Ahlfors and Beurling, [4], that predates the Bergman-Schiffer paper by a year.

Notes: By rescaling (3.1) one can deduce that if f is univalent in the plane then $Sf \equiv 0$, and hence that f is a Möbius transformation. Of course we know this, but it would be interesting to give a direct proof of the contrapositive: if f is locally univalent in the plane and Sf is nonzero at a single point then f is not univalent. And if that works, prove that if f is analytic in the disk and $|Sf(z_0)| > 6/(1-|z_0|^2)^2$ at a single point, then f is not univalent.

One other approach to the basic coefficient estimate $|b_1| \leq 1$, (3.5), is through Jenkins' General Coefficient Theorem, [54], p. 95.

A (shorter) companion paper to this one would be 'Old and New on f''/f''. Many of the results we discuss for the Schwarzian have analogues for the operator f''/f', sometimes referred to as the 'pre-Schwarzian', but we restrict ourselves to brief comments. Starting with the coefficient inequality $|a_2| \leq 2$ for normalized univalent functions $f(z) = z + a_2 z^2 + \cdots$ one derives bounds for $|f''/f'|$ by renormalizing. This leads to the sharp inequality

$$\left|\frac{f''}{f'}(z)\right| \leq \frac{6}{1-|z|^2}.$$

In [71] it was shown that if f is analytic and univalent in any simply connected domain Ω then

$$\left|\frac{f''}{f'}\right| \leq 8\lambda_\Omega.$$

This too is sharp. However, in contrast to the Beardon-Gehring theorem, there is no constant $a > 0$ such that

$$\left|\frac{f''}{f'}\right| \leq a\lambda_\Omega \tag{3.20}$$

for all multiply connected domains.

3.2 *Sufficient Conditions for Univalence and the use of Comparison Theorems*

Apart from the particular results, Nehari's main contribution in [66] was the recognition that the univalence of an analytic function can be easily related to the disconjugacy of the solutions of the differential equation associated to its Schwarzian derivative, *i.e.* that no solution vanishes more than once. Bounds on the Schwarzian such as (3.2) allow one to apply Sturm comparison theorems, and this produces sufficient conditions for univalence. It does not yield necessary conditions because the comparison theorems that apply in one direction do not have converses that apply in the other. Here we sketch an approach to the sufficient condition having the same basic ingredients as Nehari's, but following a different route. In Section 4.1 we will give more of a 'post-modern' treatment of the sufficient condition and its generalizations.

Recall that N denotes the set of univalent functions satisfying (3.2). Since $S(T \circ f) = Sf$ for any Möbius transformation T, many results for N are naturally independent of any particular normalization on f. There are instances, however, when a normalization is useful, and we let N_0 be the functions in N with

$$f(0) = 0, \quad f'(0) = 1, \quad f''(0) = 0. \tag{3.21}$$

The important conditions are on the first and second derivatives, for one then considers the initial value problem

$$u'' + qu = 0, \quad u(0) = 1, u'(0) = 0 \tag{3.22}$$

with

$$q = \frac{1}{2}Sf, \quad f(z) = \int_0^z u^{-2}(\zeta)\, d\zeta, \quad u = (f')^{-1/2}. \tag{3.23}$$

When

$$q(z) = \frac{\pm 1}{(1 - z^2)^2}$$

the corresponding normalized solutions to $Sf = 2q$ are

$$\ell(z) = \frac{1}{\sqrt{2}} \frac{(1+z)^{\sqrt{2}} - (1-z)^{\sqrt{2}}}{(1+z)^{\sqrt{2}} + (1-z)^{\sqrt{2}}}, \qquad S\ell(z) = \frac{-2}{(1-z^2)^2} \tag{3.24}$$

$$L(z) = \frac{1}{2} \log \frac{1+z}{1-z}, \qquad SL(z) = \frac{2}{(1-z^2)^2} \tag{3.25}$$

Note that while the image of L is the parallel strip $|\operatorname{Im} w| < \pi/4$, the image of l is a Jordan domain (even a quasidisk) interior to the union of two circles.

Using comparison theorems, together with some lemmas in [36], one can prove sharp growth and distortion theorems for N_0, [23]. Specifically, if $f \in N_0$ then

$$\ell(|z|) \leq |f(z)| \leq L(|z|) \tag{3.26}$$

$$\ell'(|z|) \leq |f'(z)| \leq L'(|z|). \tag{3.27}$$

In any of these inequalities, if equality holds at one point other than the origin then f is conjugate by a rotation to ℓ or L as the case may be.

A helpful consequence of the upper bound in (3.26) is that it is always possible to normalize a function in N to obtain one in N_0 without introducing a pole. For if $f(z) = z + a_2 z^2 + \cdots$ is in N then $g = f/(1 + a_2 f)$ still satisfies (3.2) and is normalized. It will satisfy the right hand inequality in (3.26) in the largest disk about the origin where it is analytic, and hence cannot have a pole in $\mathbf{D}$.

Knowing this, a helpful consequence of the lower bound in (3.26) is a quick proof of Nehari's theorem. Suppose that f satisfies (3.2) and that $f(z_1) = f(z_2)$. We want to show that $z_1 = z_2$. If $w = T(z)$ is a Möbius transformation of the disk onto itself then

$$(1 - |w|^2)^2 |Sf(w)| = (1 - |z|^2)^2 |Sf(z)|$$

by the invariant form of Schwarz's lemma and the chain rule, $S(f \circ T) = (S(f) \circ T)(T')^2$. Hence we may compose f with a suitable Möbius transformation of the disk and assume that $z_1 = 0$. We next compose f with a Möbius transformation of the range to get $f(0) = 0, f'(0) = 1$ and $f''(0) = 0$. Thus $f \in N_0$ and $0 = f(0) = f(z_2)$. It follows from $\ell(|z|) \leq |f(z)|$ that $z_2 = 0$. QED.

Comparison theorems for the initial value problem (3.22) also lead to estimates for the modulus of continuity of a function in N and are the basis of an important result of Gehring and Pommerenke [46]. See also [26], [33], and [60].

Theorem 3.4 (Gehring-Pommerenke) *A function $f \in N$ has a continuous extension to $\overline{\mathbf{D}}$. The image $f(\mathbf{D})$ is either a Jordan domain or the image of a parallel strip under a Möbius transformation.*

According to this theorem, if a function in N is not Möbius conjugate to the normalized logarithm L it must map the disk to a Jordan domain. Furthermore, as was pointed out in [25], if a function in N_0 is not Möbius conjugate to L it must be bounded on $\overline{\mathbf{D}}$. These facts are important in many phenomena for the classes N and N_0. We consider L to be *the* extremal in N_0, not ℓ even though $S\ell(z) = -2/(1-z^2)^2$. The minus sign in $S\ell$ makes all the difference in differential equations arguments.

Finally, it is not only comparison theorems, but *convexity* that plays an important role in Gehring and Pommerenke's analysis. In fact, for $f \in N_0$, let $g = f \circ L^{-1}$. Then for $\tau \in \mathbf{R}$ the function $v(\tau) = |g'(\tau)|^{-1/2}$ is convex, with $v(0) = 1$ and $v'(0) = 0$. It is not constant when f is not the log, and so it is bounded below by an affine function. This is the source of many of the estimates, and it holds in greater generality.

Notes: In his original paper [66], Nehari also proved that

$$|Sf| \leq \frac{\pi^2}{2} \tag{3.28}$$

is a sufficient condition for a function to be univalent in $\mathbf{D}$. The constant $\pi^2/2$ is sharp. A normalized extremal is

$$f(z) = \frac{1}{\pi}\tanh(\pi z).$$

Another approach to (3.2) is through the Loewner equation and subordination chains, first introduced by Pommerenke [79] and employed for this purpose by Becker [13]. Becker proved that

$$\left| z\frac{f''}{f'}(z) \right| \leq \frac{1}{1-|z|^2}$$

is a sufficient condition for univalence. The constant 1 is sharp, as was proved later by Becker and Pommerenke [14].

3.3 Quasiconformal Mappings

We recall the Ahlfors-Weill theorem, that if

$$|Sf(z) \leq \frac{2t}{(1-|z|^2)^2} \tag{3.29}$$

for some $0 \leq t < 1$ then f has a $\frac{1+t}{1-t}$-quasiconformal extension to the sphere. One of the most striking aspects of this theorem is that its original proof uses an explicit extension.

Assume at first only $f \in N$ and is not Möbius conjugate to L. Then by the Gehring-Pommerenke result f has a continuous extension to the boundary and maps the disk onto a Jordan domain. Let $\Omega = f(\mathbf{D})$. Next we define a mapping of Ω into $\mathbf{C}$ by

$$\Lambda_f(w) = w + \frac{1}{\partial_w \log \lambda_\Omega(w)} \tag{3.30}$$

$$= f(\zeta) + \frac{(1-|\zeta|^2)f'(\zeta)}{\bar{\zeta} - \frac{1}{2}(1-|\zeta|^2)\frac{f''}{f'}(\zeta)}, \quad w = f(\zeta) \tag{3.31}$$

Here

$$\partial_w = \frac{1}{2}\left(\frac{\partial}{\partial u} - i\frac{\partial}{\partial v}\right).$$

Observe that $|\partial_w \log \lambda_\Omega| = (1/2)|\nabla \log \lambda_\Omega|$. As a consequence of convexity the magnitude of the gradient tends to infinity at the boundary, so Λ_f is the identity there. Also, as a complex number, $1/\partial_w \log \lambda_\Omega$ and $\nabla \log \lambda_\Omega$ point in the same direction, so, using the Poincaré metric as a scale, $\Lambda(w)$ heads away from w as fast as possible.

One can think of the second expression for Λ_f, (3.31), in terms of the Koebe transform of f;

$$g(z) = \frac{f\left(\frac{z+\zeta}{1+\bar{\zeta}z}\right) - f(\zeta)}{(1-|\zeta|^2)f'(\zeta)} = z + \left(\frac{1}{2}(1-|\zeta|^2)\frac{f''}{f'}(\zeta) - \bar{\zeta}\right)z^2 + \cdots,$$

Now, it was shown in [25] that $-2/g''(0) \notin \overline{g(\mathbf{D})}$, and this translates to

$$\Lambda_f(w) \notin \overline{\Omega}. \tag{3.32}$$

Hence Λ_f is a *reflection* in $\partial\Omega$.

Define

$$E_f(z) = \begin{cases} f(z) & \text{for } |z| \le 1, \\ \Lambda_f(f(1/\bar{z})) & \text{for } |z| > 1. \end{cases} \tag{3.33}$$

This is the Ahlfors-Weill extension, though they did not write it in this way.

In [24] it was shown that E_f gives a *homeomorphic* extension of f to the sphere when $f \in N$ and f is not Möbius conjugate to L. To recover the original Ahlfors-Weill theorem, when f satisfies the stronger condition (3.29), compute that the complex dilatation $\mu_{E_f} = \partial_{\bar{z}}E_f/\partial_z E_f$ of E_f at a point ζ in the exterior of the disk is

$$\mu_{E_f}(\zeta) = -\frac{1}{2}\frac{z}{\bar{z}^2}(1-|z|^2)^2 Sf(z), \tag{3.34}$$

where $z = 1/\bar{\zeta}$.

It is interesting to compare this approach with the topological argument to establish injectivity that was used by Ahlfors and Weill. Without having the Gehring-Pommerenke theorem, it is first necessary to dilate to assume that f is analytic and locally univalent on the closed disk. Then if f satisfies (3.29) the formula (3.34) implies that the extension defines E_f as a local homeomorphism of the sphere onto itself (something one *cannot* conclude if $t = 1$). By the monodromy theorem E_f is injective.

There is a significant generalization of the Ahlfors-Weill theorem to quasidisks. To formulate it succinctly we first define a norm by

$$||Sf||_\Omega = \sup_{z\in\Omega} |Sf(z)|\lambda_\Omega(z)^{-2}. \tag{3.35}$$

Following Lehto [60] we then define the *inner radius of univalence* of Ω to be

$$\sigma_I(\Omega) = \sup\{a\colon |Sf| \le a\lambda_\Omega^2 \Longrightarrow f \text{ univalent in } \Omega\}.$$

Thus for the disk, $\sigma_I(\mathbf{D}) = 2$.

Is $\sigma_I(\Omega)$ positive for all domains? A complete answer is provided by two important results.

Theorem 3.5 (Ahlfors, Gehring) *Let Ω be a simply connected domain. Then $\sigma_I(\Omega) > 0$ if and only if Ω be a quasidisk.*

The sufficiency was proved by Ahlfors in 1963 [1] and the necessity by Gehring in 1977 [40].

Ahlfors' proof is to show that if $\partial\Omega$ is a K-quasicircle then there is a constant $a = a(K) > 0$ such that if $|Sf| \le a\lambda_\Omega^2$ then f has a quasiconformal extension to the sphere. This is again done by a direct construction of the extension, similar to (3.33). In this case it is necessary to use a quasiconformal reflection in $\partial\Omega$. He also uses the same topological argument based on the monodromy theorem to obtain injectivity.

It is true that if f is analytic in Ω and $|Sf| < \sigma_I(\Omega)\lambda_\Omega^2$ then f is univalent and has a quasiconformal extension to the sphere. We do not know if Ahlfors' formula provides a homeomorphic extension in the case when $|Sf| \le \sigma_I(\Omega)\lambda_\Omega^2$. One might expect that this is so except when f is an extremal function, meaning that $f(\Omega)$ is not a Jordan domain. But are there extremal functions, and are they unique?

Gehring's proof of the necessity is by contradiction and uses a stronger, quantitative version of the topological notion of local connectivity that he had introduced earlier in studying extensions of quasiconformal mappings in space, [39], see the notes, below.

Gehring and Pommerenke in [46] give a proof of the Ahlfors-Weill theorem using analytic families of injections and the "λ-lemma" of Mañé, Sad and Sullivan; reviewed in Pommerenke's book [81]. Their method of proof leads to the following theorem.

Theorem 3.6 (Gehring-Pommerenke) *Let Ω be a quasidisk and ρ any positive function on Ω. Suppose the condition $|Sf| \le a\rho^2$ implies that f is univalent in Ω. If $0 \le b < a$ and $|Sf| \le b\rho^2$ then f has a quasiconformal extension to $\overline{\mathbf{C}}$.*

Theorem 3.6 shows, in a very general way, the close connection between univalence criteria and quasiconformal extensions. There are many univalence criteria of this form, such as those of Duren and Lehto, [31], or Nehari, [68]. Other generalizations are discussed in Section 4.1.

Notes: Another construction of the Ahlfors-Weill extension is due to Thurston and uses the idea of best Möbius approximation. This approach also provides another definition of the Schwarzian. If f is analytic at a point z_0 then there is a unique Möbius transformation $M_{z_0} = M(f, z_0)$ that agrees with f to second order at z_0, that is,

$$M_{z_0}(z_0) = f(z_0),\ M'_{z_0}(z_0) = f'(z_0),\ M''_{z_0}(z_0) = f''(z_0).$$

Then

$$({M_{z_0}}^{-1} \circ f)(z) = z + \frac{1}{6} Sf(z_0)(z - z_0)^3 + O((z - z_0)^4).$$

Now Möbius transformations are defined everywhere, and if f satisfies the Ahlfors-Weill condition then the extension E_f to the sphere is defined by

$$E_f(z) = \begin{cases} f(z), \ |z| \le 1 \\ M(f, 1/\bar{z})(z), \ |z| > 1. \end{cases}$$

The idea of using Möbius approximations to prove injectivity results is in an important paper of Martio and Sarvas [61]. Thurston applied this definition of the Schwarzian in [89] to questions in Teichmüller theory, see the notes following Section 3.5. We also refer to the paper of Velling [90].

Lehto also defines an *outer radius of univalence*,

$$\sigma_O(\Omega) = \sup\{||Sf||_\Omega : f \quad \text{univalent in} \quad \Omega\}.$$

Many results can be expressed very elegantly in terms of the inner and outer radii of univalence and related domain constants. We also mention [72] which settles in the negative a question raised in [60] related to the Schwarzian distance between domains.

Gehring's local connectivity condition reads as follows:

Let $B(z_0, r) = \{z : |z - z_0| < r\}$. A set $E \subset \overline{\mathbf{C}}$ is b-locally connected, $b > 1$, if for all z_0 and r, points in $E \cap \overline{B}(z_0, r)$ can be joined in $E \cap \overline{B}(z_0, br)$, and points in $E \backslash B(z_0, r)$ can be joined in $E \backslash B(z_0, r/b)$. One says that E is *linearly locally connected* if it is b-locally connected for some b.

In [40] he proved that a simply connected domain that is linearly locally connected is a quasidisk. The converse is also true but requires several intermediate results to form a complete chain of implications, see [43]. A parallel strip is the simplest example of a domain that is *not* linearly locally connected.

The inner and outer radius, linear local connectivity, *et al* can be defined in the same manner for multiply connected domains, and this is actually the setting of Gehring's theorem: if $\sigma_I(\Omega) > 0$ then Ω is locally linearly connected and each boundary component is either a point or a quasicircle. Conversely, it was shown in [70] that $\sigma_I(\Omega) > 0$ for any *finitely* connected domain whose boundary components are quasicircles or points. There are infinitely connected circle domains for which $\sigma_I(\Omega) > 0$, [70] and for which $\sigma_I(\Omega) = 0$, [12]. See the notes at the end of the next section for some additional comments.

There are corresponding theorems on quasiconformal extensions in the multiply connnected case, but the existence of the extension follows from general theorems. What is missing is some kind of explicit construction.

There was a period when Fred Gehring made a hobby out of collecting characteristic properties of quasidisks. It is remarkable in how many areas of analysis they arise. The situation up to around 1981 is described in his lectures [43] where he has seventeen different characterizations.

At least one more recent addition to that list is a property in terms of f''/f'. It is easy to show from Ahlfors' theorem on the Schwarzian that if Ω is a K-quasidisk then there is a constant $a = a(K)$ such that $|f''/f'| \leq a\lambda_\Omega$ implies f is univalent. The converse is difficult and was established only in 1986 by Astala and Gehring, [9].

Another active area of interest has been the study of conformal mappings of the disk with quasiconformal extensions as a subclass of the class of univalent functions; see [60].

For variations and extensions of Gehring and Pommerenke's work see [25], the papers of Steinmetz [86], [87], and the paper of Chuaqui [21].

3.4 Mapping Properties

In this section we want to take a closer look at some special properties of functions satisfying Nehari's condition (3.2). Some examples mentioned earlier are the distortion and growth results (3.26), (3.27) in Section 3.2 and Theorem 3.4 of Gehring and Pommerenke on extending a map to the boundary. A much older result is a theorem of Paatero [75] from 1931, who proved that a conformal mapping of the disk onto a convex domain satisfies (3.2). (In case the reader was ever worried, this at least shows that there are plenty of functions satisfying the condition.) Paatero's result was proved in a different way by Nehari in 1976, [67]. Nehari also showed that a bounded convex map satisfies the Ahlfors-Weill condition.

We turn again to quasidisks. The Ahlfors-Weill theorem says that a function satisfying

$$|Sf(z)| \leq \frac{2t}{(1-|z|^2)^2}$$

maps the disk onto a quasidisk, and the Gehring-Pommerenke theorem says that if

$$|Sf(z)| \leq \frac{2}{(1-|z|^2)^2}$$

and if f is not conjugate to the logarithm L then, at least, f maps the disk onto a Jordan domain. What is the gap between these results? To what extent can the image of a map satisfying just Nehari's bound fail to be a quasidisk?

Some partial answers are given in [27]. To discuss them we split the definition of a quasidisk into two parts. Geometrically, there are two things a quasidisk cannot have, an inside cusp or an outside cusp. And there are two independent conditions, each one of interest, that rule these out.

A simply connected domain Ω is a *John domain* (for Fritz John) if it is bounded and there exists a constant $c > 0$ such that for any crosscut C of Ω we have

$$\operatorname{diam} H \leq c \operatorname{diam} C$$

for one of the components H of $\Omega \backslash C$. Here diam denotes the euclidean diameter. Ω is *linearly connected* (*n.b.* not linearly *locally* connected) if there is a constant $b > 0$ such that any two points z_1, z_2 in Ω can be joined by a curve $\gamma \subset \Omega$ with

$$\operatorname{diam} \gamma \leq b|z_1 - z_2|.$$

One then has that a bounded quasidisk is a linearly connected, John domain, and conversely. We refer to [81] for this result, and also to [45] for the definition of a John domain given here, which is also not the original definition.

The John condition prohibits outward pointing cusps and the linearly connected condition prohibits inward pointing cusps. The main phenomenon is that for functions satisfying Nehari's condition, one of these comes for free, [27].

Theorem 3.7 *If $f \in N$ and if $f(\mathbf{D})$ is a John domain then $f(\mathbf{D})$ is a quasidisk.*

Complementing this result are examples in [27] that establish the following.

Theorem 3.8 *There exist functions $f, g \in N_0$ such that $f(D)$ is linearly connected but not John and such that $g(\mathbf{D})$ is not linearly connected.*

It is also shown that if $\Omega = f(\mathbf{D})$, $f \in N$, is not a John domain then Ω contains an 'infinitesimal parallel strip', see also [91].

Notes: A different way of putting together conditions like linearly connected and John is the notion of a *uniform domain*, introduced by Martio and Sarvas in [61]. The definition is in terms of euclidean lengths of curves rather than their diameters. A domain Ω (not assumed to be simply connected) is *uniform* if there exist constants a and b such that each $z_1, z_2 \in \Omega$ can be joined by an arc α in Ω with

$$\text{length}(\alpha) \leq a|z_1 - z_2|,$$

and

$$\min_{j=1,2} \text{length}(\alpha_j) \leq b d(z, \partial\Omega),$$

for $z \in \alpha$, where α_1, α_2 are the components of $\alpha \backslash \{z\}$. This formulation of the definition is in [44].

If a domain is uniform then each component of $\partial\Omega$ is either a point or a quasicircle, but not conversely. Martio and Sarvas proved that if Ω is uniform then its inner radius of univalence $\sigma_I(\Omega) > 0$, but an example in [44] shows that the converse is not true. There is now an extensive literature on uniform domains.

Convexity plays an important role in the analysis in [27]. There it is shown that if a function f in the disk satisfies Nehari's condition (3.2) then $\lambda_\Omega^{1/2}$ is a hyperbolically convex function on $\Omega = f(\mathbf{D})$. Moreover, this becomes a characteristic property of functions satisfying (3.2) if one allows for shifting the range by Möbius transformations. See the paper of Kim and Minda [56] for some additional convexity properties of the Poincaré metric, and also the paper of Yamashita [92] for some connections to Bloch functions.

3.5 Teichmüller Spaces

The results in the previous sections can be seen as natural generalizations of classical theorems on univalent functions. For the most part, however, the impetus for the development came from Teichmüller theory. The complete story is quite involved, partly because there are two possible plot lines. One emphasizes Riemann surfaces and quasiconformal mappings and one emphasizes Fuchsian and Kleinian groups (and quasiconformal mappings). Fortunately, to spotlight Gehring's contributions we can stick with the universal Teichmüller space, and this cuts the prerequisites to a minimum.

In 1978, Gehring [42] settled a conjecture of Bers in the negative by proving the following theorem on conformal mappings of plane domains.

Theorem 3.9 (Gehring) *There is a simply connected domain Ω and a positive number δ such that $f(\Omega)$ is not a Jordan domain whenever f is conformal in Ω with $||Sf||_\Omega \le \delta$.*

Bers' conjecture appears in his survey paper [16] in 1972, but he had stated the problem earlier. It arises through the Bers imbedding of universal Teichmüller space into the space of Schwarzians of conformal mappings. We refer again to Lehto's book [60], and also to the article by Earle [32] for a very clear and concise coverage of the main ideas.

The domain Ω is given quite explicitly as the complement of a pair of spirals starting from $\pm i$, respectively, and spiraling into the origin. Furthermore Ω contains two disjoint spirals α, β starting at ± 1, respectively, and spiraling into the origin from opposite sides of $\partial\Omega$. Let f be a conformal mapping of Ω fixing -1, 1 and ∞. If the Schwarzian is sufficiently small, $||Sf||_\Omega \le \delta$ for a suitable δ, then the images $f(\alpha)$ and $f(\beta)$ spiral into two points w_1 and w_2, and distortion theorems for *quasiconformal* mappings force $w_1 = w_2$. Hence $f(\Omega)$ is not a Jordan domain.

Why spirals? For the class of spirals Gehring uses the curvature and its derivative are controlled, and the same is true for images under a conformal mapping with small Schwarzian. This hinges on a formula relating curvatures to the pre-Schwarzian and derivatives of curvatures to the Schwarzian. For this, see exercise 3 on page 21 in Ahlfors' book on conformal invariants [2]. Such formulas go back at least to E. Cartan [18]. They hold in greater generality, see [74].

We return to the connection with Teichmüller theory. Let $\mathbf{L}$ denote the lower half-plane. Let $B(1)$ be the space of analytic functions φ on $\mathbf{L}$ for which the norm

$$||\varphi||_{\mathbf{L}} = \sup_{z\in\mathbf{L}} 4y^2|\varphi(z)| < \infty, \quad z = x + iy.$$

We let

$$U = \{\varphi : \varphi = Sf, \quad f \text{ univalent in } \mathbf{L}\},$$

and

$$T(1) = \{\varphi : \varphi = Sf, f \quad \text{univalent in } \mathbf{L} \text{ with a quasiconformal extension to } \overline{\mathbf{C}}\}.$$

$T(1)$ is called the *universal Teichmüller space*. In all cases the 1 refers to the fact that the definitions involve the trivial group $G = \{1\}$.

Obviously $T(1) \subset U$. Ahlfors' theorem, that Ω is a quasidisk implies $\sigma_I(\Omega) > 0$, has as a consequence that $T(1)$ is an open subset of U. Gehring's converse to Ahlfors' theorem implies that $T(1)$ is the *interior* of U. Bers asked if it is true that U is the closure of $T(1)$, and Gehring's theorem above says that this is not so. We give the argument for this, following Gehring's own exposition in his survey paper [41].

Suppose Theorem 3.9 holds. Let $\varphi = Sg$ where g is a conformal mapping of $\mathbf{L}$ onto the domain Ω in the theorem. Take any $\psi = Sh \in U$ with $||\varphi - \psi||_{\mathbf{L}} \leq \delta$, where δ is as in the theorem. The function $f = h \circ g^{-1}$ is a conformal mapping of Ω. The chain rule (1.8) for the Schwarzian together with the conformal invariance of the Poincaré metric imply that

$$||Sf||_{\Omega} = ||Sh - Sg||_{\mathbf{L}} = ||\psi - \varphi||_{L} \leq \delta.$$

Hence $h(\mathbf{L}) = f(\Omega)$ is not a Jordan domain, in particular not a quasidisk, and h does not have a quasiconformal (or even a homeomorphic) extension to the plane. Therefore $\psi \notin T(1)$, that is, the distance from φ to $T(1)$ is at least δ and so U is not the closure of $T(1)$.

Notes: Notice that Gehring's example of a mapping h with $Sh \notin \overline{T(1)}$ has $h(\mathbf{L})$ not a Jordan domain. In [38], Flinn gave an example where $h(\mathbf{L})$ is Jordan and $Sh \notin \overline{T(1)}$.

Especially interesting is Thurston's approach to the problem in [89]. In a much different way he proves that U has components that reduce to a point. Another proof of Thurston's theorem is in Astala's paper [8].

For recent work on the connections between univalent functions, Teichmüller theory and other areas of complex analysis, see [9], [10], and [49]. The Astala-Gehring paper [9] on f''/f' gives another model of universal Teichmüller space.

4 The Post-Modern, Abstract Expressionist, Neo-Classical Period

We want to discuss a generalization of the Schwarzian in the setting of an n-dimensional Riemannian manifold (M, g), $n \geq 2$. Let ∇ be the Riemannian connection, or covariant derivative, for the metric $g = \langle\ ,\ \rangle$. For a smooth function φ we define a symmetric, traceless 2-tensor

$$B_g(\varphi) = \text{Hess}(\varphi) - d\varphi \otimes d\varphi - \frac{1}{n}\left\{\Delta\varphi - ||\text{grad}\varphi||^2\right\} g,$$

where $\text{Hess}(\varphi)$ is the Hessian of φ. The Hessian, gradient, and Laplacian are taken with respect to the metric g. In terms of vector fields X, Y on M,

$$\begin{aligned} B_g(\varphi)(X,Y) &= X(Y\varphi) - (\nabla_X Y)(\varphi) - (X\varphi)(Y\varphi) \\ &\quad - \frac{1}{n}\left\{\Delta\varphi - ||\text{grad}\varphi||^2\right\}\langle X, Y\rangle. \end{aligned}$$

We always consider $B_g(\varphi)$ in connection with a conformal metric $\hat{g} = e^{2\varphi}g$ and we call it the *Schwarzian tensor* [74]. If $f\colon(M,g) \to (M',g')$ is a conformal local diffeomorphism with $f^*g' = e^{2\varphi}g$, $\varphi = \log\|df\|$, then we define the Schwarzian of f to be

$$\mathcal{S}(f) = B_g(\varphi).$$

Before going on, if f is an ordinary analytic function and g is the euclidean metric, then $\mathcal{S}(f) = B(\log|f'|)$. Computing in standard coordinates one easily finds

$$\begin{pmatrix} \operatorname{Re}\mathcal{S}(f) & -\operatorname{Im}\mathcal{S}(f) \\ -\operatorname{Im}\mathcal{S}(f) & -\operatorname{Re}\mathcal{S}(f) \end{pmatrix},$$

when we represent the tensor as 2 x 2 matrix.

The tensor $B_g(\varphi)$ has a 'warped' additivity property when there is a conformal change in the background metric, namely

$$B_g(\varphi+\sigma) = B_g(\varphi) + B_{\hat{g}}(\sigma), \tag{4.1}$$

where, as above, $\hat{g} = e^{2\varphi}g$. This is exactly the transformation formula:. If

$$(M,g) \xrightarrow{h} (M',g') \xrightarrow{f} (M'',g''),$$

then, suppressing the dependence on the several metrics,

$$\mathcal{S}(f\circ h) = h^*\mathcal{S}(f) + \mathcal{S}h. \tag{4.2}$$

One can see why the *square* of the derivative appears in the classical formula

$$S(f\circ h) = (S(f)\circ h)(h')^2 + Sh,$$

because when the action of the pullback by h is computed on the tensor $\mathcal{S}(f)$ we apply the pushforward h_* on a *pair* of vector fields. However, there is even a better reason why the transformation formula reads as it does which we will discuss later.

There are two things about the generalization to emphasize. First, it makes sense to talk about the Schwarzian of a conformal metric as well a conformal mapping. We say that a conformal mapping of a manifold to itself, or between manifolds, is Möbius if its Schwarzian vanishes, and we use the same term for conformal metrics on a single manifold. On the sphere S^n with its standard metric $g_0 = (1+|x|^2)^{-2}|dx|^2$ the Möbius metrics $\hat{g} = e^{2\varphi}g_0$ are obtained by solving the equation

$$B_g(\varphi) = 0.$$

The solutions are easily found to be

$$\varphi(x) = -\log\left(\frac{A|x|^2 + B\cdot x + C}{|x|^2+1}\right). \tag{4.3}$$

It is extremely helpful to have such a simple and explicit family of metrics on the sphere.

Second, because the Schwarzian is computed with respect to a background metric, there is a new role for the transformation formula when there is a conformal change in the background metric. For example, the Poincaré metric on the disk is a Möbius metric (with respect to the euclidean metric). Hence for a conformal mapping f of the disk into the euclidean plane

$$\mathcal{S}_{hyp} = \mathcal{S}_{euc}.$$

There is a second order ordinary differential equation associated with the Schwarzian tensor. Just as in the classical case we regard

$$B_g(\varphi) = q$$

as a Riccati equation, where q is a traceless 2-tensor. The substitution $v = e^{-\varphi}$ turns this into the linear equation

$$\operatorname{Hess} v + uq = \frac{1}{n}(\Delta v)g.$$

Along a unit speed geodesic $\gamma(t)$ this becomes

$$v'' + vq(\gamma', \gamma') = \frac{1}{n}\Delta v.$$

Notes: The Schwarzian tensor has connections to basic notions in differential geometry through the Ricci curvature. It determines how the trace free part of the Ricci tensor changes when there is a conformal change of metric. So, for instance, if (M, g) is an Einstein manifold and if $\hat{g} = e^{2\varphi}g$, then $(M, \hat{g})$ is an Einstein manifold if and only if $B(\varphi) = 0$.

There are also many formulas in higher dimensions analogous to those relating the Schwarzian derivative to curvature of 1-dimensional curves. All of this is covered in [74], along with a discussion of the Möbius group of a Riemannian manifold.

A generalization of the Schwarzian based on the idea of Möbius approximation has been given by Carne in [17]. In [3], Ahlfors considered separate generalizations of the real and imaginary parts of the Schwarzian to non-conformal diffeomorphisms of $\mathbf{R}^n$. The basis for his generalization of the imaginary part is the connection with curvature and for the real part it is an interesting relation between the Schwarzian and cross ratio. If f is analytic and a, b, c, d are distinct points in $\mathbf{C}$ then

$$(f(z+ta), f(z+tb), f(z+tc), f(z+td)) = \\ (a,b,c,d)\{1 + \frac{1}{6}(a-b)(c-d)Sf(z)t^2 + O(t^3)\}.$$

Thus to terms of order t^2 the Schwarzian measures the infinitesimal distortion of cross ratio. This is one of the ways that the Schwarzian finds its way into one-dimensional dynamics.

For functions of several complex variables, FitzGerald and Gong proposed a definition in [37]. Molzon and Mortenson have developed a different approach in several interesting papers [65], [62], [64], [63]. We also mention two papers of Yoshida [93], [94] on reducing a system of differential equations to a canonical form, similar in spirit to the birth of the Schwarzian in the Classical Period.

4.1 Generalizations of Sufficient Conditions for Univalence

There is a natural generalization of Nehari's univalence criterion in terms of $\mathcal{S}f$, [73].

Theorem 4.1 *Let $(M.g)$ be a Riemannian manifold of dimension $n \geq 2$ and $f\colon (M,g) \longrightarrow (\mathbf{R}^n, |dx|^2)$ be a conformal local diffeomorphism. Suppose that the scalar curvature of g is bounded above by $n(n-1)K$, for some $K \in \mathbf{R}$ and that any two points in M can be joined by a geodesic of length $< \delta$, for some $0 < \delta \leq \infty$. If*

$$\|\mathcal{S}(f)\| \leq \frac{2\pi^2}{\delta^2} - \frac{1}{2}K \tag{4.4}$$

then f is injective.

The norm of the Schwarzian is computed by regarding $\mathcal{S}(f)$ as a bilinear form on each tangent space. One can also state the theorem for a conformal local diffeomorphism from M into the sphere S^n with its standard metric.

An interesting feature is that there is both a curvature and a diameter term in the bound for the Schwarzian. When the metric g is complete one can take $\delta = \infty$, in which case it makes sense to consider only metrics with nonpositive scalar curvature.

We show how the theorem can be applied to obtain other univalence criteria. Recall first the Nehari conditions

$$|\mathcal{S}f(z)| \leq \frac{2}{(1-|z|^2)^2} \quad \text{or} \quad |\mathcal{S}f(z)| \leq \frac{\pi^2}{2}.$$

We apply the theorem with $M = \mathbf{D}$. First take the background metric to be the Poincaré metric. Then

$$g = \frac{|dz|^2}{(1-|z|^2)^2}, \quad K = -4,\ \delta = \infty.$$

Because the Poincaré metric is a Möbius metric we get the euclidean Schwarzian, and in computing the norm we get two factors of $(1-|z|^2)$.

The condition becomes $(1-|z|^2)^2|Sf(z)| \le 2$. Next we take the background metric to be the euclidean metric. Then $K = 0$ and $\delta = 2$, and the condition is $|Sf| \le \pi^2/2$.

Essentially all of the known univalence criteria for functions analytic in the disk follow from the general theorem by choosing different background metrics and appealing to the rules (4.1) or (4.2). This has been studied by M. Chuaqui in [22]. As an example, one can obtain the very general criterion of Epstein, [33]:

$$\left|Sf(z) - 2(\tau_{zz} - \tau_z^2)(z) + \frac{4\bar{z}\tau_z(z)}{1-|z|^2}\right| \le 2\frac{(1+(1-|z|^2)^2)\tau_{z\bar{z}}(z)}{(1-|z|^2)^2}. \tag{4.5}$$

In this case the metric to take in Theorem 4.1 is $e^{2\tau}|dz|^2/(1-|z|^2)^2$, where τ is a real-valued function satisfying some mild extra conditions.

The proof of Theorem 4.1 again uses comparison theorems for differential equations, but only very simple ones. The basic fact one needs is that an upper bound for the Schwarzian tensor leads to a lower bound for the Hessian of an associated function. Recall that $f\colon (M,g) \longrightarrow (\mathbf{R^n}, |dx|^2)$. Let $f^*g_0 = e^{2\varphi}g$ and set

$$u = e^{-\varphi/2}. \tag{4.6}$$

A calculation shows that if Sf satisfies (4.4) then

$$\operatorname{Hess} u + \frac{\pi^2}{\delta^2}ug \ge 0. \tag{4.7}$$

Injectivity can be deduced by using this equation along geodesics and ruling out the possibility that solutions vanish twice, but we refer to [73] for an explanation on how the initial value problem is set up.

In the 2-dimensional case of a conformal mapping of a plane domain, $\varphi = \log|f'|$ and $u = (f')^{-1/2}$, which is the substitution (3.23) we used in Section 3.2. Notice also that in the case of a complete metric, so $\delta = \infty$, (4.7) is the statement that u is convex.

It is interesting to see how much more of the geometric function theory from the 'Modern Period' carries over. As a first step, the paper [26] studies some of the geometry of functions f satisfying (4.4) on the disk $\mathbf{D}$.

Notes: Special mention must be made of Epstein's methods [33], [34], [35] in bringing many new geometric ideas to the original 2-dimensional case. He regards the disk and its image under a map f as lying on the boundary of hyperbolic 3-space $\mathbf{H}^3$, and constructs surfaces in $\mathbf{H}^3$ over these regions as envelopes of horospheres. The function τ in (4.5) is related to the support function of the horospheres, and the condition (4.5) with a $2t$ instead of a 2 on the right hand side implies that the principal curvatures of these surfaces are less than 1. From this one knows enough about how the surfaces are embedded to define a hyperbolic Gauss map and use it to reflect in the surfaces to obtain quasiconformal extensions. For versions of Epstein's constructions in higher dimensions see [20].

Pommerenke gave a self-contained classical approach to the Epstein criterion in [80]. Anderson and Hinkkanen [6] were likewise able to sidestep the geometry and prove even a more general version of the criterion allowing for complex values of some of the parameters. Theirs is truly the most general criterion. It is curious that neither Epstein's method itself nor the way of deriving (4.5) from Theorem 4.1 seem to be able to incorporate such complex parameters.

Becker's method of using subordination chains for proving univalence was considered by Pfaltzgraff for mappings of $\mathbf{C}^n$ in [76] and [77]. He also uses subordination chains to derive further criteria for quasiconformal extensions in the plane case, [78].

We know almost nothing about possible necessary conditions for injectivity of conformal mappings of Riemannian manifolds. However, some work of D. Stowe and the author suggests that there is an approach via a generalizatiuon of the area theorem.

4.2 The Möbius Connection

So much for some of the Post-Modern news. We end by venturing briefly into the deep and still largely uncharted waters of Abstract Expressionism. We have now come to feel that the most geometric setting for the Schwarzian, in two dimensions and up, is as a measurement of the difference between two natural connections (in the differential geometric sense) when there is a conformal change of metric. To say everything very precisely requires quite a bit of preparation, but I would like to try to get across some of the geometric ideas. An exposition of conformal connections, as a special case of Cartan connections, is presented in Kobayashi's book, [57].

This section represents work with D. Stowe, and it is still very much 'in progress'. For more details, but certainly not the final word, we refer to a write-up of the author's lecture [69].

The setting is the bundle of conformal 2-jets from S^n with its standard metric g_0 to a manifold of the same dimension. A 2-jet carries the information: source point, target point, first and second derivatives at the source point. This much makes sense for any smooth manifold. When the manifold is Riemannian one can speak of conformal 2-jets. First, a 2-jet is invertible if it is the 2-jet of a local diffeomorphism. Next, we say an invertible 2-jet $f^{[2]}$ from (S^n, g_0) into (M, g) is conformal if $f^*g = e^{2\varphi}g_0$ on the level of 1-jets. Consider the fiber of 2-jets over a point $y \in M$. The Möbius group of the sphere, $\text{Möb}(S^n)$, acts on a 2-jet on the right by composition to move the source points, but it leaves the fiber over y invariant. So the bundle P of invertible conformal 2-jets is a principal $\text{Möb}(S^n)$-bundle, and we want to talk about connections in P. This only depends on the conformal structure of M.

A connection allows one to define horizontal vectors and horizontal lifts to P of curves in M. Take a curve β in M and a horizontal lift to P. Such a horizontal lift is a 1-parameter family of conformal 2-jets from S^n to M, and so we have a corresponding curve α in S^n consisting of the source points. One says that α *develops* onto β. Think of this as a way of rolling the sphere on M along α in S^n and β in M. One of the defining conditions of a conformal connection is that the sphere should roll without slipping.

One naturally looks for conformal connections with special properties by making restrictions, perhaps on how curves develop, or as is more commonly done, on the curvature form of the connection. One notable example is the Weyl conformal connection which depends on the conformal structure on M and not on the particular Riemannian metric. This connection results from asking that certain contractions of the curvature form vanish (Ricci contraction of the linear part of the curvature form), and it only works in dimensions 3 and up.

We were quite surprised that one gets very nice conditions by asking for the decidedly Neo-Classical condition that circles develop onto circles. A circle on M is a submanifold whose second fundamental form has vanishing covariant differential.

Theorem 4.2 *There is a unique connection in the principal bundle P for which circles develop to circles and for which the scalar contraction of the linear part of the curvature form is zero.*

We call this connection the *Möbius connection.* If there is a conformal change of metric the Möbius connection changes. The Schwarzian measures this change.

Theorem 4.3 *Let ω be the Möbius connection for (M,g) and ω' the Möbius connection for (M',g'). If $f\colon(M,g,\omega) \longrightarrow (M',g',\omega')$ is a conformal local diffeomorphism then $f^*\omega' - \omega = \mathcal{S}f$.*

This is an abuse of notation. A more careful statement of the theorem involves use of a special section constructed from the 2-jet of the exponential map to pullback connections from P to M.

The general chain rule (4.2) now comes with no calculation. If

$$(M,g,\omega) \xrightarrow{h} (M',g',\omega') \xrightarrow{f} (M'',g'',\omega'')$$

then

$$\begin{aligned} \mathcal{S}(f\circ h) &= (f\circ h)^*\omega'' - \omega = h^*(f^*\omega'') - \omega \\ &= h^*(\mathcal{S}(f) + \omega') - \omega \\ &= h^*\mathcal{S}(f) + h^*\omega' - \omega \\ &= h^*\mathcal{S}(f) + \mathcal{S}h. \end{aligned}$$

Further applications of the Möbius connection are given in [69], but there is much more to do.

Notes: In his lectures notes on differential equations, [29], Deligne gets the classical Schwarzian in a way similar to Theorem 4.3 by considering projective connections. Part of our purpose is to stick with the Riemannian setting.

In two other interesting papers, [82], [50], Schiffer and Hawley also want to consider the Schwarzian as a kind of connection, but they use the term somewhat informally. They seek to define covariant derivatives on a Riemann surface and they speak of ' ... an entity which transforms under a linear but non-homogeneous law if the coordinates are changed.' They have in mind the chain rule for the Schwarzian. We repeat that from our point of view the Schwarzian is not a connection, but it is, exactly, a difference of connections.

Following up on the Schiffer and Hawley papers is a paper of Hejhal [51], which brings back differential equations and monodromy groups.

For our part, our fondest hope is to work our way back through the Modern Period into the Classical Period.

REFERENCES

[1] L. Ahlfors. Quasiconformal reflections. *Acta math.*, 109:291–301, 1963.

[2] L. Ahlfors. *Conformal Invariants: Topics in Geometric Function Theory.* McGraw Hill, New York, 1973.

[3] L. Ahlfors. Cross-ratios and Schwarzian derivatives in $\mathbf{R}^n$. In *Complex Analysis:Articles Dedicated to Albert Pfluger on the Occasion of his 80th Birthday*, pages 1–15, Boston, 1989. Birkhauser.

[4] L. Ahlfors and A. Beurling. Conformal invariants and function theoretic null sets. *Acta Math.*, 83:101–129, 1950.

[5] L. Ahlfors and G. Weill. A uniqueness theorem for Beltrami equations. *Proc. Amer. Math. Soc.*, 13:975–978, 1962.

[6] J. M. Anderson and A. Hinkkanen. Univalence criteria and quasiconformal extensions. *Trans. Amer. Math. Soc.*, 324:823–842, 1991.

[7] V. I. Arnold. *Geometrical methods in the theory of ordinary differential equations.* Springer-Verlag, New York, 1983.

[8] K. Astala. Self-similar zippers. In *Holomorphic functions and moduli, vol. 1, Math. Sci. Res. Inst. Pub. 10*, pages 61–73, New York, 1988. Springer.

[9] K. Astala and F. W. Gehring. Injectivity, the BMO norm, and universal Teichmüller space. *Jour. d'Analyse Math.*, 46:16–57, 1986.

[10] K. Astala and M. Zinsmeister. Teichmüller spaces and BMOA. *Math. Ann.*, 289:613–625, 1991.

[11] F. Baldassari and B. Dwork. On second order linear differential equations with algebraic solutions. *Amer. Jour. Math.*, 101:42–76, 1979.

[12] A. Beardon and F. W. Gehring. Schwarzian derivatives, the Poincaré metric and the kernel function. *Comment. Math. Helv.*, 55:50–64, 1980.

[13] J. Becker. Löwnersche Differentialgleichung und quasiconform fortsetzbare schlichte Funktionen. *Jour. für die Reine und Angew. Math.*, 255:23–43, 1972.

[14] J. Becker and Ch. Pommerenke. Schlichtheitskriterien und Jordangebiete. *Jour. für die Reine und Angew. Math.*, 354:74–94, 1984.

[15] S. Bergman and M. Schiffer. The kernel function and conformal mapping. *Compositio Math.*, 8:205–249, 1951.

[16] L. Bers. Uniformization, moduli, and Kleinian groups. *Bull. London Math. Soc.*, 4:257–300, 1972.

[17] K. Carne. The Schwarzian derivative of a conformal mapping. *Jour. für die Reine und Angew. Math.*, 408:10–33, 1990.

[18] E. Cartan. *Lecons sur la Theorie des Espaces à connexion Projective.* Gauthier-Villars, Paris, 1937.

[19] A. Cayley. On the Schwarzian derivative and the polyhedral functions. *Trans. Camb. Phil. Soc.*, 13, 1880.

[20] M. Chuaqui. The Schwarzian derivative and quasiconformal reflections on S^n. *Ann. Acad. Sci. Fenn.*, 17:315–326, 1992.

[21] M. Chuaqui. On a theorem of Nehari and quasidiscs. *Ann. Acad. Sci. Fenn.*, 18:117–124, 1993.

[22] M. Chuaqui. A unified approach to univalence criteria in the disk and simply connected domains. *Proc. Amer. Math. Soc.*, 123:441–453, 1995.

[23] M. Chuaqui and B. Osgood. Sharp distortion theorems associated with the Schwarzian derivative. *Jour. London Math. Soc.(2)*, 48:289–298, 1993.

[24] M. Chuaqui and B. Osgood. Ahlfors-Weill extensions and critical points of the Poincaré metric. *Comment. Math. Helv.*, 69:659–668, 1994.

[25] M. Chuaqui and B. Osgood. An extension of a theorem of Gehring and Pommerenke. *Israel Jour. Math.*, pages 393–407, 1995.

[26] M. Chuaqui and B. Osgood. General univalence criteria in the disk: extensions and extremal functions. to appear.

[27] M. Chuaqui, B. Osgood, and Ch. Pommerenke. John domains, quasidisks, and the Nehari class. *Jour. für die Reine und Angew. Math.*, to appear.

[28] W. de Melo and S. van Strien. A structure theorem in one dimensional dynamics. *Annals of Math.*, 129:519–546, 1989.

[29] P. Deligne. *Equations differentielles à points singuliers réguliers*, volume 163 of *Springer Lecture Notes.* Springer Verlag, New York, 1970.

[30] P. Duren. *Univalent Functions.* Springer-Verlag, New York, 1983.

[31] P.L. Duren and O. Lehto. Schwarzian derivatives and homeomorphic extensions. *Ann. Acad. Sci. Fenn. Ser. A 1 Math.*, 477:3–11, 1970.

[32] C. Earle. Teichmüller theory. In *Discrete groups and automorphic forms.* Academic Press, New York, 1977.

[33] C. Epstein. The hyperbolic Gauss map and quasiconformal reflections. *J. für die reine u. angew. Math.*, 372:96–135, 1986.

[34] C. Epstein. The asymptotic boundary of a surface in $\mathbf{H}^3$ with nonnegative curvature. *Michigan Math. J.*, 34:227–239, 1987.

[35] C. Epstein. Univalence criteria in hyperbolic space. *J. für die reine u. angew. Math.*, 380:196–214, 1987.

[36] M. Essén and F. Keogh. The Schwarzian derivative and estimates of functions analytic in the unit disk. *Math. Proc. Camb. Phil. Soc.*, 78:501–511, 1975.

[37] C. FitzGerald and S. Gong. The Schwarzian derivative in several complex variables. *to appear.*

[38] B. Brown Flinn. Jordan domains and the universal Teichmüller space. *Trans. Amer. Math. Soc.*, 282:603–610, 1984.

[39] F. W. Gehring. Extensions of quasiconformal mappings in three space. *Jour. d'Analyse*, 14:171–182, 1965.

[40] F. W. Gehring. Univalent functions and the Schwarzian derivative. *Comment. Math. Helv.*, 52:561–572, 1977.

[41] F. W. Gehring. Remarks on the universal Teichmüller space. *L'Enseigment Math.*, 24:173–178, 1978.

[42] F. W. Gehring. Spirals and the universal teichmüller space. *Acta Math.*, 141:99–113, 1978.

[43] F. W. Gehring. *Characteristic properties of quasidosks*, volume 84. Presses Univ. Montreal, Montreal, 1982.

[44] F. W. Gehring and B. Osgood. Uniform domains and the quasihyperbolic metric. *Jour. d'Analyse Math.*, 36:50–74, 1979.

[45] F.W. Gehring, K. Hag, and O. Martio. Quasihyperbolic geodesics in John domains. *Math. Scand.*, 65:72–92, 1989.

[46] F.W. Gehring and Ch. Pommerenke. On the Nehari univalence criterion and quasicircles. *Comment. Math. Helv.*, 59:226–242, 1984.

[47] E. Goursat. *Lecons sur les series hypergeometriques et sur quelques fonctions qui s'y rattachent, vol. 2 Intégrales Algébrique.* Hermann, Paris, 1938.

[48] J. J. Gray. Fuchs and the theory of differential equations. *Bull. Amer. Math. Soc.*, 10:1–26, 1984.

[49] D. H. Hamilton. The closure of Teichmüller space. *Jour. d'Analyse Math.*, 55:40–50, 1990.

[50] N. Hawley and M. Schiffer. Connections and conformal mapping. *Acta Math.*, 107:175–274, 1962.

[51] D. Hejhal. Monodromy groups and linearly polymorphic functions. *Acta Math*, 135:1–55, 1975.

[52] E. Hille. Remarks on a paper by Zeev Nehari. *Bull. Amer. Math. Soc.*, 55:552–553, 1949.

[53] E. Hille. *Ordinary differential equations in the complex domain.* John Wiley and Sons, New York, 1976.

[54] J. A. Jenkins. *Univalent functions and conformal mapping.* Springer-Verlag, New York, 1965.

[55] N. Katz. Algebraic solutions of differential equations (p curvature and hodge filtration). *Invent. math.*, 18:1–118, 1972.

[56] S.-A. Kim and D. Minda. The hyperbolic and quasihyperbolic metrics in convex regions. *J. Analysis*, 1:109–118, 1993.

[57] S. Kobayashi. *Transformation groups in differential geometry.* Springer Verlag, New York, 1972.

[58] W. Krauss. Über den Zusammenhang einiger Charakteristiken eines einfach zusammenhängenden Bereiches mit der Kreisabbildung. *Mitt. Math. Sem. Giessen*, 21:1–28, 1932.

[59] E. Kummer. Über die hypergeometrische Reihe. *Crelle*, 15:39–83, also 127–172, 1836.

[60] O. Lehto. *Univalent functions and Teichmüller spaces.* Springer-Verlag, New York, 1987.

[61] O. Martio and J. Sarvas. Injectivity theorems in plane and space. *Ann. Acad. Sci. Fenn.*, 4:383–401, 1978.

[62] R. Molzon and K. Pinney Mortenson. Differential operators associated with holomorphic mappings. *Ann. of Global Analysis and Geom.*, to appear.

[63] R. Molzon and K. Pinney Mortenson. The Schwarzian derivative for maps between manifolds with complex projective connections. *Trans. Amer. Math. Soc.*, to appear.

[64] R. Molzon and K. Pinney Mortenson. Univalence of holomorphic mappings. *Manuscripta math.*, to appear.

[65] R. Molzon and K. Pinney. A characterization of complex projective space up to biholomorphic isometry. *J. of Geometric Analysis*, to appear.

[66] Z. Nehari. The Schwarzian derivative and schlicht functions. *Bull. Amer. Math. Soc.*, 55:545–551, 1949.

[67] Z. Nehari. A property of convex conformal maps. *Jour. d'Analyse Math.*, 30:390–393, 1976.

[68] Z. Nehari. Univalence criteria depending on the Schwarzian derivative. *Illinois J. of Math.*, 23:345–351, 1979.

[69] B. Osgood. The Möbius connection in the bundle of conmformal 2-jets. In *Proceedings of Idaho State conference on Vector Bundles in Complex Differential Geometry*, Pocatello, ID.

[70] B. Osgood. Univalence criteria in multiply connected domains. *Trans. Amer. Math. Soc.*, 260:459–473, 1980.

[71] B. Osgood. Some properties of f''/f' and the Poincaré metric. *Indiana Univ. Math. Jour.*, 31:449–461, 1982.

[72] B. Osgood and D. Stowe. The Schwarzian distance between domains: a question of O. Lehto. *Ann. Acad. Sci. Fenn.*, 12:313–318, 1987.

[73] B. Osgood and D. Stowe. A generalization of Nehari's univalence criterion. *Comment. Math. Helv.*, 65:234–242, 1990.

[74] B. Osgood and D. Stowe. The Schwarzian derivative and conformal mapping of Riemannian mainfolds. *Duke Math. Journal*, 67:57–97, 1992.

[75] V. Paatero. Über die konforme Abbildung von Gebieten deren Ränder von beschränkter Drehung sind. *Ann. Acad. Sci. Fenn. Ser. A XXXIII*, 9:1–79, 1931.

[76] J. Pfaltzgraff. Subordination chains and univalence of holomorphic maps in $\mathbf{C}^n$. *Math. Ann.*, 210:55–68, 1974.

[77] J. Pfaltzgraff. Subordination chains and quasiconformal extension of holomorphic maps in $\mathbf{C}^n$. *Ann. Acad. Sci. Fenn.*, 1:13–25, 1975.

[78] J. Pfaltzgraff. k-Quasiconformal extension criteria in the disk. *Complex Variables*, 21:293–301, 1993.

[79] Ch. Pommerenke. *Univalent Functions.* Vanderhoeck and Ruprecht, Göttingen, 1975.

[80] Ch. Pommerenke. On the Epstein univalence criterion. *Results. Math.*, 10:143–146, 1986.

[81] Ch. Pommerenke. *Boundary Behavior of Conformal Maps.* Springer-Verlag, Berlin, 1992.

[82] M. Schiffer and N. Hawley. Half-order differentials on Riemann surfaces. *Acta Math.*, 115:199–236, 1966.

[83] H. Schwarz. Über diejenigen Fälle in welchen die Gaussische hypergeometrische Reihe eine algebraissche Function ihres vierten Elemntes darstellt. *J. Math.*, 75:292–335 (=Abhandlungen II, pp. 211–259), 1873.
[84] H. Schwarz. *Mathematische Abhandlungen.* Springer, Berlin, 1890.
[85] D. Singer. Stable orbits and bifurcations of maps of the interval. *SIAM J. Appl. Math.*, 35:260–267, 1978.
[86] N. Steinmetz. Locally univalent functions in the unit disk. *Ann. Acad. Sci. Fenn.*, 8:325–332, 1983.
[87] N. Steinmetz. Homeomorphic extensions of univalent functions. *Complex Variables*, 6:1–9, 1985.
[88] D. Sullivan. Bounds, quadratic differentials, and renormalization conjectures. In *Mathematics into the Twenty-First Century, Volume 2*, Providence, RI, 1991. American Math. Soc.
[89] W. Thurston. Zippers and univalent functions. In *The Bieberbach Conjectuire*, Providence, RI, 1986. American Math. Soc.
[90] J. Velling. A geometric interpretation of the Ahlfors-Weill mapping and an induced foliation on $\mathbf{H}^3$. In *Holomorphic functions and moduli, vol. 2, Math. Sci. Res. Inst. Pub. 10*, New York, 1988. Springer.
[91] J. Velling. Degenerations of quasicircles: inner and outer radii of teichmüller spaces. *Ann. Acad. Sci. Fenn.*, 18:191–202, 1993.
[92] S. Yamashita. The Schwarzian derivative and local maxima of the Bloch derivative. *Math. Japonica*, 37:1117–1128, 1992.
[93] M. Yoshida. Canonical forms of some systems of linear partial differential equations. *Proc. Japan Acad*, 52:473–476, 1976.
[94] M. Yoshida. Orbifold-uniformizing differential equations. *Math. Ann.*, 267:125–142, 1984.

Department of Mathematics
Stanford University
Stanford, CA 94305-2125

The Gehring-Hayman Inequality in Conformal Mapping

Ch. Pommerenke and S. Rohde

The purpose of this paper is twofold: First we want to give an overview of results related to the Gehring-Hayman inequality, and second we shall present some new results related to it.

1 Simply connected domains

In 1962, F.W. Gehring and W.K. Hayman proved a conjecture of G. Piranian.

Theorem 1.1. [GeHa62] *Let f map the unit disk $\mathbb{D}$ conformally onto G. If $-1 \leq x_1 < x_2 \leq 1$, then*

$$\ell(f([x_1, x_2])) \leq K_1 \ell(f(\Gamma)) \tag{1.1}$$

for any curve $\Gamma \subset \overline{\mathbb{D}}$ from x_1 to x_2.

Here ℓ denotes the euclidean length, and by $K_1, K_2, \ldots$ we denote absolute constants. The best value of K_1 in (1.1) is still unknown. S. Jaenisch [Ja68] has shown that $4.56 \leq K_1 \leq 17.45$.

The celebrated and important Gehring-Hayman theorem has been extended in many directions. The diameter version

$$\operatorname{diam} f([x_1, x_2]) \leq K_2 \operatorname{diam} f(\Gamma) \tag{1.2}$$

is implicit in their proof; see ([Po92], p.88).

Now we turn to the conformally invariant formulation of the Gehring-Hayman theorem. In the simply connected domain $G \subsetneqq \mathbb{C}$, let d_G denote the hyperbolic (non-euclidean) metric and furthermore

$$d_G^*(w_1, w_2) = \min_C \int_C \frac{|dw|}{\operatorname{dist}(w, \partial G)} \tag{1.3}$$

[1]The authors acknowledge support through the HCM network "Conformal geometry and geometric function theory".

the quasi-hyperbolic metric; here C runs through all curves $C \subset G$ from w_1 to w_2.

Theorem 1.2. [GeHa62][HeNa94] *Let G be simply connected. If S is the hyperbolic geodesic from $w_1 \in G$ to $w_2 \in G$ and S^* the quasi-hyperbolic geodesic, then*

$$\ell(S) \leq K_1 \ell(C) \quad , \quad \operatorname{diam} S \leq K_2 \operatorname{diam} C, \tag{1.4}$$

$$\ell(S^*) \leq K_3 \ell(C) \quad , \quad \operatorname{diam} S^* \leq K_4 \operatorname{diam} C \tag{1.5}$$

for any curve $C \subset G$ from w_1 to w_2.

Thus these geodesics are essentially also shortest in the euclidean sense. The statements (1.5) about quasi-hyperbolic geodesics are due to J. Heinonen and R. Näkki [HeNa94].

The Gehring-Hayman theorem has been extended to quasiconformal mappings and to higher dimensions.

Theorem 1.3. [HeNa94][HeRo93] *Let f be a K- qc map of the unit ball $\mathbb{B}^n$ $(n \geq 2)$ onto $G \subset \mathbb{R}^n$. If S^* is the quasi-hyperbolic geodesic from $w_1 \in G$ to $w_2 \in G$ then*

$$\ell(S^*) \leq c\, \ell(C) \quad , \quad \operatorname{diam} S^* \leq c \operatorname{diam} C \tag{1.6}$$

for any curve $C \subset G$ from w_1 to w_2. Here c depends only on K and n.

Since quasiconformal maps need not be differentiable, the original form (1.1) does not carry over. Furthermore [HeRo93], for every $n \geq 2$ there is a qc-selfmap of $\mathbb{B}^n$ and a compact set $E \subset \partial\mathbb{B}^n$ of Hausdorff dimension $n-1$ such that

$$\ell(f([0, z])) = \infty \quad \text{for } z \in E. \tag{1.7}$$

This is in striking contrast to Beurling's theorem about conformal mapping ([Po92], p.215) that $\ell(f([0, z])) < \infty$ for $z \in \partial\mathbb{D} \setminus E$ where the exceptional set E has capacity zero and thus dimension zero.

For an overview of results related to the quasihyperbolic metric we refer to the article [Ko95] of P. Koskela in the same volume.

2 Multiply connected domains

Now we turn to multiply connected domains G in $\mathbb{C}$ with at least two finite boundary points. Then there is a universal covering map f of $\mathbb{D}$ onto G, that is, an analytic map with $f'(z) \neq 0$ such that, for $z, z' \in \mathbb{D}$,

$$f(z') = f(z) \quad \Longleftrightarrow \quad z' = \gamma(z) \text{ for some } \gamma \in F \tag{2.1}$$

where F is a Fuchsian group in $\mathbb{D}$. Thus we have $G \cong \mathbb{D}/F$.

The hyperbolic (Poincaré) metric in G has the length element

$$\rho_G(w)|dw| \quad , \quad \rho_G(w) = \frac{1}{(1-|z|^2)|f'(z)|} \quad , \quad w = f(z) \in G, z \in \mathbb{D}. \tag{2.2}$$

Its geodesics are the images of non-euclidean geodesics (arcs on orthocircles) in $\mathbb{D}$. They have locally shortest hyperbolic length and every homotopy class contains exactly one geodesic. We denote the hyperbolic distance again by d_G and the hyperbolic length by λ_G while ℓ is the euclidean length.

We say that ∂G is uniformly perfect if the doubly connected domains in G that separate ∂G have bounded moduli; see [Po79][Po84]. This holds if and only if [BePo78]

$$\rho_G(w) \geq c_1 / \operatorname{dist}(w, \partial G) \qquad (w \in G) \tag{2.3}$$

for some constant $c_1 > 0$, thus if and only if the hyperbolic and quasihyperbolic metrics are comparable. Furthermore ∂G is uniformly perfect if and only if

$$\inf\{|\text{trace } \gamma| : \gamma \in F\} > 2 \tag{2.4}$$

where again $G \cong \mathbb{D}/F$. See e.g., [Fe89][Go92] for further properties.

Now the Gehring-Hayman theorem can be extended to geodesics in given homotopy classes if and only if ∂G is uniformly perfect [Po79]:

Theorem 2.1. *The domain $G \subset \mathbb{C}$ has uniformly perfect boundary if and only if there is a constant $c_2 < \infty$ such that*

$$\ell(S) \leq c_2 \ell(C) \tag{2.5}$$

for every hyperbolic geodesic S of G and every curve $C \subset G$ with the same endpoints as S that is homotopic to S.

In the proof given in [Po79] the case that G has isolated boundary points had been overlooked. In this case it is easy to construct geodesics near the isolated boundary point for which (2.5) is violated.

We will now show that, if there are no short closed geodesics in G, then (2.5) can fail only if S stays near an isolated boundary point. To make this precise, represent G again as $\mathbb{D}/F$ and denote by $r(w)$ the injectivity radius

$$r(w) = \min\{d_{\mathbb{D}}(w, \gamma(w)) : \gamma \in F, \gamma \neq id\}$$

for $w \in \mathbb{D}$, where $d_{\mathbb{D}}$ is the hyperbolic distance in $\mathbb{D}$.

It is well known ([Be83], Theorem 11.7.1) that every isolated boundary point $\zeta \in \partial G$ has a neighborhood $N(\zeta) \subset G \cup \{\zeta\}$ bounded by an analytic Jordan curve of hyperbolic length 1 such that $f^{-1}(N(\zeta))$ consists of disjoint horodisks that are fixed by some parabolic $\gamma \in F$. Moreover, horodisks corresponding to different boundary points are also disjoint. Denote those neighborhoods by N_k and the horodisks by H_j.

Finally, there is a universal constant ε_0 such that, for every $\varepsilon < \varepsilon_0$, every $z \in G$ with $r(w) < \varepsilon$ $(f(w) = z)$ belongs to some N_k if there are no closed geodesics of hyperbolic length $\leq \varepsilon$. For $\varepsilon > 0$ set

$$G_\varepsilon = \{z \in G : r(w) > \varepsilon, f(w) = z\}.$$

The decomposition $G = (G \setminus G_{\varepsilon_0}) \cup G_{\varepsilon_0}$ is sometimes called the thin-thick decomposition of G.

Theorem 2.2. *Let $G \subset \mathbb{C}$ and assume that there are no closed geodesics of hyperbolic length $\leq \varepsilon$ in G. Assume further that ∞ is not an isolated boundary point. If S is a geodesic in G with $S \cap G_\varepsilon \neq \emptyset$, then*

$$\ell(S) \leq c(\varepsilon)\ell(C) \tag{2.6}$$

for every curve $C \subset G$ with the same endpoints as S that is homotopic to S.

This result is definitive in the sense that none of the assumptions can be omitted. It follows from (2.4) that ∂G is uniformly perfect if and only if $G = G_\varepsilon$ for some $\varepsilon > 0$. In this case, we again obtain (2.5). If G is a domain with uniformly perfect boundary and $\zeta_1, ..., \zeta_k \in G$, then $G \setminus \{\zeta_1, ..., \zeta_k\}$ satisfies the assumption of the theorem for suitable ε.

Proof of Theorem 2.2. Choose a lift $\hat{S}$ of S under f and set

$$S_1 = \{z \in \hat{S} : d_{\mathbb{D}}(z, f^{-1}(G_\varepsilon)) \leq 1\},$$

$$S_2 = \hat{S} \cap (\mathbb{D} \setminus f^{-1}(G_\varepsilon))$$

so that $\hat{S} = S_1 \cup S_2$. Let $\hat{C}$ be the lift of C that has the same endpoints as $\hat{S}$. Note that in general $S_1 \cap S_2 \neq \emptyset$.

The idea of the proof is to show $\ell(f(S_1)) \leq c\ell(C)$ and $\ell(f(S_2)) \leq c\ell(C)$ separately. The first inequality is obtained by subdividing S_1 into arcs A_j of hyperbolic length about 1, associating a subarc C_j of $\hat{C}$ to each A_j, and proving $\ell(f(A_j)) \leq c\ell(f(C_j))$. The second inequality will be proven by a corresponding estimate for each *component* B_j of S_2. By the definition of S_1 and S_2, components B_j of hyperbolic length ≤ 1 are contained in S_1 and need not be considered again.

We may assume that $\lambda_G(S) \geq 1$, otherwise (2.6) is easy. We will also assume that both endpoints of S are in G_ε. The general case needs only minor modifications. Next, we may assume that $\ell(C) = \Lambda_1(C)$, where Λ_1 denotes the linear measure of the point set. Finally we assume that $\varepsilon = \varepsilon_0$ where ε_0 is the constant described above. For the general case, one has to replace the neighborhoods N_k by smaller neighborhoods so that the hyperbolic length of the boundary is ε instead of 1. The rest of the proof is the same, up to different constants.

Let $H'_k \subset H_k$ denote the horodisk whose boundary has hyperbolic distance 1 from the boundary of H_k. Let $\Gamma \subset \mathbb{D}$ be the curve obtained from $\hat{C}$ by replacing every component of $\hat{C} \cap \bigcup H_k$ by the corresponding arc on ∂H_k. It is easy to verify

$$\operatorname{diam} f(\Gamma') \le K\ell(C') \tag{2.7}$$

for every subarc Γ' of Γ whose endpoints are in $\overline{f^{-1}(G_\varepsilon)}$ and the corresponding arc C' on $\hat{C}$. Roughly speaking, (2.7) says that walking around N_k does not increase the diameter of a curve by much, compared to passing through N_k.

Our proof rests on the following fact: If $A \subset \mathbb{D}$ is a geodesic of hyperbolic length $\ge 1/2$, if $B \subset \mathbb{D}$ is any other arc whose non-euclidean orthogonal projection onto the geodesic line through A contains A, and if $f(A)$ and $f(B)$ are contained in G_ε, then

$$\ell(f(A)) \le c \operatorname{diam} f(B). \tag{2.8}$$

This follows from the fact that $f(\mathbb{D} \setminus \bigcup_k H'_k)$ has uniformly perfect boundary, together with ([Po64], Satz 1.8 and Satz 1.2).

We show first that

$$\ell(f(S_1)) \le c\ell(C). \tag{2.9}$$

Decompose S_1 into arcs A_j of hyperbolic lengths between 1/2 and 1.Then $A_j \subset D = \mathbb{D} \setminus \bigcup_k H'_k$.

For each j, let C_j be a subarc of $\hat{C}$ whose non-euclidean orthogonal projection onto $\hat{S}$ is A_j. If one (or both) endpoint(s) of C_j is not in $\overline{f^{-1}(G_\varepsilon)}$, consider the minimal arc of $\hat{C}$ that contains C_j and has both endpoints in $\overline{f^{-1}(G_\varepsilon)}$. Call the resulting arc again C_j and let Γ_j be the subarc of Γ corresponding to C_j.

From (2.8) and (2.7) it follows that

$$\ell(f(A_j)) \le c\ell(C_j)$$

and (2.9) follows since every point of C is contained in at most two arcs $f(C_j)$.

To show

$$\ell(f(S_2)) \le c\ell(C) \tag{2.10}$$

we only need to consider those components B_j of S_2 that are not contained in S_1. For each B_j there exists an $H_k \supset B_j$, and $H'_k \cap B_j \ne \emptyset$. In particular $\lambda_{\mathbb{D}}(B_j) \ge 1$. Let $C_j \subset \hat{C}$ be such that the projection of C_j onto $\hat{S}$ is B_j, and as above extend C_j (if necessary) so that its endpoints are in $\overline{f^{-1}(G_\varepsilon)}$. Denote by Γ_j the subarc of Γ corresponding to C_j. Choose a geodesic $A_j \subset \mathbb{D}$ of hyperbolic length 1 joining two points of ∂H_k between the endpoints of B_j. Then $f(A_j) \subset G_{\varepsilon/2}$ and the projection of Γ_j onto the geodesic line through A_j contains A_j. Using (2.8) and (2.7) we obtain

$$\Lambda_1(\partial N_j) \leq K\ell(f(A_j)) \leq c\,\mathrm{diam}\, f(\Gamma_j) \leq c'\ell(f(C_j)).$$

It is not hard to show that $\ell(f(B_j)) \leq K\Lambda_1(\partial N_j)$. Roughly speaking, the reason is that $f(B_j)$ approaches the isolated boundary point in N_k exponentially fast in the number of turns around it. Summing over j we obtain (2.10) and are finished.

We now turn to an inequality that does not require any information about G, except hyperbolicity of course, but makes a topological assumption about S and C instead.

Theorem 2.3. *There is a universal constant K_1 such that*

$$\ell(S) \leq K_1\ell(C)$$

whenever $G \subset \mathbb{C}$ has at least two boundary points, $S \subset G$ is a geodesic and C is any curve homotopic to S such that $S \cup C$ is a Jordan curve.

Unfortunately, the assumption of C being a Jordan arc in G does not suffice to even guarantee that S is a Jordan arc if S is the geodesic homotopic to C. An example is the geodesic arc $\exp(2\pi i e^{it}), \pi/4 \leq t \leq 3\pi/4$, in the punctured disk. But it seems plausible that some curves C, such as line segments, do have the property that the corresponding geodesic is a Jordan arc for *every* domain G containing C. We also conjecture that every chord-arc curve C satisfies (2.5), with c_2 not depending on G.

Proof of Theorem 2.3. Throughout the proof, K denotes various absolute constants. Since S and C are homotopic there is a Jordan domain $\Omega \subset G$ with $\partial\Omega = S \cup C$. Let a, b denote the endpoints of S and let Γ be the geodesic for the hyperbolic metric d_Ω of Ω from a to b. We may and will assume that C is smooth. By (1.4) we have $\ell(\Gamma) \leq K\ell(C)$ so that it suffices to show

$$\ell(S) \leq K\ell(\Gamma). \tag{2.11}$$

For a point $z \in \Gamma$ denote by R_z the geodesic ray in Ω (that is, the ray geodesic for d_Ω) perpendicular to Γ that ends at a point $e_z \in S$. Write $\delta_\Omega(z) = \mathrm{dist}(z, \partial\Omega)$ and d_Ω resp. λ_Ω for the hyperbolic distance resp. length in Ω. Using ([Po92], Prop. 4.17) it is not difficult to find K_2 and a sequence $z_n \in \Gamma$ $(-\infty < n < \infty)$ with $1 \leq d_\Omega(z_n, z_{n+1}) \leq K_2$ such that

$$\ell(R_{z_n}) \leq K_2\delta_\Omega(z_n), \tag{2.12}$$

$$|e_{z_n} - e_{z_{n+1}}| \geq \frac{1}{K_2}\delta_\Omega(z_n), \tag{2.13}$$

$$|z - e_{z_n}| \leq K_2\delta_\Omega(z) \quad \text{for} \quad z \in R_{z_n}. \tag{2.14}$$

Write R_n instead of R_{z_n} etc., and let S_n be the arc of S between e_n and e_{n+1}, Γ_n the arc of Γ between z_n, z_{n+1}.

Below we will show the existence of constants $K_{3,4} > 1$ with the properties

$$\lambda_G(R_n) \le K_3 \lambda_G(S_n) \tag{2.15}$$

and

$$\text{if} \quad \ell(S_n) \ge K_4 \ell(\Gamma_n) \qquad \text{then} \quad \lambda_G(S_n) \ge 4K_2K_3. \tag{2.16}$$

From this we obtain (2.11) as follows: If $\ell(S_n) \le K_4\ell(\Gamma_n)$ for all n then we only need to sum over n. Now assume that $\ell(S_{n_0}) > K_4\ell(\Gamma_{n_0})$ and let $n_1 < n_0 < n_2$ be the largest resp. smallest integer with $\lambda_G(S_{n_{1,2}}) \le K_2$. The existence is clear since $\lambda_G(S_n) \to 0$ as $|n| \to \infty$. Now by (2.16)

$$\lambda_G(S_{n_1} \cup S_{n_1+1} \cup ... \cup S_{n_2-1}) \ge K_2(n_2 - n_1 - 2) + 4K_2K_3$$

whereas by (2.15)

$$\lambda_G(R_{n_1} \cup \Gamma_{n_1} \cup \Gamma_{n_1+1} \cup ... \cup \Gamma_{n_2-1} \cup R_{n_2}) \le 2K_2K_3 + K_2(n_2 - n_1).$$

The second union is a curve homotopic to the first union, of smaller hyperbolic length than the first. This contradicts the minimality of S, thus proving (2.11).

It remains to verify (2.15) and (2.16). The proof of (2.15) is standard: Subdivide R_n into arcs A_j with $\lambda_G(A_j) = 1$ and decompose S_n into disjoint arcs B_j with $\operatorname{diam} B_j \ge \frac{1}{K} \operatorname{diam} A_j$, $\operatorname{dist}(A_j, B_j) \le K \operatorname{dist}(A_j, \partial\Omega)$. Using ([BePo78], Theorem 1) it is straightforward to estimate $\lambda_G(A_j) \le K\lambda_G(B_j)$ and (2.15) follows from summation.

To prove (2.16) we fix n and may assume that $\delta_\Omega(z_n) = 1$ and $z_n = 0$. Then $\ell(\Gamma_n) \ge 1/5$ and we need to show that $\ell(S_n) \ge K_4/5$ implies $\lambda_G(S_n) \ge 4K_2K_3$, if K_4 is large enough (depending on K_2, K_3 only).

Let $r > 1$. If $\partial G \cap D_r(0)$ contains two points of euclidean distance at least $1/r$, then the existence of K_4 (depending on K_2, K_3 and r only) easily follows from (2.13) and the monotonicity of the hyperbolic metric. If $D_{\sqrt{r}}(0) \cap \partial G = \emptyset$, then ρ_G is essentially constant in $D_{\sqrt{r}/2}(0)$ (in the sense that $\max_{\{|z|<\sqrt{r}/2\}} \rho_G(z) \le K \min_{\{|z|<\sqrt{r}/2\}} \rho_G(z)$), hence $\ell(S_n) < K_4\ell(\Gamma_n)$ for large r and K_4 as otherwise $R_n \cup \Gamma_n \cup R_{n+1}$ would be hyperbolically shorter than S_n by (2.12).

Finally, let us assume the existence of a closed disk $B = \overline{D_\varepsilon(b)}$ with $|b| < \sqrt{r}$ and $\partial G \cap D_r(0) \subset B$. Let ε be minimal with this property, then $0 \le \varepsilon < 2/r$. Let $r' = \min\{|z| : z \in \partial G \setminus D_r(0)\}$, $r' = \infty$ if $\partial G \subset D_r(0)$. Set $m = \max\{\sqrt{\varepsilon}, 1/\sqrt{r'}\}$, then $m \le 1/\sqrt{r}$.

Consider $z \in G$ with $|z - b| \ge m$, $|z| \le 1/m$. Using ([BePo78], Theorem 1), we find that

$$\frac{1}{K} \le \rho_G(z)|z - b| \log(\frac{1}{m}) \le K.$$

Hence ρ_G is essentially constant in $G' = \{z \in G : |z-b| \geq 1/(4K_2), |z| \leq 4K_2\}$ (provided K_4 and r are large enough). If $R_n \cup \Gamma_n \cup R_{n+1} \subset G'$, then $\ell(S_n) < K_4/5$ for large K_4 since S_n is geodesic. Otherwise we may assume that $|e_k - b| \geq 1/(2K_2)$ for $k = n-1, n+1$ by (2.13) and

$$R_{n-1} \cup \Gamma_{n-1} \cup \Gamma_n \cup R_{n+1} \subset \{z \in G : |z-b| \geq \frac{1}{K_2 K}, |z| \leq K_2 K\}$$

follows using (2.12) and (2.13). Now the fact that $S_{n-1} \cup S_n$ is geodesic implies $\ell(S_n) \leq K_4/K$ for large K_4 as before. We have proven (2.16).

3 Further results on hyperbolic geodesics

We cannot resist to mention in passing the beautiful theorem of V. Jørgensen.

Theorem 3.1. [Jo56] *Let G be any domain in $\hat{\mathbb{C}}$ with at least three boundary points. If S is a shortest hyperbolic geodesic of G then $D \cap S$ is connected for every disk D in $\hat{\mathbb{C}}$.*

A shortest geodesic is the curve of shortest hyperbolic length between two given points regardless of homotopy class. The disks in $\hat{\mathbb{C}}$ of the theorem cannot be replaced by any more general class of objects [BF83]. D. Minda [Mi87] has further generalized Theorem 3.1.

Finally we present two new results where the sharp bounds can be determined. Let $\mathbb{T}^{\pm} = \{z \in \mathbb{T} : \ \operatorname{Im} z \gtrless 0\}$. The Gehring-Hayman theorem implies

$$\ell(f([-1,1])) \leq K_1 \Lambda_1(f(\mathbb{T}^+)) \tag{3.1}$$

where $f(A)$ for $A \subset \mathbb{T}$ is the set of angular limits $f(\zeta)$ with $\zeta \in A$. Conversely Theorem 1.1 can be deduced from (3.1) by considering a conformal map of $\mathbb{D}$ onto the domain(s) bounded by Γ and the reflection of Γ upon $[x_1, x_2]$.

Theorem 3.2. *Let f map $\mathbb{D}$ conformally onto $G \subset \mathbb{C}$.*
(i) *If* $f(\mathbb{T}^+) \subset \overline{\mathbb{D}}$ *then* $|f(x)| \leq 3 + \sqrt{8}$ *for* $-1 < x < 1$.
(ii) *If* $f(\mathbb{T}^+) = [-1,1]$ *then* $|f(x)| \leq \sqrt{3}$ *for* $-1 < x < 1$.

Equality in (i) for $x = 0$ holds for the function f_1 defined by

$$f_1(z) - \frac{1}{f_1(z)} = 2i + 4i\left(\frac{i-z}{i+z}\right)^2 \quad , \quad |f_1(z)| > 1 \tag{3.2}$$

with $\mathbb{C} \setminus f_1(\mathbb{D}) = \overline{\mathbb{D}} \cup [-i\infty, -i]$, and equality in (ii) for $x = 0$ holds for

$$f_2 = \frac{1}{2}(g + \frac{1}{g}) \quad , \quad g(z) - \frac{1}{g(z)} = 4i\frac{z^2-1}{(i+z)^2} \quad , \quad |g(z)| > 1 \tag{3.3}$$

with $\mathbb{C} \setminus f_2(\mathbb{D}) = [-1, +1] \cup [-i\infty, 0]$.

It is easy to deduce results about hyperbolic geodesics from Theorem 3.2. We write $\overline{D}(c,r) = \{w : |w-c| \le r\}$. Let S be a geodesic from a to b in the simply connected domain G.

(i) *If there is a curve* $C \subset G \cap \overline{D}(c,r)$ *from* a *to* b *then*

$$S \subset \overline{D}(c, (3+\sqrt{8})r).$$

(ii) *If* $[a,b] \subset G$ *then*

$$S \subset \overline{D}(\frac{a+b}{2}, \frac{\sqrt{3}}{2}|a-b|).$$

Since any plane set of diameter d lies in a suitable closed disk of radius $d/\sqrt{3}$ (with equality for the Reuleaux triangle) it follows from (i) that the constant in (1.2) can be chosen to be $K_2 = 2(3+\sqrt{8})/\sqrt{3} < 6.74$.

The proof of Theorem 3.2 (i) is similar to that of Theorem 3.3 below and is not difficult.

Proof of Theorem 3.2 (ii). We may assume that $x = 0$. We determine $b = re^{it}$ with $0 < r < 1$ such that $f(0) = (b+b^{-1})/2$. Since $G \subset \mathbb{C} \setminus [-1,1]$ we can write

$$f = \frac{1}{2}\left(\frac{g+b}{1+\bar{b}g} + \frac{1+\bar{b}g}{g+b}\right)$$

where g is a conformal map of $\mathbb{D}$ into $\mathbb{D} \setminus \{-b\}$ with $g(0) = 0$. It is easy to check that

$$\Lambda_1(g(\mathbb{T}^+)) = \arg g(-1) - \arg g(1) = \pi + 2\phi, \tag{3.4}$$

$$\phi = \arg\frac{1+b}{1-b} = \arctan\frac{2r\sin(t)}{1-r^2}. \tag{3.5}$$

An estimate of Beurling (see e.g., ([Po92], p.217/18)) shows that

$$\frac{1}{\sqrt{2}} = \sin\frac{\ell(\mathbb{T}^+)}{4} \le \frac{2\sqrt{r}}{1+r}\sin\frac{\Lambda_1(g(\mathbb{T}^+))}{4} = \frac{2\sqrt{r}}{1+r}\sin\left(\frac{\pi}{4}+\frac{\phi}{2}\right).$$

We conclude that $r \ge r_0 = 3 - \sqrt{8}$ and furthermore that

$$\frac{(1+r)^2}{4r} \le 2\sin^2\left(\frac{\pi}{4}+\frac{\phi}{2}\right) = 1 + \frac{\tan\phi}{\sqrt{1+\tan^2\phi}}.$$

Using (3.5) we deduce that

$$4|f(0)|^2 = |b+\frac{1}{b}|^2 = r^2 + r^{-2} + 2 - 4\sin^2 t$$

$$\le \frac{(1+r^2)^2(16r^2-(1-r)^4)-(1-r^2)^2(1-r)^4}{16r^4-r^2(1-r)^4}$$

and it is elementary to show that the right-hand side is ≤ 12 for $r_0 \leq r < 1$; the maximum is attained for $r = 2 - \sqrt{3}$.

The last theorem shows that a complete geodesic lies roughly in the middle between the two parts of the boundary determined by its endpoints.

Theorem 3.3. *If f maps $\mathbb{D}$ conformally into $\mathbb{C}$ then*

$$3 - \sqrt{8} \leq \frac{\operatorname{dist}(f(x), f(\mathbb{T}^+))}{\operatorname{dist}(f(x), f(\mathbb{T}^-))} \leq 3 + \sqrt{8} \quad \textit{for} \quad -1 < x < 1$$

and the bounds are sharp.

Proof. To prove the left-hand inequality, we may assume that $x = 0$, $f(0) = 0$ and

$$\operatorname{dist}(f(0), f(\mathbb{T}^+)) = 1 > \operatorname{dist}(f(0), f(\mathbb{T}^-)) = |w|$$

with $w \in f(\mathbb{T}^-)$. Then $f(\mathbb{T}^+) \subset \mathbb{C} \setminus \mathbb{D}$. Let H be the component of $\mathbb{D} \cap G$ containing 0. Then the harmonic measure satisfies

$$\omega(0, H \cap \partial G, H) \leq \omega(0, H \cap \partial G, G) \leq \frac{1}{2}$$

and since $\partial H \setminus (H \cap \partial G) \subset \mathbb{T}$ we obtain from Beurling's estimate ([Po92], p.217/18) that

$$\frac{1}{\sqrt{2}} \leq \sin \frac{\omega(0, \partial H \setminus (H \cap \partial G), H)}{4} \leq \frac{2\sqrt{|w|}}{1 + |w|}$$

which implies $|w| \geq 3 - \sqrt{8}$. The right-hand inequality follows by symmetry.

The function f_1 defined in (3.2) satisfies $f_1(0) = ia$ with $a = 3 + \sqrt{8}$. Hence $f_3 = (f_1 - ia)/(1 + iaf_1)$ maps $\mathbb{D}$ conformally onto $\mathbb{D} \setminus [-i/a, -i]$ such that $f_3(0) = 0$ and $f_3(\mathbb{T}^+) = \mathbb{T}$, $f_3(\mathbb{T}^-) = [-i/a, -i]$. Therefore $\operatorname{dist}(0, f_3(\mathbb{T}^+)) = 1$, $\operatorname{dist}(0, f_3(\mathbb{T}^-)) = 1/a = 3 - \sqrt{8}$.

References

[Be83] A.F. Beardon, *The geometry of discrete groups*, Springer, Berlin, 1983.

[BePo78] A.F. Beardon and Ch. Pommerenke, *The Poincaré metric of plane domains*, J. London Math. Soc. (2) **18** (1978), 475–483.

[BF83] B. Brown Flinn, *Hyperbolic convexity and level sets of analytic functions*, Indiana Univ. Math. J. **32** (1983), 831–841.

[Fe89] J. L. Fernández, *Domains with strong barrier*, Rev. Mat. Iberoamericana **5** (1989), 47–65.

[GeHa62] F.W. Gehring and W.K. Hayman, *An inequality in the theory of conformal mapping*, J. Math. Pures Appl. (9) **41** (1962), 353–361.

[Go92] M. J. González, *Uniformly perfect sets, Green's function, and fundamental domains*, Rev. Mat. Iberoamericana **8** (1992), 239–269.

[HeNa94] J. Heinonen and R. Näkki, *Quasiconformal distortion on arcs*, J. Analyse Math. **63** (1994), 19–53.

[HeRo93] J. Heinonen and S. Rohde, *The Gehring-Hayman inequality for quasihyperbolic geodesics*, Math. Proc. Camb. Phil. Soc. **114** (1993), 393–405.

[Ko95] P. Koskela, *Old and new on the quasihyperbolic metric*, in Quasiconformal Mappings and Analysis, A Collection of Papers Honoring F.W. Gehring (P. Duren, J. Heinonen, B. Osgood, B. Palka, eds.), Springer-Verlag, 1997, pp. 211–226.

[Ja68] S. Jaenisch, *Length distortion of curves under conformal mappings*, Michigan Math. J. **15** (1968), 121–128.

[Jo56] V. Jørgensen, *On an inequality for the hyperbolic measure and its applications to the theory of functions*, Math. Scand. **4** (1956), 113–124.

[Mi87] D. Minda, *A reflection principle for the hyperbolic metric and applications to geometric function theory*, Complex Variables **8** (1987), 129–144.

[Po64] Ch. Pommerenke, *Linear-invariante Familien analytischer Funktionen I*, Math. Ann. **155** (1964), 108–154.

[Po79] Ch. Pommerenke, *Uniformly perfect sets and the Poincaré metric*, Arch. Math. **32** (1979), 192–199.

[Po84] Ch. Pommerenke, *On uniformly perfect sets and Fuchsian groups*, Analysis **4** (1984), 299–321.

[Po92] Ch. Pommerenke, *Boundary behaviour of conformal maps*, Springer, Berlin, 1992.

Fachbereich Mathematik
TU Berlin
10623 Berlin, Germany

Extremal Extensions from the Circle to the Disk

Edgar Reich

0 Introduction

Let $\Delta = \{|z| < 1\}$, $\Gamma = \{|z| = 1\}$, $\partial = (\partial_x - i\partial_y)/2$, $\bar{\partial} = (\partial_x + i\partial_y)/2$. In this exposition we shall be concerned with two kinds of extension problems from Γ to $\Delta \cup \Gamma$, namely, firstly, the *least non-analytic* (abbreviated LNA) extensions of *continuous* complex-valued boundary values, and, secondly, the *extremal quasiconformal* (abbreviated EQC) extensions of *homeomorphic* boundary values. We formulate the first problem as follows. Let $g(z)$, $z \in \Gamma$, be a continuous complex-valued function. It will often be convenient to normalize g by the conditions

$$Re[\bar{z}g(z)] = 0, \; z \in \Gamma, \tag{0.1}$$

$$g(1) = g(i) = g(-1) = 0. \tag{0.2}$$

Let $\mathcal{V}[g]$ denote the class of continuous complex-valued extensions G of g to $\Delta \cup \Gamma$, possessing bounded generalized $\bar{\partial}$-derivatives, $\|\bar{\partial}G\|_\infty < \infty$. If $\mathcal{V}[g]$ is non-empty it contains at least one element G_0 such that

$$\|\bar{\partial}G_0\|_\infty = m_0[g] = \inf\{\|\bar{\partial}G\|_\infty : G \in \mathcal{V}[g]\}.$$

Such a G_0 is then a LNA extension of g. Evidently, the quantity $m_0[g]$ measures the minimal non-analyticity with which g can be extended; g possesses an analytic extension if and only if $m_0[g] = 0$. $\mathcal{V}[g]$ is non-empty if and only if $m_0[g] < \infty$.

For a quasiconformal mapping F of Δ, denote by $\mu = \mu[F]$ the complex dilatation $\mu[F] = \bar{\partial}F/\partial F$. If F is onto Δ and keeps 1, i, and -1 fixed, it is denoted by F^μ. For the second problem, let $f(z)$, $z \in \Gamma$, be a continuous, one-to-one, complex-valued function on Γ. We assume that f is normalized (by an auxiliary conformal or anti-conformal map) so that

$$f \text{ is a sense-preserving self-homeomorphism of } \Gamma, \tag{0.3}$$

---

[1]Work done in part with support of an SERC grant at the University of York.

$$f(1) = 1,\ f(i) = i,\ f(-1) = -1. \tag{0.4}$$

Let $Q[f]$ denote the class of quasiconformal extensions F of f to $\Delta \cup \Gamma$. If $Q[f]$ is non-empty it contains at least one element F_0 such that

$$\|\mu[F_0]\|_\infty = k_0[f] = \inf\{\|\mu[F]\|_\infty : F \in Q[f]\}.$$

Such an F_0 is referred to as an EQC extension of f.

As our review will show, facts about LNA extensions give an indication of parallel facts that hold [Re4] (possibly in a modified "souped up" form) for EQC extensions. The reason for the relationship between LNA and EQC is, roughly, that when F is nearly the identity, and $G(z) = F(z) - z$, then, having $\mu[F]$ small, is equivalent to having $\bar{\partial}G$ small. Now, it is easier to study the LNA extensions than the EQC extensions, because there is a classical explicit linear integral representations of G in terms of $\bar{\partial}G$, while, by contrast, obtaining F from $\mu[F]$ requires the solution of a partial differential equation. Contrary to following the historical order in which developments occurred, it therefore makes sense in an exposition to look at the LNA problem before proceeding to the EQC case.

It is easy to replace the disk by other Jordan domains or, say, the half-plane, and to state all results in conformally invariant form. Furthermore, essentially all the results for the EQC problem are known to generalize to arbitrary, not necessarily simply-connected regions. Restricting consideration in this review to the disk is for the double purpose of simplifying the exposition and staying within the main expertise of the writer.

For a review of the literature on EQC we refer the reader to Strebel's survey [S3]. The article by Earle and Li Zhong [E-L] in the current volume is also very relevant.

1 LNA Extensions

1.1 A class of linear transforms.

Suppose $g \in C(\Gamma)$ is given. When looking for an extension G of g with $\bar{\partial}G \in L^\infty(\Delta)$, if one exists, it is natural to first try a linear transform of g. It turns out that if such an extension does exist it can always be found by any of a class of linear transforms.

Corresponding to the real parameter $\alpha, (\alpha \geq 1)$, let

$$(T_\alpha g)(z) = \frac{(1-|z|^2)^{\alpha-1}}{2\pi i} \int_\Gamma \frac{g(\zeta)d\zeta}{(1-\bar{z}\zeta)^{\alpha-1}(\zeta - z)},\quad z \in \Delta, \tag{1.1}$$

where $(1-\bar{z}\zeta)^{\alpha-1}$ is the branch equalling 1 at $\zeta = 0$. According to (1.1),

$$(T_\alpha g)(z) = \frac{1}{2\pi} \int_\Gamma \left(\frac{1-|z|^2}{1-\bar{z}\zeta}\right)^{\alpha-2} \frac{1-|z|^2}{|\zeta - z|^2} g(\zeta)|d\zeta|. \tag{1.2}$$

One identifies $T_1 g$ as the Cauchy integral of g, T_2 as the Poisson integral, while T_4 is the integral transform encountered by Earle [E] on the basis of certain invariance properties (See also [Re3]). If g is holomorphic in Δ and continuous in $\Delta \cup \Gamma$, then $T_\alpha g = g$, but of course $T_1 g$ is not continuous in $\Delta \cup \Gamma$ if $m_0[g] \neq 0$. On the other hand, if $\alpha \geq 2$, then for any $g \in C(\Gamma)$, $T_\alpha g$ provides an extension of g to $\Delta \cup \Gamma$.

We next consider $\bar{\partial}(T_\alpha g)$. By (1.1),

$$\bar{\partial}(T_\alpha g)(z) = \frac{\alpha - 1}{2\pi i}(1 - |z|^2)^{\alpha - 2} \int_\Gamma \frac{g(\zeta)}{(1 - \bar{z}\zeta)^\alpha} d\zeta. \tag{1.3}$$

Suppose, now, that $m_0[g] < \infty$. Let $G_0(z)$ be a LNA extension of g, and $q_0(z) = \bar{\partial} G_0(z)$. By Green's formula, we can rewrite the right side of (1.3) as

$$\bar{\partial}(T_\alpha g)(z) = \frac{\alpha - 1}{\pi}(1 - |z|^2)^{\alpha - 2} \iint_{|\zeta|<1} \frac{q_0(\zeta)}{(1 - \bar{z}\zeta)^\alpha} d\xi d\eta, \ (\zeta = \xi + i\eta).$$

Changing variable of integration, for fixed z, from ζ to w according to $w = (\zeta - z)/(1 - \bar{z}\zeta)$, gives

$$\bar{\partial}(T_\alpha g)(z) = \frac{\alpha - 1}{\pi} \iint_{|w|<1} \frac{(1 + \bar{z}w)^{\alpha - 2}}{(1 + z\bar{w})^2} q_0 \left(\frac{w + z}{1 + \bar{z}w} \right) dudv, (w = u + iv).$$

Therefore,

$$\|\bar{\partial} T_\alpha g\|_\infty \leq C_\alpha m_0[g], \tag{1.4}$$

where[2]

$$\begin{aligned} C_\alpha &= \frac{\alpha - 1}{\pi} \sup_{|z|<1} \iint_{|w|<1} |1 + \bar{z}w|^{\alpha - 4} dudv \\ &= \frac{\alpha - 1}{\pi} \iint_{|w|<1} \frac{dudv}{|1 + w|^{4 - \alpha}} = \frac{(\alpha - 1)\Gamma(\alpha - 2)}{[\Gamma(\alpha/2)]^2}, (\alpha > 2). \end{aligned} \tag{1.5}$$

In particular,

$$C_2 = \infty, \ C_3 = \frac{2}{\pi} \iint_{|w|<1} \frac{dudv}{|1 + w|} = \frac{8}{\pi}, \ C_4 = 3. \tag{1.6}$$

A numerical computation indicates that $2.52710 < \min C_\alpha < 2.52711$; there is a unique minimum at $\alpha = 3.1407...$.

---

[2]The expression of C_α in terms of the Gamma function in (1.5) also follows from [Fo-Ru, Proposition 2.7, Part (c)]. I would like to thank Frank Bonsall for calling this reference to my attention as well as supplying much information regarding the transformations T_α. See also [Bo-W].

In view of the foregoing, we have

$$m_0[g] \le \|\bar\partial T_{3.1407} g\|_\infty \le 2.52711 m_0[g], \quad m_0[g] \le \|\bar\partial T_3 g\|_\infty \le \frac{8}{\pi} m_0[g], \tag{1.7}$$

while for Earle's transform,

$$m_0[g] \le \|\bar\partial T_4 g\|_\infty \le 3 m_0[g]. \tag{1.8}$$

We see that, e.g. either T_3 or T_4 will provide a member of $\mathcal{V}[g]$ whenever this class is non-empty. While the fact that e.g. C_3 is less than C_4 indicates that $\|\bar\partial T_3 g\|_\infty$ tends to be closer to $m_0[g]$ than $\|\bar\partial T_4 g\|_\infty$, T_4 is technically easier to handle.

1.2 For which g is $m_0[g]$ finite?

The relevant condition turns out to hinge on the Zygmund class $\Lambda_*(\Gamma)$ of functions $g(z)$ with the property that

$$|g(z+h) - 2g(z) + g(z-h)| \le A|h| \text{ whenever } z, z+h, z-h \in \Gamma. \tag{1.9}$$

Theorem 1.2. [Re-C] *Suppose $g \in \Lambda_*(\Gamma)$. Then $m_0[g] < \infty$. Conversely, if g satisfies* (0.1), (0.2), *and if $m_0[g] < \infty$ then $g \in \Lambda_*(\Gamma)$.*

Outline of proof. Assuming (1.9), one can derive an explicit bound on $\|\bar\partial T_4 g\|_\infty$ in terms of the constant A. The converse follows from the [A] representation (1.10) of $G(z)$ in terms of $q(z) = \bar\partial G(z)$, when $G\,|_\Gamma$ satisfies (0.1), (0.2):

$$G(z) = -\frac{1}{\pi}\iint_\Delta \left[\frac{q(\zeta)}{\zeta - z} + \frac{z^3\overline{q(\zeta)}}{1 - z\bar\zeta} - \beta(\zeta)z^2 - B(\zeta)z + \overline{\beta(\zeta)}\right] d\xi d\eta, \tag{1.10}$$

where

$$\beta(\zeta) = \Big(\frac{1}{i-\zeta} + \frac{i+\zeta}{1-\zeta^2}\Big)\frac{q(\zeta)}{2} + \Big(\frac{i+\bar\zeta}{1-\bar\zeta^2} - \frac{1}{i+\bar\zeta}\Big)\frac{\overline{q(\zeta)}}{2}, \quad B(\zeta) = \frac{\overline{q(\zeta)}}{1-\bar\zeta^2} - \frac{q(\zeta)}{1-\zeta^2}.$$

If we do not assume that g is normalized then we can only conclude that g belongs to $\Lambda_*(\Gamma)$ up to an additive function h for which $m_0[h] = 0$. The function h need not necessarily belong to $\Lambda_*(\Gamma)$.

1.3 Evaluation of $m_0[g]$. Characterization of LNA extensions.

Let $\mathcal{B}$ denote the Banach space[3] of functions $\phi(z)$ holomorphic in Δ with finite L^1 norm

$$\|\phi\|_1 = \iint_\Delta |\phi(z)| dx dy < \infty.$$

[3] Also frequently denoted as $L^1_a(\Delta)$.

Let G be any member of $\mathcal{V}[g]$, and consider the linear functional

$$\Lambda_g[\phi] = \iint_\Delta \phi \bar{\partial} G \, dxdy, \; \phi \in \mathcal{B}. \tag{1.11}$$

If ϕ is, say, a polynomial, one sees by Green's formula that $\Lambda_g[\phi]$ depends only on the boundary values of G, i.e. on g. Since polynomials are dense in $\mathcal{B}$ we see that Λ_g depends only on g, justifying the notation.

Lemma 1.3. *Suppose $\nu \in L^\infty(\Delta)$ has the property that*

$$\iint_\Delta \nu \phi dxdy = 0 \text{ for all } \phi \in \mathcal{B}.$$

Then there exists $H \in C(\Delta \cup \Gamma)$ such that $\bar{\partial} H = \nu$ a.e. and $H(z) = 0$ for $z \in \Gamma$.

Proof. Form H using (1.10) with $q = \nu$.

Theorem 1.31. *Suppose $m_0[g] < \infty$. Then*

$$m_0[g] = \|\Lambda_g\| = \sup \left\{ \frac{|\Lambda_g[\phi]|}{\|\phi\|_1} : \phi \in \mathcal{B} \right\}.$$

Moreover, if

$$\iint_\Delta q_1 \phi \, dxdy$$

is a Hahn-Banach extension of Λ_g from $\mathcal{B}$ to $L^1(\Delta)$, then $q_1 = \bar{\partial} G_1$ where G_1 is a LNA extension of g.

Proof. By the remark following Theorem 1.2, there is no loss of generality in assuming that g is normalized by (0.1), (0.2). Using a LNA extension G_0 in the representation (1.11), we see that

$$\Lambda_g[\phi] = \iint_\Delta q_0 \phi dxdy, \; q_0 = \bar{\partial} G_0, \; \|q_0\|_\infty = m_0[g]. \tag{1.12}$$

Thus, $\|\Lambda_g\| \leq m_0[g]$. In the other direction, by the Hahn-Banach theorem there exists $q_1 \in L^\infty(\Delta)$, such that $\|q_1\|_\infty = \|\Lambda_g\|$ and

$$\iint_\Delta q_1 \phi dxdy = \Lambda_g[\phi] \text{ for all } \phi \in \mathcal{B}.$$

Therefore, by Lemma 1.3, there exists $G_1 \in \mathcal{V}[g]$ with $\bar{\partial} G_1 = q_1$. Therefore, $\|q_1\|_\infty \geq m_0[g]$, and the assertions follow.

As a special case, one can assert:

Theorem 1.32. *A necessary and sufficient condition that G is a LNA extension of $G|_\Gamma$ is that*

$$\sup\left\{\frac{|\int\int_\Delta \phi\bar{\partial}G dx dy|}{\|\phi\|_1} : \phi \in \mathcal{B}\right\} = \|\bar{\partial}G\|_\infty. \tag{1.13}$$

1.4 Uniqueness theorems.

We have the following immediate corollary of Theorem 1.31:

Theorem 1.41. *The boundary values g have a unique LNA extension to $\Delta \cup \Gamma$ if and only if Λ_g has a unique Hahn-Banach extension from $\mathcal{B}$ to $L^1(\Delta)$.*

While Theorem 1.41 is theoretically interesting its successful applications have been more in the direction of providing examples [Re1] of the non-uniqueness of Hahn-Banach extensions than the other way around. Two types of useful sufficient conditions for uniqueness of LNA extensions are known. The first type is based on a-priori information about a single LNA extension. The second type involves a hypothesis on g that guarantees the possiblity of inward extensions G with sufficiently small values of $\bar{\partial}G$ near Γ. The simplest prototype of the first type is the following trivial application of Theorem 1.41:

Theorem 1.42. *Suppose $G_0 \in \mathcal{V}[g]$ and*

$$\bar{\partial}G_0(z) = \|\bar{\partial}G_0\|_\infty \frac{\overline{\phi_0(z)}}{|\phi_0(z)|} \quad \text{a.e. in } \Delta \tag{1.14}$$

for some $\phi_0 \in \mathcal{B}$. Then G_0 is the unique LNA extension of g.

The hypothesis of Theorem 1.42 can be refined considerably in the way of weakening the assumption that $\phi_0 \in \mathcal{B}$. In Section 2.3 more delicate conditions for the EQC case that would also work here are given.

For the simplest example of the second type let us consider the Zygmund class $\lambda_*(\Gamma)$ of continuous complex-valued functions g satisfying

$$g(z+h) - 2g(z) + g(z-h) = o(h) \tag{1.15}$$

uniformly with respect to z, whenever $z, z+h, z-h \in \Gamma$.

Theorem 1.43. [Re-C] *Suppose $g \in \lambda_*(\Gamma)$. Then there is a unique LNA extension G_0, and it must be of the form* (1.14).

Outline of proof. We form the extension

$$G = T_4 g$$

as in the proof of Theorem 1.2, and then consider

$$q(z) = (\bar{\partial}G)(z) = \frac{3}{2\pi i}(1-|z|^2)^2 \int_\Gamma \frac{g(\zeta)}{(1-\bar{z}\zeta)^4} d\zeta, \ z \in \Delta. \tag{1.16}$$

Following up the consequences of the hypothesis (1.15), one finds from (1.16) that

$$\lim_{|z|\to 1} q(z) = 0. \tag{1.17}$$

If G_0 is any LNA extension of $g, q_0 = \bar{\partial}G_0$, $\|q_0\|_\infty = m_0[g]$, we can write

$$\Lambda_g[\phi] = \iint_\Delta q_0 \phi dx dy = \iint_\Delta q\phi dx dy. \tag{1.18}$$

By Theorem 1.32, there is a sequence $\phi_n \in \mathcal{B}$, $\|\phi_n\|_1 = 1$, such that

$$\lim_{n\to\infty} \Lambda_g[\phi_n] = m_0[g]. \tag{1.19}$$

We can assume that $m_0[g] > 0$ to avoid triviality. Since $\{\phi_n\}$ is a normal family there is a subsequence, also to be denoted as $\{\phi_n\}$, such that

$$\lim \phi_n(z) = \phi_0(z), \tag{1.20}$$

uniformly on compact subsets of Δ. While it is clear that $\phi_0 \in \mathcal{B}$, in general (1.20) would of course only imply that $\|\phi_0\|_1 \leq 1$. In view of (1.17), however, it is not difficult to show, with the help of (1.18) and (1.19) and the assumption $m_0[g] > 0$, that actually $\|\phi_0\|_1 = 1$ and $|\Lambda_g[\phi_0]| = m_0[g]$. Thus, G_0 must be of the form (1.14), and uniqueness follows by Theorem 1.42.

For related examinations of the consequences of local properties of g we may refer the reader to [Re5] which is an adaptation of earlier work of Fehlmann's [Fe] from EQC to LNA.

1.5 An Example.

We consider the boundary values

$$g(z) = \bar{z}^n, \ z \in \Gamma, \tag{1.21}$$

where n is a positive integer. Since $g \in \lambda_*(\Gamma)$, the LNA extension G_0 is unique; by Theorem 1.43, it is of the form (1.14). Evidently, $\|\bar{\partial}G_0\|_\infty = m_0[g] > 0$ as there is no holomorphic extension of g to $\Delta \cup \Gamma$. In view of the uniqueness of G_0, the symmetry of g dictates that $G_0(z)$ has the form

$$G_0(z) = p(r^2)\bar{z}^n, \ (z = re^{i\theta}), \tag{1.22}$$

where

$$G_0(0) = \lim_{r\to 0} r^n p(r^2) = 0, \ \lim_{r\to 1} p(r^2) = 1. \tag{1.23}$$

Since ϕ_0 has at most finitely many zeroes in any compact subset of Δ, one sees that p is a C^∞ function of r^2, $0 < r < 1$. By, (1.22),

$$|\bar{\partial} G_0(z)| = nr^{n-1}p(r^2) + r^{n+1}p'(r^2).$$

So, in view of (1.14),

$$nr^{n-1}p(r^2) + r^{n+1}p'(r^2) \equiv \text{const} = K = m_0[g].$$

The only solution of this differential equation satisfying (1.23) is

$$p(t) = \frac{2}{n+1}Kt^{\frac{1-n}{2}}.$$

Hence

$$K = m_0[g] = \frac{n+1}{2}, \ G_0(re^{i\theta}) = re^{-in\theta}. \tag{1.24}$$

For $\bar{\partial} G_0(z)$ we obtain

$$\bar{\partial} G_0(z) = \frac{n+1}{2} \frac{\overline{\phi_0(z)}}{|\phi_0(z)|}, \ \phi_0(z) = z^{n-1}. \tag{1.25}$$

2 EQC Extensions

2.1 *Two fundamental lemmas.*

Analogously to (1.11), we associate the linear functional

$$\Lambda_\mu[\phi] = \iint_\Delta \mu(z)\phi(z)dxdy, \ \phi \in \mathcal{B}, \tag{2.1}$$

to a given complex dilatation function $\mu \in L^\infty(\Delta)$, $\|\mu\|_\infty < 1$. In contrast with Λ_g, Λ_μ is not determined by merely $F^\mu|_\Gamma$ alone. In particular, a literal translation of Lemma 1.3 does not hold. There are, however, two important partial substitutes. The first, Bers' Lemma [Ber1], says that a condition on Λ_μ resembling that of Lemma 1.3 implies that $F^\mu|_\Gamma$ is "approximately" the identity. The second, Lemma 2.12, goes in the opposite direction.

Bers' Lemma. *Suppose*

$$\Lambda_\mu[\phi] = 0 \ \textit{for all} \ \phi \in \mathcal{B}. \tag{2.2}$$

Let $f^{t\mu}$ denote the restriction of $F^{t\mu}$ to Γ, $0 \le t \le 1$. Then $k_0[f^{t\mu}] = o(t)$ as $t \to 0$.

Outline of proof. [Re3] The heart of the matter is to show that one has $o(t)$ rather than just $0(t)$ which would be trivial. This can be done by constructing an extension of $f^{t\mu}$ that, as $t \to 0$, is much more like an EQC extension than $F^{t\mu}$ itself is liable to be. (Finding an EQC extension constructively is unfortunately not practical, of course.) The construction in [Re3] makes use of the fact [A] that F^{μ} can be decomposed into mappings close to the identity. In the limiting version (cf [G-Re]) one generates $F^{t\mu}(z)$ as $w = F(z,t)$, $0 \le t \le 1$, as the solution of a differential equation of the form

$$\begin{aligned} \frac{dw}{dt} &= G(w,t), \qquad 0 \le t \le 1,\ w \in \Delta \cup \Gamma, \\ F(z,0) &= z, \qquad z \in \Delta \cup \Gamma \end{aligned} \tag{2.3}$$

with $\|\bar\partial G\|_\infty$ uniformly bounded with respect to t. Basically, an elementary computation shows that if in a differential equation of type (2.3) one has $\|\bar\partial G\|_\infty \le M$, then the solution has a complex dilatation bounded by $\sinh(Mt)$. Thus, in order to bring the solution of (2.3) closer to an EQC extension of the boundary values we should reduce $\|\bar\partial G\|_\infty$ without changing the boundary values of G. So we are brought back to the LNA extension problem! The discussion in Section 1.1 suggests that we should replace G by $T_\alpha G$ for some $\alpha \ge 3$. In particular, replacing G by $G^* = T_4 G$, it is technically possible to bring in the hypothesis (2.2) to obtain a sufficiently good estimate of $\|\bar\partial G^*\|_\infty$ to prove the Lemma.

Lemma 2.12. [Re-S1] *Suppose the restriction of F^μ to Γ is the identity. Then*

$$\mathcal{R}e \iint_\Delta \frac{\mu}{1-|\mu|^2}\phi\, dx dy \le \iint_\Delta \frac{|\mu|^2}{1-|\mu|^2}|\phi| dx dy, \quad \text{for all } \phi \in \mathcal{B}. \tag{2.4}$$

There have been two proofs. The first [Re-S1] is a "geometric" one, using the classical length-area method with the Teichmüller metric [4]

$$\int |\phi(z)|^{1/2}|dz|$$

An alternative proof is due to Bers [Ber2].

2.2 *For which f is $k_0[f]$ finite? Characterization of EQC extensions.*

The basic condition on finiteness of $k_0[f]$ is the famous condition of Beurling and Ahlfors [Beu-A]. Formulated originally for extensions from the real axis to the upper half-plane it states:

---

[4]Question: Does Lemma 2.12 already follow from the special case of single-valued $\sqrt{\phi}$? (In the case of Lemma 1.3 the answer is yes.)

Theorem of Beurling & Ahlfors. *Let f be an increasing homeomorphism of $\mathbf{R}$ onto itself. There exists a quasiconformal extension of f to the upper half-plane if and only if there exists $\rho > 0$, such that*

$$\frac{1}{\rho} \leq \frac{f(x+h)-f(x)}{f(x)-f(x-h)} \leq \rho, \text{ whenever } x, x+h, x-h \in \mathbf{R}, h \neq 0. \quad (2.5)$$

Evidently, Theorem 1.2 is the LNA analogue of the Beurling-Ahlfors condition. However, there is no known procedure for determining $k_0[f]$ from a knowledge of Λ_μ, analogous to Theorem 1.31. But just as Lemma 1.3 is used to derive the necessary and sufficient condition, Theorem 1.32, for LNA, so can (but in a less direct manner) Bers' Lemma and Lemma 2.12 be used to derive a necessary and sufficient condition for EQC. The interesting fact is that the result has precisely the same form as Theorem 1.32, namely,

Theorem 2.2. [Re-S2] *A necessary and sufficient condition that F^μ is an EQC extension of its restriction to Γ is that*

$$\sup\left\{\frac{|\Lambda_\mu[\phi]|}{\|\phi\|_1} : \phi \in \mathcal{B}\right\} = \|\mu\|_\infty \quad (2.6)$$

Necessity was first proved by Hamilton [H]. Alternatively, a technique due to Krushkal [K] using Bers' Lemma and the Hahn-Banach theorem can be used to prove necessity analogously to the way Lemma 1.3 is used to prove half of Theorem 1.31.

To prove sufficiency one needs something to take the place of the way the representation (1.11) is used. Of course, nothing as simple as (1.11) involving complex dilatations holds, but Lemma 2.12 provides the key to the following substitute.

Lemma 2.2. *Suppose F_1, F_2 are quasiconformal self-mappings of Δ agreeing on Γ. Let*

$$\kappa = \mu[F_1], \alpha = \mu[F_1^{-1}] \circ F_1, \ \beta = \mu[F_2^{-1}] \circ F_1, \ \tau = \frac{\kappa}{\alpha} = -\frac{\overline{\partial F_1}}{\partial F_1}. \quad (2.7)$$

Then, for any $\phi \in \mathcal{B}$,

$$\begin{aligned} &\mathcal{R}e \iint_\Delta \frac{(\alpha-\beta)(1-\alpha\bar{\beta})\tau}{(1-|\alpha|^2)(1-|\beta|^2)} \phi\, dxdy \\ &\leq \iint_\Delta \frac{|\alpha-\beta|^2}{(1-|\alpha|^2)(1-|\beta|^2)} |\phi|\, dxdy. \end{aligned} \quad (2.8)$$

To prove Lemma 2.2 we let $F = F_1^{-1} \circ F_2$ and rewrite (2.4) using the quantities defined by (2.7). The result is equivalent to (2.8). Relation (2.8) should be thought of as a souped-up version of (1.11) because, formally, (2.8) becomes

$$\iint_{\Delta} (\bar{\partial}F_1 - \bar{\partial}F_2)\phi dxdy \approx 0$$

when F_1, F_2 are close to the identity.

Now to prove that (2.6) is sufficient in Theorem 2.2 one manipulates [Re-S2] (2.8) in apparently tricky ways the details of which shall not be dwelled on in this exposition. What was the idea and why did it work? The answer is that we may regard (2.8) as merely a technically refined version of the length-area principle already successfully applied in the pioneering work of Grötzsch and Teichmüller to obtain sharp extremal results for quasiconformal mappings subject to various boundary restrictions. As the simplest example of a complex dilatation satisfying (2.6), one has the Teichmüller type:

$$\mu(z) = \|\mu\|_\infty \frac{\overline{\phi_0(z)}}{|\phi_0(z)|} \text{ for some } \phi_0 \in \mathcal{B}. \tag{2.9}$$

Note that if $\bar{\partial}G$ were of form (2.9), then, by Theorem 1.32, G would be a LNA extension of $G|_\Gamma$.

2.3 Non-decreasable dilatations. Uniqueness theorems.

In view of the fact that the conditions for LNA and EQC are identical when formulated in terms of the respective functionals Λ_g and Λ_μ it is natural to ask whether Theorem 1.41 carries over similarly:

Question. *Is the condition that*

$$\iint_{\Delta} \mu(z)\phi(z)dxdy$$

is the unique Hahn-Banach extension of Λ_μ from $\mathcal{B}$ to $L^1(\Delta)$ necessary and/or sufficient for the condition that F^μ is the unique EQC extension of its restriction to Γ?

While all examples known to me speak in favor of an affirmative answer, as far as I know the question is open in both directions.[5]

In Theorem 2.32 we will illustrate one way in which a sufficient condition for uniqueness of EQC can be formulated.[6] Theorem 2.33 [S2] is

[5] Added in proof: N. Lakić, V. Marković, and M. Mateljević have announced that the answer is yes, both as regards necessity and sufficiency.

[6] There is also a more refined sufficient condition of the same type [Re2]. Either condition is also sufficient for uniqueness of the Hahn-Banach extension. Whether either is necessary for uniqueness of the EQC or Hahn-Banach extension is unknown.

another way of formulating a sufficient condition, but it is definitely not simultaneously a necessary condition.

To approach Theorem 2.32, instead of uniqueness the more general concept of non-decreasable dilatation turns out to be appropriate. If F_1 is a quasiconformal self-mapping of Δ let us say that F_1 *has non-decreasable dilatation if the hypothesis that* F_2 *is also a quasiconformal self-mapping of* Δ *together with the conditions,*

$$F_1|_\Gamma = F_2|_\Gamma,\ |\mu[F_2](z)| \le |\mu[F_1](z)| \text{ a.e. in } \Delta, \tag{2.10}$$

imply that $F_2 = F_1$.

Theorem 2.31. *Let* $F(z)$ *be a quasiconformal self-mapping of* Δ, $\kappa(z) = \mu[F](z) \neq 0$ *a.e. Suppose there exists a sequence* $\phi_n \in \mathcal{B}$ *satisfying the following conditions* (2.11), (2.12), (2.13)*:*

$$\begin{array}{l}\phi_0(z) = \lim \phi_n(z) \textit{ exists}^6 \textit{ a.e. in } \Delta, \textit{ is locally} \\ \textit{integrable and non-vanishing a.e. in } \Delta;\end{array} \tag{2.11}$$

$$\operatorname{sgn} \kappa(z) = \operatorname{sgn} \overline{\phi_0(z)},\ z \in \Delta; \tag{2.12}$$

$$\lim_{n\to\infty} \iint_\Delta [|\phi_n| - \mathcal{R}e(\phi_n \operatorname{sgn} \overline{\phi_0})]dxdy = 0. \tag{2.13}$$

Then F^{-1} *has non-decreasable dilatation.*

Proof. We will apply (2.8) with our current F in the role of the F_1 of Lemma 2.2. In order to group $|\phi| - \phi \operatorname{sgn} \kappa$ together we rewrite (2.8) as

$$\begin{aligned}\iint_\Delta \frac{|\alpha|^2 - |\beta|^2 + (1-|\alpha|)[|\alpha| - \mathcal{R}e(\bar\beta \operatorname{sgn}\alpha)]}{(1+|\alpha|)(1-|\beta|^2)}|\phi|dxdy \le \\ \le \mathcal{R}e \iint_\Delta \frac{\bar\alpha(\alpha-\beta)(1-\alpha\bar\beta)}{|\alpha|(1-|\alpha|^2)(1-|\beta|^2)}(|\phi| - \operatorname{sgn}\kappa)dxdy.\end{aligned} \tag{2.14}$$

Postulating that $|\beta(z)| \le |\alpha(z)|$, and setting $k = \|\alpha\|_\infty = \|\kappa\|_\infty$, one finds that a lower bound on the coefficient of $|\phi|$ on the left side of (2.14) is

$$\begin{aligned}\frac{1-|\alpha|}{1+|\alpha|}[|\alpha| - \mathcal{R}e(\bar\beta \operatorname{sgn}\alpha)] \ge \frac{1-k}{1+k}\frac{|\alpha|}{k}[|\alpha| - \mathcal{R}e(\bar\beta \operatorname{sgn}\alpha)] \ge \\ \ge \frac{1-k}{2(1+k)k}[|\alpha|^2 + |\beta|^2 - 2\mathcal{R}e(\alpha\bar\beta)] = \frac{1-k}{2(1+k)k}|\alpha-\beta|^2.\end{aligned}$$

[6] An important point is that ϕ_0 does not necessarily belong to $\mathcal{B}$. One may,however, conjecture (?) that (2.11) and (2.13) together imply that ϕ_0 must be holomorphic in Δ.

An upper bound for the integrand on the right side of (2.14) is

$$|\alpha-\beta|\frac{1+|\alpha|^2}{(1-|\alpha|^2)(1-|\beta|^2}||\phi|-\phi\operatorname{sgn}\kappa|\le\frac{1+k^2}{(1-k^2)}|\alpha-\beta|\cdot|\,|\phi|-\phi\operatorname{sgn}\kappa|.$$

Therefore,

$$\begin{aligned}&\iint_\Delta|\alpha-\beta|^2|\phi|dxdy\\&\le\frac{2k(1+k^2)}{(1+k)(1-k)^3}\iint_\Delta|\alpha-\beta|\,|\,|\phi|-\phi\operatorname{sgn}\kappa|dxdy= \qquad (2.15)\\&=C\iint_\Delta|\alpha-\beta||\phi|^{1/2}[|\phi|-\mathcal{R}e(\phi\operatorname{sgn}\kappa)]^{1/2}dxdy,\ \text{for all }\phi\in\mathcal{B},\end{aligned}$$

with $C=\frac{2\sqrt{2}k(1+k^2)}{(1+k)(1-k)^3}$.

Hence, by Schwarz's Inequality,

$$\iint_\Delta|\alpha-\beta|^2|\phi|dxdy\le C^2\iint_\Delta[|\phi|-\mathcal{R}e(\phi\operatorname{sgn}\kappa)]^{1/2}dxdy,\ \ \phi\in\mathcal{B}.\quad(2.16)$$

In particular, setting $\phi=\phi_n$, we have, by Fatou's Lemma,

$$\iint_\Delta|\alpha-\beta|^2|\phi_0|dxdy=0.$$

Thus, $\alpha=\beta$, completing the proof.

If a particular complex dilatation $\kappa(z)$ satisfies the hypotheses (2.11)-(2.13) of Theorem 2.31, then this evidently remains true for any determination of $|\kappa(z)|$. So a simple example is the following:

Example. *If*

$$\mu[F](z)=k(z)\frac{\overline{\phi_0(z)}}{|\phi_0(z)|},\qquad(2.17)$$

where $k(z)>0$, $\phi_0\in\mathcal{B}$, *then* F^{-1} *has non-decreasable dilatation.*

Now if F is an EQC extension and at the same time possesses non-decreasable dilatation, then, obviously, F is a *unique* EQC extension. When μ is of the Teichmüller type (2.9) then one knows [S1] that F^{-1} is also of Teichmüller type. Thus we have a roundabout way of confirming the classical fact [S1] that for (2.9) one has *unique* extremality. More generally, however, ϕ_0 need not belong to $\mathcal{B}$, namely:

Theorem 2.32. [Re2] *Suppose* $\kappa(z)$ *is a complex dilatation of the form,*

$$\kappa(z)=\|\kappa\|_\infty\frac{\overline{\phi_0(z)}}{|\phi_0(z)|},\ \ z\in\Delta,\qquad(2.18)$$

satisfying (2.11) – (2.13). *Then* F^{κ} *is the unique EQC extension of its boundary restriction.*

Proof. In this case we have $|\mu[F](z)|$ a.e. constant. Hence $|\mu[F^{-1}](z)|$ is also a.e. constant. Since, by Theorem 2.31, F^{-1} has non-decreasable dilatation, it follows that F^{-1} is the unique EQC extension of $F^{-1}|_{\Gamma}$. Hence F is the unique EQC of $F|_{\Gamma}$.

Examples of complex dilatations of type (2.18) with ϕ_0 not belonging to $\mathcal{B}$ have been known since [S1].

It is also feasible to relate the possibility of unique EQC extensions of boundary values f directly to properties of f. As Strebel [S2] observed, Lemma 2.2 immediately implies, roughly speaking, that if $k_0[f]$ is determined by the global nature of f rather than by its local behavior (for example, if f is so smooth as to have locally conformal extensions without having a globally conformal extension to $\Delta \cup \Gamma$) then there is just one EQC extension, namely one of type (2.9). We saw with Theorem 1.43 that this also happens in the LNA case. The corresponding result for the EQC case is the following.

Theorem 2.33. [S2] *Suppose* f *is a sense-preserving homeomorphism of* Γ *onto itself with the property that*

$$\lim_{h\to 0} \frac{f(e^{i(\theta+h)}) - f(e^{i\theta})}{f(e^{i\theta}) - f(e^{i(\theta-h)})} = 1.$$

Then f *has a unique EQC extension* F_0, *and* $\mu[F_0]$ *is of Teichmüller type* (2.9).

References

[A] Lars V. Ahlfors, *Lectures on Quasiconformal Mappings*, Van Nostrand Math. Studies N° 10, Princeton, 1966.

[Ber1] Lipman Bers, *A non-standard integral equation with applications to quasiconformal mappings*, Acta Math. **116** (1966), 113–134.

[Ber2] Lipman Bers, *A new proof of a fundamental inequality for quasiconformal mappings*, J. d'Analyse Math. **36** (1979), 15–30.

[Beu-A] A. Beurling and L. Ahlfors, *The boundary correspondence under quasiconformal mappings*, Acta Math. **96** (1956), 125–142.

[Bo-W] F.F. Bonsall and D. Walsh, *Symbols for trace class Hankel operators with good estimates for norms*, Glasgow Math. J. **28** (1986), 47–54.

[E] Clifford J. Earle, *Conformally natural extensions of vector fields from* S^{n-1} *to* B^n, Procs. Am. Math. Soc. **102** (1988), 145–149.

[E-L] Clifford J. Earle and Li Zhong, *Extremal quasiconformal mappings in plane domains*, [*this volume*].

[Fe] R. Fehlmann, *Über extremale quasikonforme Abbildungen*, Comm. Math. Helv. **56** (1981), 558–580.

[Fo-Ru] Frank Forelli and Walter Rudin, *Projections on spaces of holomorphic functions in balls*, Indiana Univ. Math. J. **24** (1974), 593–602.

[G-Re] F.W. Gehring and E. Reich, *Area distortion under quasiconformal mappings*, Ann. Acad. Sci. Fenn. **388** (1966), 1–15.

[H] Richard S. Hamilton, *Extremal quasiconformal mappings with prescribed boundary values*, Trans Am. Math. Soc. **138** (1969), 399–406.

[K] S.L. Krushkal', *Extremal quasiconformal mappings*, Siberian Math. J. **10** (1969), 411–418.

[Re1] Edgar Reich, *Uniqueness of Hahn-Banach extensions from certain spaces of analytic functions*, Math. Zeitschr. **167** (1979), 81–89.

[Re2] Edgar Reich, *On criteria for unique extremality of Teichmüller mappings*, Ann. Acad. Sci. Fenn. AI Math. **6** (1981), 289–301.

[Re3] Edgar Reich, *A quasiconformal extension using the parametric representation*, J. d'Analyse Math. **54** (1990), 246–258.

[Re4] Edgar Reich, *On some related extremal problems*, Rev. Roum. Math. Pures Appl. **39** (1994), 613–626.

[Re5] Edgar Reich, *A theorem of Fehlmann-type for extensions with bounded $\bar{\partial}$-derivatives*, Complex Variables **26** (1995), 343–351.

[Re-C] Edgar Reich and Chen Jixiu, *Extensions with bounded $\bar{\partial}$-derivative*, Ann. Acad. Sci. Fenn. AI Math **16** (1991), 377–389.

[Re-S1] Edgar Reich and Kurt Strebel, *On quasiconformal mappings which keep the boundary points fixed*, Trans. Am. Math. Soc. **138** (1969), 211–222.

[Re-S2] Edgar Reich and Kurt Strebel, *Extremal quasiconformal mappings with given boundary values*, Contributions to Analysis, A Collection of Papers dedicated to Lipman Bers, Academic Press, New York, 1974, pp. 375–391.

[S1] Kurt Strebel, *Zur Frage der Eindeutgkeit extremaler quasikonformer Abbildungen des Einheitskreises (I); II*, Comm. Math. Helv. **36** (1961/1962), 306-323; **39** (1964), 77–89.

[S2] Kurt Strebel, *On the existence of extremal Teichmüller mappings*, J. d'Analyse Math. **30** (1976), 464–480.

[S3] K. Strebel, *Extremal quasiconformal mappings*, MSRI Report N° 09419-86 (August 1986) = Resultate der Mathematik **10** (1986), 168–210.

Department of Mathematics
127 Vincent Hall
206 Church Street S.E.
University of Minnesota
Minneapolis, MN 55455

Construction of Quasiregular Mappings

Seppo Rickman

1 Introduction

Quasiregular mappings were introduced and studied by Yu. G. Reshetnyak in a series of articles that began to appear in 1966. Reshetnyak used the term "mappings with bounded distortion" and defined them by means of the so called analytic definition of quasiconformal mappings leaving out the homeomorphism requirement. One of Reshetnyak's main result in the theory is that, if not constant, a quasiregular mapping is discrete and open. This means that such mappings are branched covers in a general sense.

One of the most interesting features of quasiregular mappings is that they give a natural generalization of the geometric aspects of the theory of complex analytic functions in the plane to Euclidean n-space R^n, or more generally, to Riemannian n-manifolds. For example, a Picard type theorem on omitted values is valid in the following form: a quasiregular mapping of R^n into itself is constant if it omits at least q points, where q is an integer that depends only on n and the dilatation coefficient. General results in value distribution are known too. The main result there is a defect relation in the spirit of Ahlfors' theory of covering surfaces.

The potential theoretic side of the theory of quasiregular mappings is based on extremals of certain variational integrals which generalize the Dirichlet integral and which are invariant under conformal change of metric. The Euler-Lagrange equations for these integrals are second order quasilinear partial differential equations. As a consequence, there is a strong interaction between PDE theory and quasiregular mappings.

One of the basic questions in quasiregular theory is the existence of maps of desired properties. In dimension two this question often reduces to the study of analytic functions and to solving a Beltrami equation. In higher dimension the only known way to show existence is by concrete construction. More precisely, this means that the map is defined by splitting the domain into pieces where the map is required to be quasiconformal, or the map is obtained as a locally uniform limit of maps of the said type. This state of affairs leads often to laborious constructions which are hard to accomplish, and moreover, are difficult for the reader. The proofs of the sharpness of the Picard type theorem [R2] and the defect relation [R5] are examples.

The purpose of this paper is to give an overview of the principles behind various constructions of quasiregular maps. The main theme will be the description of how the so called Alexander map principle comes into play in the subject.

2 Quasiregular mappings

Let G be a domain in R^n, $n \geq 2$. A continuous map $f: G \to R^n$ is called *quasiregular* (qr) if f belongs locally to the Sobolev class W_n^1 and there exists K, $1 \leq K < \infty$, such that

$$|f'(x)|^n \leq K J_f(x) \text{ a.e.} \tag{2.1}$$

Here $f'(x)$ is the formal derivative of f, $|f'(x)|$ its operator norm, and $J_f(x)$ the Jacobian determinant. In this article we call a qr map $K - qr$ if (2.1) is satisfied. The definition extends immediately to the case $f: M \to N$ where M and N are connected oriented Riemannian n-manifolds. A qr homeomorphism is called *quasiconformal* (qc). For the general theory of qr maps we refer to [Re2], [MRV], [V], [R4], and [HKM].

Quasiregular theory has recently been generalized to various non-Rieman-nian settings. J. Heinonen and I. Holopainen [HH] have established a theory on Carnot groups with respect to the Carnot-Caratheodory metric. They have shown that Reshetnyak's theorem is valid for 2-step Carnot groups, i.e., for Carnot groups of Heisenberg type. They also prove a Picard type theorem on omitted values. The quasiconformal theory on such spaces has been studied extensively by A. Koranyi and H.M. Reimann [KR1], [KR2], G.D. Mostow [M1], [M2], P. Pansu [Pa], J. Heinonen and P. Koskela [HK].

In the Euclidean case $1 - qr$ maps are exactly the complex analytic functions for $n = 2$. For $n \geq 3$ $1 - qr$ maps are restrictions of Möbius transformations or constants. This is the content of a generalized Liouville theorem by Yu. G. Reshetnyak [Re1] and F.W. Gehring [G]. Locally homeomorphic qr maps are in a sense rigid when $n \geq 3$. This is clearly seen from Zorich's theorem [Z]: If $f: R^n \to R^n$ is qr, locally homeomorphic, and $n \geq 3$, then f is homeomorphic, that is, qc.

3 Alexander maps and products

Let the 2-sphere S^2 be given a finite triangulation. For example, by passing to the barycentric subdivision we obtain a triangulation with the property that each vertex belongs to an even number of 2-simplicies. The new triangulation can then be used to define an orientation preserving discrete open map $f: S^2 \to S^2$ by the method of Alexander, namely, each 2-simplex is mapped alternatively onto the upper and lower hemisphere of S^2. If the 2-simplices of the triangulation have a regularity property, for example, if each of them can be mapped by an L-bilipschitz map onto a standard 2-simplex up to some scaling factor, then the map f can be made $K - qr$ with K depending only on L. We understand that an Alexander map is bilipschitz in each simplex of the given triangulation.

This principle applies to any connected oriented Riemannian n-manifold M with a given triangulation. If the triangulation satisfies a regularity property as above, the existence of a nonconstant qr map $f: M \to S^n$ follows. The construction of a triangulation of the desired kind on an arbitrary M was done by K. Peltonen in [Pe]. For special cases see also [MS].

Let us return to the case of an Alexander map $f: S^2 \to S^2$ defined by a triangulation with the regularity property. A point x in $R^3 \setminus \{0\}$ has a spherical representation $x = (t, y)$, where $y = x/|x| \in S^2$ and $t = |x|$. The map f has a trivial extension to a map $g: R^3 \to R^3$ by

$$\begin{aligned} &g(t,y) = (t, f(y)), \quad t > 0 \\ &g(0) = 0. \end{aligned}$$

The extension is also qr but the dilatation bound K in (2.1) tends to ∞ when the sizes of the triangles in S^2 tend to zero. By a suitable change in the first factor we may keep the dilatation under a universal bound. For example, if the triangles have approximately diameter δ, then such an extension h is obtained by setting

$$\begin{aligned} &h(t,y) = (t^{1/\delta}, f(y)), \quad t > 0, \\ &h(0) = 0. \end{aligned} \tag{3.1}$$

The map h in (3.1) gives a $K - qr$ counterpart in R^3 of the planar power map $z \mapsto z^k$ in complex notation and the bound K is independent of δ.

Suppose R^2 is given a triangulation with a regularity property as described above and such that Alexander's method applies to give a qr map $f: R^2 \to S^2$. We may now ask again: How do we extend f to a qr map of R^3? Suppose that the 2-simplices in the triangulation of R^2 have approximately diameter δ. Then a $K - qr$ extension h to R^3 is obtained by setting

$$h(x_1, x_2, x_3) = (e^{x_3/\delta}, f(x_1, x_2)). \tag{3.2}$$

The map h omits the origin, it can be regarded as a counterpart for the exponential function $z \mapsto e^z$, and K is again independent of δ. This type of map was given by Zorich [Z].

Both in (3.1) and (3.2) the second component is a fixed map f. We now ask what conditions are required if we replace f in the case (3.2) by maps

f_{x_3} depending on the parameter x_3. For example, given two Alexander maps $f_0: R^2 \times \{0\} \to S^2$ and $f_1: R^2 \times \{1\} \to S^2(e^{1/\delta})$ with diameter approximately δ of the 2-simplexes of their triangulations, is it possible to extend $f_0 \cup f_1: R^2 \times \{0,1\} \to S^2 \cup S^2(e^{1/\delta})$ to a $K - qr$ map of a neighborhood of $R^2 \times [0,1]$? In a very general sense this is possible, and furthermore, the construction can be performed in a concrete geometric way. It amounts to exhibit a suitable deformation from f_0 to f_1 through discrete open maps $f_t: R^2 \times \{t\} \to S^2(e^{t/\delta})$. Precise conditions for such a deformation theory is given in [R2, Section 5]. Observe, however, that the maps f_t are in general not Alexander maps for $0 < t < 1$. In Section 5 we will describe an application of this.

4 Prisms and quasiregularity

As we saw in Section 3, a natural method to produce *qr* maps in dimension three is to form a product of a 2-dimensional Alexander map and a 1-dimensional homeomorphism. This idea, developed a little further, can be used to produce several types of examples to answer specific questions. Among other things this method has given a counter example to Lindelöf's theorem on asymptotic values [R1], [R4, VII.7.1], nonremovability of sets with small Hausdorff dimension [R6], and a counter example to Denjoy's theorem on the order and asymptotic values [HoR].

There is a common feature in the proofs of the results mentioned above. Let $P = T \times L \subset R^3$ be a prism, where T is a rectilinear triangle in R^2 and $L = [a, b]$ a line segment. We map P onto the upper half of a spherical ring $\overline{B}^3(r_2) \setminus B^3(r_1)$ by a map h, quasiconformal in int P, so that $h(T \times \{a\})$ is the upper hemisphere $S^2_+(r_1), h(T \times \{b\})$ the upper hemisphere $S^2_+(r_2)$, and $h(\dot{T} \times L)$ is contained in the plane $R^2 \times \{0\}$ ($\dot{T}$ is the relative boundary of T). Let $P' = T' \times L$ be another prism with $P \cap P' = (\dot{T} \times L) \cap (\dot{T}' \times L)$ one common face. Let $h': P' \to R^3$ be a map, *qc* in int P', so that h and h' coincide on $P \cap P'$. The maps h and h' together define a *qr* map in $\operatorname{int}(P \cup P')$. The trick in the constructions depends on the observation that the map h' can be defined somewhat freely outside the common face $P \cap P'$. For example, the image $h'P'$ can be a slightly deformed half of a spherical ring. As a consequence, if the domain of a map to be constructed is decomposed into prisms in some suitable way, then passing through a sequence of prisms we are able to "deform" the behavior of the map in a desirable manner. Limits for such deformation is given by the dilatation bound of the map. Putting this in another way, we perform a map which is a deformation of the type (3.2) in the sense that the image sphere in the Alexander map varies.

Let us describe more precisely the proof of the following nonremovability theorem in [R6]: *For each $\lambda > 0$ there is a self-similar Cantor set E in*

R^3 with Hausdorff dimension $\dim_H E \leq \lambda$ *and a bounded* $K(\lambda) - qr$ *map* $f: R^3 \setminus E \to R^3$ *that does not extend continuously to any point of* E.

The Cantor set E is obtained as follows. Let Q be the rectangular box $[0,1] \times [0,\gamma] \times [0,\gamma^2]$ in R^3, where

$$\gamma = (\frac{1 - m\delta}{m+1})^{1/3},$$

$m > 20$ is a positive integer and $\delta > 0$ such that $1/10 \leq m\delta \leq 1/3$. First we delete from Q m slices

$$\{x \in Q: j\gamma^3 + (j-1)\delta < x_1 < j\gamma^3 + j\delta\}, \quad j = 1, \ldots, m,$$

which are parallel to the x_2x_3-plane. Let the union of these slices be D_1. We are left with boxes $Q_{1k} = I_k \times [0,\gamma] \times [0,\gamma^2]$, $k = 1, \ldots, m+1$, where the length of I_k is γ^3. From each box Q_{1k} we delete similarly m slices parallel to the x_1x_3-plane etc. If the union of boxes in the ith step is denoted by E_i, the Cantor set is

$$E = \bigcap_{i=1}^{\infty} E_i.$$

Without loss of generality we may consider an unbounded f omitting a ball. Start with the identity map g_0 outside Q. To extend g_0 to the first generation of slices we decompose each such slice into prisms. More precisely, if the slice is $A = M \times L$, where M is a 2-dimensional rectangle and L a line segment (here we shifted the order of coordinates), we triangulate M so that the 2-simplices satisfy a regularity condition and their sizes have a fixed upper bound. In each prism $T \times L$, T a 2-simplex in the triangulation, we let the map be qc onto a somewhat deformed half of a spherical ring as described above. It is possible to arrange all this so that we get a qr extension of g_0 to $(R^3 \setminus Q) \cup A$ so that the map omits a ball and that the size of the image of a prism increases gradually when we approach a middle part of the slice. When such an extension is performed to all slices of the first generation, we get a qr extension g_1 of g_0 to $(R^3 \setminus Q) \cup D_1$. The next step is to extend g_1 to the second generation of slices. The boundary correspondence for the extension, call it g_2, is inherited from the original identity map and from the values of g_1 in the first generation slices. Here we arrange things so that the intersection of the second generation slices with the first generation slice is contained in a part where g_2 is locally a diffeomorphism. This way (together with suitable choice of the parameters) we are able to avoid a cumulative effect on the dilatation when passing from g_1 to g_2. Again we increase gradually the image of a prism when we approach a middle part of a slice of the second generation. We continue similarly to subseqent generations of slices. We can keep a uniform lower bound on the image of prisms of the middle part of the slices of all generations. We also take care that the extensions omit a fixed ball. The limit map has then the required properties.

Only recently results for *qr* maps giving removability of sets with positive Hausdorff dimension were proved for $n \geq 3$. By ideas originating from [DS] T. Iwaniec and G. Martin proved such removability for even dimension in [IM]. In [I] Iwaniec extended the results to all dimensions.

5 The Picard example

The sharpness of the Picard type theorem mentioned in the introduction is known in dimension three. The result is the following: *For every positive integer p there exists a nonconstant $K(p) - qr$ map $f\colon R^3 \to R^3$ omitting p points.* The construction is given in [R2]. Modifications of it has been applied to several other problems. First, it was shown in [R3] that the local index can attain an arbitrary large value, independent of the dilatation, in a tamely embedded Cantor set. Second, the method provides a starting point to prove the sharpness of a defect relation [R5]. Third, the method can be used to show that there is radical difference in the local growth behavior between *qc* and *qr* maps in dimension three [HeR], contrary to the case in dimension two. This follows as a corollary to a result where arbitrarily high local index is attained on a wildly embedded Cantor set, namely, on the Antoine necklace. In [HeR] we also apply similar ideas to show that an example by M.H. Freedman and R. Skora [FS] of a *qc* group acting on S^3 and not topologically conjugate to a Möbius group is in fact semiconjugate by a *qr* map to a Möbius group. In this section the main ideas of the construction in [R2] will be outlined.

In (3.2) we gave the construction of a *qr* map $h\colon R^3 \to R^3 \setminus \{0\}$ (called the Zorich map), which is a counterpart of the exponential function. The construction extends to any dimension. Let us look at the map h from a particular viewpoint. Let the given triangulation of R^2 be G as a simplicial 2-complex. If A is any 2-simplex of G, then the cylinder $\text{int}A \times]0, \infty[$ is mapped alternatively either onto $H_+ \setminus \overline{B}^3$ or $H_- \setminus \overline{B}^3$, where $H_\pm = \{x \in R^3 \colon x_3 \gtrless 0\}$ and B^3 is the open unit ball. Similarly $\text{int}A \times]-\infty, 0[$ is mapped onto $H_+ \cap B^3$ or $H_- \cap B^3$. Write $U_1 = R^3 \setminus \overline{B}^3$, $U_2 = B^3 \setminus \{0\}$, $u_1 = \infty$, $u_2 = 0$, $W_j = h^{-1}U_j$, $j = 1, 2$. We observe $W_1 = H_+$, $W_2 = H_-$. The unit sphere S^2, which separates u_1 and u_2, has the preimage $R^2 = h^{-1}S^2$. Hence the common boundary R^2 of W_1 and W_2 has a simple structure.

Suppose we are given a nonconstant *qr* map $f\colon R^3 \to R^3 \setminus \{u_2, u_3\}$, where $u_2 = -e_3/2$, $u_3 = e_3/2$ (e_1, e_2, e_3 is the standard base in R^3). Set $u_1 = \infty$ and let U_1, U_2, U_3 be the components of $R^3 \setminus (S^2 \cup B^2 \cup \{u_2, u_3\})$ such that $u_j \in \overline{U}_j$, $j = 2, 3$. Now the situation is necessarily much more complicated than in the Zorich map. In particular, the preimages $W_j = f^{-1}U_j$ become exceedingly complicated when they approach ∞. The argument to see this is the following. First, since f is discrete open and since the points u_j are omitted, any component of W_j is unbounded. Since f is *qr* and omits at least one point in R^3, the lower order of f is positive (see [R4, VII. 6.16]). This implies (see [R4, V.(8.4), V.(8.10)]) that there is a sequence of balls $B^3(x_i, r_i)$ with $x_i \to \infty$, $r_i/|x_i| \to 0$, and the average covering number $\nu(\overline{B}^3(x_i, r_i), S^2)$ of the ball $\overline{B}^3(x_i, r_i)$ over the sphere S^2 tending to ∞. Here $\nu(E, S^2)$ is the average of the counting number

$$n(E, y) = \sum_{x \in f^{-1}(y) \cap E} i(x, f)$$

over S^2 (with respect to the 2-measure in S^2), E a Borel set, and $i(x, f)$ the local topological index at x. From the comparison lemma in [R4, IV. 1.1] it follows that, for i large, each ball $B(x_i, 2r_i)$ meets some component of W_j for all $j = 1, 2, 3$. This means that the sets W_j have the peculiar property that in many places of a common boundary of two of them the third must be present in a near neighborhood, the size of the neighborhood being determined by the local covering behavior. Furthermore, the narrowness of a component of some W_j in the Phragmén-Lindelöf sense (see [GLM] or [R4, VII. 6.19]) has the tendency to increase the covering growth. So, there is a balance between the growth behavior and complexity of the sets W_j (more growth implies more balls $B(x_i, r_i)$ of the type described above), and this balance requires a certain opening behavior of every component of each W_j. All these properties impose severe restrictions on the sets W_j.

The main ideas in the construction of a qr map $f: R^3 \to R^3$ omitting two points in R^3 are the following, We build the sets W_j with few components (in fact one component for $j = 1, 2$ and six components for $j = 3$) so that the components have "conelike" opening to ∞ and so that "near" any common boundary point of two of the W_j's the third W_j appears. This is accomplished by a certain procedure of digging tunnels. More precisely, tunnels for, say W_j, has the purpose of making connections to common boundary points of W_k and W_l, $k, l \neq j$. In [R2] this process is called cave refinement.

At a specific distance from $\partial W_1 \cup \partial W_2 \cup \partial W_3$ we define the first level surfaces in each W_j, say N_1 in W_1. N_1 is homeomorphic to R^2 and it will be given a triangulation G_1 with a regularity property so that the Alexander map method applies. This level surface is mapped by an Alexander map onto a sphere $S^2(s_0)$ where $s_0 > 1$. Other level surfaces in W_1 are constructed by similarity from N_1. Namely, we fix one similarity map φ_1 and obtain the ith level surface as $\varphi_1^{i-1} N_1$. For larger i the level surface is at a larger distance from ∂W_1. The similarity transfers the triangulation on N_1 to $\varphi_1^{i-1} N_1$. The obtained triangulation $\varphi_1^{i-1} G_1$ is canonically subdivided so that we get the same order of size of the triangles as in the original G_1. By means of the subdivided triangulation we define an Alexander map from $\varphi_1^{i-1} N_1$ onto $S^2(s_{i-1})$, where $s_{i-1} > s_0$ is chosen in a specified way.

The next step is to extend these Alexander maps to the layers between the level surfaces (cf. the end of Section 3). The problem here is that the triangulations on two successive level surfaces are combinatorially different. The solution to this problem is a general deformation theory of discrete open maps in dimension two. Without the deformation theory there would be a cumulative effect on the dilatation. As a result we are able to perform these extensions with a fixed bound on the dilatation. The sets W_2 and W_3 are treated similarly.

It remains to extend the obtained map to the parts between the first level surfaces and $\partial W_1 \cup \partial W_2 \cup \partial W_3$. This requires a technique which differs somewhat from the Alexander map method. On the other hand, there is no danger of a cumulative effect on the dilatation in this part.

The construction for omitting p points in R^3 is in principle very similar, it only involves more combinatorics.

References

[DS] S.K. Donaldson and D.P. Sullivan, *Quasiconformal 4- manifolds*, Acta Math. **163** (1990), 181–252.

[FS] M.H. Freedman and R. Skora, *Strange actions of groups of spheres*, J. Diff. Geom. **25** (1987), 75–98.

[G] F.W. Gehring, *Rings and quasiconformal mappings in space*, Trans. Amer. Math. Soc. **103** (1962), 353–393.

[GLM] S. Granlund, P. Lindqvist, and O. Martio, *Phragmén–Lindelöf's and Lindelöf's theorem*, Ark. Mat. **23** (1985), 103–128.

[HH] J. Heinonen and I. Holopainen, *Quasiregular mappings on Carnot groups*, J. Geometric Analysis (To appear).

[HK] J. Heinonen and P. Koskela, *Definitions of quasiconformality*, Invent. Math. **120** (1995), 61–79.

[HKM] J. Heinonen, T. Kilpeläinen, and O. Martio, *Nonlinear Potential Theory*, Oxford University Press, 1993.

[HeR] J. Heinonen and S. Rickman, *Quasiregular maps $S^3 \to S^3$ with wild branch set*, Topology (To appear).

[HoR] I. Holopainen and S. Rickman, *Failure of the Denjoy theorem for quasiregular maps in dimension $n \geq 3$*, Proc. Amer. Math. Soc. **124** (1996), 1783–1788.

[I] T. Iwaniec, *p-harmonic tensors and quasiregular mappings*, Ann. Math. **136** (1992), 589–624.

[IM] T. Iwaniec and G. Martin, *Quasiregular mappings in even dimensions*, Acta Math. **170** (1993), 29–81.

[KR1] A. Korányi and H.M. Reimann, *Quasiconformal mappings on the Heisenberg group*, Invent. Math. **80** (1985), 309–338.

[KR2] A. Korányi and H.M. Reimann, *Foundations of the theory of quasiconformal mappings on the Heisenberg group*, Adv. in Math. **111** (1995), 1–87.

[MRV] O. Martio, S. Rickman, and J. Väisälä, *Definitions for quasiregular mappings*, Ann. Acad. Sci. Fenn. Ser. AI Math. **448** (1969), 1–40.

[MS] O. Martio and U. Srebro, *On the existence of automorphic quasimeromorphic mappings in R^n*, Ann. Acad. Sci. Fenn. Ser. AI Math. **3** (1977), 123–130.

[M1] G.D. Mostow, *Strong Rigidity of Locally Symmetric Spaces*, Annals of Math. Studies 78, Princeton University Press, 1973.

[M2] G.D. Mostow, *A remark on quasiconformal mappings on Carnot groups*, Michigan Math. J. **41** (1994), 31–37.

[Pa] P. Pansu, *Métriques de Carnot–Carathéodory et quasiisométries des espaces symétriques de rang un*, Ann. Math. **129** (1989), 1–60.

[Pe] K. Peltonen, *On the existence of quasiregular mappigns*, Ann. Acad. Sci. Fenn. Ser. AI Math. Diss. **85** (1992), 1–48.

[Re1] Yu. G. Reshetnyak, *The Liouville theorem with minimal regularity conditions*, Sibirsk. Math. Zh. **8** (1967), 835–840. (Russian)

[Re2] Yu. G. Reshetnyak, *Space Mappings with Bounded Distortion*, Transl. of Math. Monographs Vol. 73, Amer. Math. Soc., Providence, R.I., 1989.

[R1] S. Rickman, *Asymptotic values and angular limits of quasiregular mappings of a ball*, Ann. Acad. Sci. Fenn. Ser. AI Math. **5** (1980), 185–196.

[R2] S. Rickman, *The analogue of Picard's theorem for quasiregular mappings in dimension three*, Acta Math. **154** (1985), 195–242.

[R3] S. Rickman, *Sets with large local index of quasiregular mappings in dimension three*, Ann. Acad. Sci. Fenn. Ser. AI Math. **10** (1985), 493–498.

[R4] S. Rickman, *Quasiregular Mappings*, Ergebnisse der Mathematik und ihrer Grenzgebiete, 3. Folge, Band 26, Springer-Verlag, 1993.

[R5] S. Rickman, *Defect relation and its realization for quasiregular mappings*, Ann. Acad. Sci. Fenn. Ser. AI Math. **20** (1995), 207–243.

[R6] S. Rickman, *Nonremovable Cantor sets for bounded quasiregular mappings*, Ann. Acad. Sci. Fenn. Ser. AI Math. **20** (1995), 155–165.

[V] M. Vuorinen, *Conformal Geometry and Quasiregular Mappings*, Lecture Notes in Math. Vol. 1319, Springer-Verlag, Berlin-Heidelberg, 1988.

[Z] V.A. Zorich, *The theorem of M. A. Lavrent'ev on quasiconformal mappings in space*, Mat. Sb. **74** (1967), 417–433. (Russian)

Department of Mathematics
P.O. Box 4 (Yliopistonkatu 5)
FIN–00014 University of Helsinki
Finland
E-mail address: rickman@cc.helsinki.fi

Mappings and Spaces

Stephen Semmes

ABSTRACT In these notes we look at mappings which respect some geometric structure and the problem of being able to produce such mappings in order to study geometry.

1 Basic concepts

Let M and N be sets. We shall consider various kinds of structure on M and N and various ways that a mapping $f : M \to N$ can respect these structures. We would like to understand how rigid such mappings need to be, and how easy it is to produce them.

If M and N are topological spaces, then we can look at mappings f which are continuous, or which are embeddings (homeomorphisms onto their images).

If M and N are smooth manifolds, then we can look at smooth mappings, or diffeomorphisms from M onto some submanifold in N.

If M and N are polyhedra, then we can look at piecewise linear mappings and embeddings.

To be complete we ought to consider also algebraic and analytic categories, but for simplicity we shall not.

We shall be interested in the distinction between finite and infinite complexity of mappings and spaces. This distinction will be an informal one, in part an emotional one. Finite polyhedra and piecewise linear mappings between them will be considered to have finite complexity. They are after all described by finite sets of data, they are obtained by combining finitely many linear pieces, and we can think of the linear pieces as being very simple. Compact smooth manifolds and diffeomorphisms between them will be considered as having finite complexity as well. This is slightly questionable, since they are not exactly described by finitely many parameters. However the deviation from finiteness is very small and easily controlled. They can be decomposed into finitely many pieces which are very close to being linear, for instance. Homeomorphisms, by contrast, can have infinite amounts of twisting, even homeomorphisms between compact smooth man-

[1]The author is supported by the U.S. National Science Foundation.

ifolds. They can contain infinitely many copies of some basic model. By our reckoning homeomorphisms between compact smooth manifolds can have infinite complexity.

Spaces or mappings with infinite complexity in the informal sense used here should involve some kind of nontrivial infinite process.

Now suppose that $(M, d(x,y))$ and $(N, \rho(u,v))$ are metric spaces. This is a natural intermediate level of structure between topological spaces and the much more rigid smooth manifolds and polyhedra. Infinite complexity is permitted, but it makes sense to talk about bounds too. For metric spaces we can look at Lipschitz mappings and bilipschitz embeddings. A mapping $f : M \to N$ is *Lipschitz* if there is a constant $C > 0$ so that

$$\rho(f(x), f(y)) \leq C\, d(x,y) \tag{1.1}$$

for all $x, y \in M$, and f is *bilipschitz* if there is a constant C so that

$$C^{-1}\, d(x,y) \leq \rho(f(x), f(y)) \leq C\, d(x,y) \tag{1.2}$$

for all $x, y \in M$. The first condition means that f cannot expand distances by more than a bounded factor, the second condition means that f cannot shrink distances by more than a bounded factor either. Two metric spaces are said to be *bilipschitz equivalent* if there is a bilipschitz mapping from one onto the other. (In this case the inverse mapping is well-defined and the bilipschitz condition is equivalent to asking that both the map and its inverse be Lipschitz.)

There is another nice class of embeddings between metric spaces, the quasisymmetric mappings ([TV1]). A mapping $f : M \to N$ is *quasisymmetric* if it is not constant and if there exists a homeomorphism $\eta : [0, \infty) \to [0, \infty)$ such that for any points $x, y, z \in M$ and $t > 0$ we have that

$$d(x,y) \leq t\, d(x,z) \quad \text{implies} \quad \rho(f(x), f(y)) \leq \eta(t)\, \rho(f(x), f(z)). \tag{1.3}$$

Thus if y is much closer to x than z is, then we can take t to be very small, and we conclude that $f(y)$ is much closer to $f(x)$ than $f(z)$ is, because $\eta(t)$ will also be very small. The requirement that η be a homeomorphism is not truly serious, just conveniently concise, all that really matters is that $\eta(t) \to 0$ as $t \to 0$ and that $\eta(t)$ remain bounded on bounded intervals.

The quasisymmetry condition implies that the mapping f distorts distances in roughly the same way in all directions. If $d(x,y)$ and $d(x,z)$ are approximately the same, then $\rho(f(x), f(y))$ and $\rho(f(x), f(z))$ should also be approximately the same, even if y and z lie in very different directions from x.

When M and N are both Euclidean spaces of the same dimension quasisymmetric mappings are the same as quasiconformal mappings. The concept of quasisymmetry is more convenient for dealing with embeddings and mappings between general metric spaces.

Bilipschitz mappings roughly preserve distances, quasisymmetric mappings roughly preserve relative distances. Quasisymmetric mappings may distort actual distances enormously though. For instance if $f : \mathbf{R}^n \to \mathbf{R}^n$ is quasisymmetric, then so is the composition of f with any dilation, and with the same choice of η. If a quasisymmetric mapping does distort distances in a severe way, it should do so in all directions, and by roughly the same amount. On reasonable metric spaces, like Euclidean spaces, the distortion of quasisymmetric mappings is never more than polynomial in a certain sense. (As distances in the first space double, the distances in the image can increase or decrease by only a bounded factor.)

The mapping $f : \mathbf{R}^n \to \mathbf{R}^n$ given by $f(x) = |x|^a\, x$ is quasisymmetric when $a > -1$, but it is bilipschitz only when $a = 0$.

Bilipschitz and quasisymmetric mappings can have infinite complexity, in the sense of infinite spiralling, for instance. To see this let $R(t)$, $-\infty < t < \infty$, be a one-parameter family of rotations on $\mathbf{R}^n$, and assume that $R(t)$ is Lipschitz as a mapping from the real line into the group of rotations with its usual metric. (The rotation group is a compact smooth manifold, any reasonable metric will do.) Define $f : \mathbf{R}^n \to \mathbf{R}^n$ by $f(x) = R(\log|x|)x$ when $x \neq 0$, $f(0) = 0$. That is, we take a point $x \in \mathbf{R}^n$, we set $t = \log|x|$, we get a rotation $R(t)$ on $\mathbf{R}^n$, and we apply this rotation to x to get $f(x)$. One can check that this always gives a bilipschitz mapping, even though it permits f to spin around a lot near the origin. We can compose this mapping with the quasisymmetric mappings mentioned in the previous paragraph to get quasisymmetric maps which both spin around and distort distances in a strong way.

To understand these examples we should think about what happens in other categories. In the topological category we can take any continuous family of rotations $R(t)$ on $\mathbf{R}^n$ and get a homeomorphism f by the recipe above. In order to get a mapping which is differentiable at the origin we would need $R(t)$ to have a limit as $t \to -\infty$. To get higher smoothness we would have to impose stronger conditions.

If we define $f : \mathbf{R}^n \to \mathbf{R}^n$ by the recipe $f(x) = r(|x|) \cdot x/|x|$, $f(0) = 0$, then in order to get a homeomorphism it is enough to ask that $r : [0, \infty) \to [0, \infty)$ be a homeomorphism. We do not have to impose any condition on the growth of $r(t)$ as $t \to \infty$ nor on the behavior at the origin. To get a quasisymmetric mapping we do need to impose additional conditions, including the requirements that $r(t)$ grow at infinity no faster than a power (so that exponentials are not permitted) and to vanish at the origin no faster than a power also (but different powers are permitted). To get a bilipschitz mapping we have to impose more stringent conditions on $r(t)$, and to get smoothness we need stronger conditions still.

These examples are quite simple but they illustrate the nonsmooth behavior which can occur. Of course bilipschitz and quasisymmetric mappings can be much more complicated, with much worse than isolated singularities.

I like bilipschitz and quasisymmetric mappings. They permit infinite complexity, part of life, but with strong constraints, part of life too. They are defined by geometric conditions that are close enough to the earth that one can hope to be able to produce them under nontrivial circumstances. Circumstances under which one cannot expect smoothness or piecewise linearity. They provide a way to try to ask for something without necessarily asking for too much.

That is the idea anyway. In the next sections we shall look more closely at the question of how easy it is to make bilipschitz and quasisymmetric mappings, and how they can be used to study the geometry of spaces. But first we shall take a look at the structure they do have, structure perhaps beyond what is obvious in the definitions.

2 Rigidity properties of mappings

A fantastic theorem in analysis states that a Lipschitz function $f : \mathbf{R}^n \to \mathbf{R}$ is differentiable at almost all points (with respect to Lebesgue measure). I think that people are often too complacent about this theorem, too old, too familiar, the proof too easy for analysts using modern technology. But it really is quite amazing. It says that if you pick a random point $x \in \mathbf{R}^n$ and blow up f there then you get a limit and the limit is linear. This is remarkable because nothing in the definition of a Lipschitz mapping explicitly suggests that that should be true. The Lipschitz condition does explicitly require the blow-ups to be bounded, but not that the limit should exist, let alone belong to such a special class of mappings (linear mappings).

Of course one has stronger rigidity properties of smooth and piecewise linear mappings, but these rigidity properties are made explicitly part of the definition, unlike the case of Lipschitz mappings.

One can try to formulate similar statements in other contexts but they usually do not work. For instance, the concept of a real-valued Lipschitz function makes sense on any metric space, and there are plenty of metric spaces for which there is a natural measure to use to interpret the phrase "almost everywhere", but there is generally nothing like the rigidity that one has for Lipschitz mappings on Euclidean spaces. One has to be careful in trying to formulate this idea of rigidity on metric spaces which lack the structure to make it easy to talk about derivatives, but there are natural ways to do it. For instance one can show that the behavior of Lipschitz functions on Cantor sets can vary from scale to scale in a nearly arbitrary manner (with no simple limiting behavior at almost all points). And there is no nice (finite-dimensional!) class of local models to capture frequent behavior of Lipschitz functions, in the way that affine functions do for $\mathbf{R}^n$.

Similarly, if one merely requires that $f : \mathbf{R}^n \to \mathbf{R}$ be Hölder continuous of some order $0 < \alpha < 1$, so that

$$|f(x) - f(y)| \leq C\,|x - y|^{\alpha} \tag{2.1}$$

for some $C > 0$ and all $x, y \in \mathbf{R}^n$, then one can also try to formulate rigidity properties for such functions, in terms of the limiting behavior of suitably normalized "blow-ups", but there is no nice theorem like this. One can make such functions behave in very different ways from scale to scale, location to location, with no limiting rigidity. This is a well-known phenomenon in Fourier analysis, integer orders of smoothness being special and enjoying rigidity properties that are not present for fractional orders of smoothness or for metric spaces in general.

A lot of Fourier analysis can be formulated for very general spaces, as in [CW1], [CW2], but differentiability is more delicate and requires more special structure. It does work in some non-Euclidean geometries, however, as in [Pa]. It is not at all clear for which geometries there is such a kind of rigidity for Lipschitz functions, and it is not even clear how best to formulate the question. One approach is given in [DS4] through the property called "GWALA", for "Generalized Weak Approximation of Lipschitz functions by Affine functions". This is an abstract version of the idea that a Lipschitz function on $\mathbf{R}^n$ is approximately affine at most scales and locations.

The theorem of differentiability almost everywhere says that Lipschitz functions on Euclidean spaces have a lot more rigidity, are closer to being smooth, than is immediately apparent or explicit in the definition. There are other manifestations of this phenomenon. If $f : \mathbf{R}^n \to \mathbf{R}$ is Lipschitz and $\epsilon > 0$, for instance, there is a subset of $\mathbf{R}^n$ on which f is continuously differentiable and whose complement has measure $< \epsilon$. There are subtle quantitative aspects of the existence of derivatives which come from Littlewood-Paley theory and Carleson's corona construction. (See [DS4] and the references therein.)

Of course bilipschitz mappings are Lipschitz and so similar considerations apply, with the extra information that the derivatives cannot be small.

What about quasisymmetric mappings? Again on general metric spaces, like Cantor sets, they can be pretty wild. But on Euclidean spaces they cannot be, and they too have to be differentiable almost everywhere. A wonderful theorem of Gehring [Ge] states that if $f : \mathbf{R}^n \to \mathbf{R}^n$ is quasisymmetric and B is a ball in $\mathbf{R}^n$, then there is a subset E of B with at least 90% of the points in B (as measured by Lebesgue measure) on which f is bilipschitz. One can even get a bound on the bilipschitz constant if one first normalizes f by composing it with a suitable dilation. There is nothing special about 90% here, anything less than 100% would work, one has only to pay for increasing the proportion by increasing the bound on the bilipschitz constant. Thus quasisymmetric mappings on Euclidean spaces are very close to being bilipschitz. They *are* bilipschitz on large pieces of the Euclidean space. This is another rigidity property which fails for metric spaces in general.

3 How many mappings, how easy are they to construct?

So we have these interesting classes of mappings for geometry, we should think about how easy it is to build them. Let us keep things simple and think about mappings between Euclidean spaces.

Let us begin with the "linear" question. Consider mappings from $\mathbf{R}^n$ into $\mathbf{R}^n$ which are continuous, or Lipschitz, or smooth, or piecewise linear. These are all linear conditions, in the sense that sums and scalar multiples of such mappings have the same property. (Quasisymmetric mappings do not linearize in quite such a simple manner.) If we impose suitable bounds we can get Banach or Frechet spaces of functions.

Let us take these linear spaces of functions as being reasonably well understood. This is not to say that there is no difficulty; to be able to take apart a function in some class like these into "simple" pieces or to build functions in some class by combining pieces in a certain way is often far from trivial. But there is a lot of technology in harmonic analysis for doing this, and many basic issues are understood.

The real issue comes with the nonlinear spaces of functions, the homeomorphisms, diffeomorphisms, piecewise-linear homeomorphisms, bilipschitz mappings, and quasisymmetric mappings. Or more generally the various kinds of embeddings.

For the purposes of this discussion let us consider only local phenomena. Locally a diffeomorphism is not very interesting, it is just a small perturbation of a nondegenerate affine mapping, nothing more. Globally it can be very different, practically anything can happen at infinity. Locally it is very simple. Similarly for piecewise-linear mappings.

Homeomorphisms and bilipschitz mappings and quasisymmetric mappings, by contrast, can be as complicated locally as at infinity. Whatever behavior is present at infinity can be rescaled to occur near a given point, even to repeat infinitely often closer and closer to the given point. This cannot happen for diffeomorphisms and piecewise-linear homeomorphisms.

As an illustration, consider the mapping $f : \mathbf{R}^n \to \mathbf{R}^n$ associated to a one-parameter family $R(t)$ of rotations on $\mathbf{R}^n$ as in Section 1. To get a mapping f which is differentiable at the origin one has to choose $R(t)$ so that it has a limit as $t \to -\infty$. This is not necessary to get a homeomorphism or a bilipschitz mapping or a quasisymmetric mapping. In these cases one can take $R(t)$ to be periodic.

So diffeomorphisms and piecewise-linear homeomorphisms are fairly simple locally, we can practically say exactly how they look and how they can be constructed. Homeomorphisms and bilipschitz and quasisymmetric mappings can be much more complicated. But exactly how complicated, and how can we construct them?

The answer to this is not so clear.

It is not hard to build plenty of examples, starting with simple examples with some finite complexity and then putting copies of them on top of themselves infinitely often. Mappings which spiral at infinitely many points, for instance.

This is not exactly a crisp statement though, it does not give a clear idea of what is possible. We shall return to this point in Section 4.

What about building new mappings from old ones? Suppose we have a mapping from the standard n-sphere $\mathbf{S}^n$ in $\mathbf{R}^{n+1}$ to itself, when can we extend it to a mapping from the closed ball in $\mathbf{R}^{n+1}$ to itself, a mapping of the same class?

For homeomorphisms this is trivial, one can simply extend radially to the ball. This also works for bilipschitz mappings.

In the piecewise-linear category the question does not really make good sense, one should use a polyhedral version of the sphere and ball (like a simplex) instead of the standard versions. If one does this then one can just use a radial extension again and the matter is trivial.

The smooth category is more tricky. One cannot just extend radially, for then the cone point at the center would not normally be a smooth point. Although this simple-minded method does not work, that does not mean that it is not true however, that one cannot extend a diffeomorphism from the sphere to a diffeomorphism on the ball. The answer is that one can in some dimensions and not in others, and that this is essentially the same as the existence of exotic spheres.

The question for quasisymmetric homeomorphisms turns out to be difficult. The answer is yes, because of results of Beurling-Ahlfors, Ahlfors, Carleson, and Tukia-Väisälä. (See [TV3].)

There are some results which suggest that the classes of quasiconformal and bilipschitz mappings should be practically as rich as the class of homeomorphisms. Sullivan [Su] has proved that in all dimensions except 4 every topological manifold can be given a unique quasiconformal or bilipschitz structure. One of the steps in the proof is to show that homeomorphisms on open subsets of $\mathbf{R}^n$ can be approximated by quasiconformal and bilipschitz mappings. Further approximation results were obtained in [TV2], which in particular allow a quasisymmetric mapping on a domain in $\mathbf{R}^n$ to be approximated well on the inside by one which is locally bilipschitz while still having approximately the same behavior at the boundary. (It is not true that Lipschitz or quasiconformal structures on topological manifolds have to exist or be unique in dimension 4, because of the results of [DoS]. Some of the results on approximating homeomorphisms by bilipschitz or quasisymmetric mappings also fail in dimension 4.)

So, in conclusion, we cannot say exactly what homeomorphisms or bilipschitz or quasiconformal mappings look like locally, but in some sense we can say that there are a lot of them.

Note that the topologists have another view on these matters. One can look at the topology of the space of germs of homeomorphisms, diffeomorphisms, etc., which fix the origin in $\mathbf{R}^n$ and ask how the topologies of these spaces differ, if at all. (See [KS2].) This analysis would be used to study questions of when a topological manifold can be made smooth, when a homeomorphism between smooth manifolds can be deformed to a diffeomorphism, etc.

4 Nervous worries and nagging doubts

In the previous section we saw how we could say that there are plenty of quasisymmetric and bilipschitz mappings. But is this really correct? Were we cheating ourselves?

Let us start with the assertion that we can build plenty of complicated bilipschitz and quasisymmetric mappings with infinite complexity by taking a simple example with finite complexity and putting infinitely many copies of it on top of themselves. This is true, but there is a delicate point that one has to remember. While making these infinite constructions, one still has to maintain some bounds, on the bilipschitz constant or the quasiconformal dilatation. When putting maps on top of themselves infinitely often one has to make sure that these dilatations do not become unbounded. This is a subtle condition. In harmonic analysis there is a good understanding of how to pile things on top of each other with good L^p bounds when $p < \infty$, or even BMO bounds, but L^∞ bounds are much more subtle. (See [J], [GJ].) And indeed in the basic examples of complicated mappings that one typically constructs one would normally arrange them so that at each stage of the construction one is adding something new to a place that was "flat" at the previous stage. So that there can be infinite twisting, but the twisting is layered in a fairly simple way.

That sounds pretty reasonable, but suppose we want to go in the other direction and start with some definite geometric problem in which one wants to construct a bilipschitz or quasisymmetric mapping with prescribed properties. Think of being given a surface in $\mathbf{R}^3$ and being asked to produce a quasisymmetric or bilipschitz parameterization of it. Or being given a metric space, like $(\mathbf{R}^2, |x-y|^{.9})$, and being asked to embed it bilipschitzly into the standard $\mathbf{R}^3$. If one tries to build such mappings through an iterative construction, one usually finds that the problem is not so well organized as one would like. That one cannot arrange the construction so that at each stage new twisting is added only at places which were pretty flat before. It is not hard to generate intricate examples with that sort of organization built into them, but in applications where one has to build a mapping with specific properties one typically does not have that kind of good structure. Everything is more mixed up, unless one makes additional assumptions that are not natural. That is what happened to me, anyway, in [Se1] and [Se2].

So yes one can build complicated examples, but the constructions are usually much tamer than what one would like to be able to handle in applications.

Note that this problem of having the dilatation pile up does not arise when one tries to construct homeomorphisms through infinite processes. For homeomorphisms there are no associated quantities which are required to be bounded, as there are for bilipschitz and quasisymmetric mappings.

The results of extension and topology should not be viewed without suspicion either. Consider Sullivan's theorem, to the effect that a topological manifold in dimension $\neq 4$ can be given a unique quasiconformal or bilipschitz structure. In dimensions ≤ 3 one has in fact the existence and uniqueness of a smooth structure for the manifold. In dimensions ≥ 5 this is not the case, and topological manifolds may not admit piecewise-linear structures either ([Si], §2, [KS2], p.311), so that it really may be advantageous to use bilipschitz or quasiconformal mappings. But in high dimensions one knows from the Kirby-Siebenmann theory that topological manifolds can be smoothed off a set of codimension 4. So that even though it may be natural to use quasisymmetric or bilipschitz mappings to accommodate the singularities, the mappings that need to arise for these topological purposes ought to be much nicer than generic quasisymmetric or bilipschitz mappings.

Note that the Kirby-Siebenmann theory also implies that a homeomorphism between piecewise-linear manifolds in high dimensions is isotopic to a piecewise-linear homeomorphism if the 3-dimensional cohomology with coefficients in $\mathbf{Z}_2$ vanishes. (See [KS1], [KS2], p.299.)

I don't know of any meaningful treatment of the nature of the singularities that are forced by the topology. I mean what type of maps does one really have to take, what kind of singularities they ought to have.

Similarly, in the extension and approximation results, one should expect to be able to obtain mappings which are better behaved than generic quasisymmetric or bilipschitz mappings.

In short, these results of topology and extension probably do not lead to quasisymmetric or bilipschitz mappings of the general type.

5 Facing the music

There is practically no existence theory for quasisymmetric and bilipschitz mappings.

I mean a theory that can produce a mapping with required properties, like parameterizing a space. And produce it from scratch, not as a modification of some given mapping. This is not a question of constructibility, there is not even a decent nonconstructive existence theory.

There are some positive results, but they are much more limited than the scope of the problem. In dimension 1 one can do pretty much what one would like to do, but dimension 1 is very special. Parameterizations by arclength have no counterpart in dimensions larger than 1. It is very helpful to always know which direction is forward and which is backward, with no intermediate sideways directions.

Dimension 2 is also special, because of the uniformization theorem and the measurable Riemann mapping theorem (so that all conformal structures on the 2-sphere are equivalent to each other). These can be used to give some reasonable criteria for 2-dimensional spaces to admit quasisymmetric parameterizations. See [Se1], [DS3], and [HK].

Already in dimension 2 the situation for bilipschitz mappings is much more difficult. The only general criteria for the existence of a bilipschitz parameterization that I know comes from the work of Toro [T1], [T2]. The results of [T2] also provide criteria in higher dimensions, but in dimension 2 there is a particularly nice formulation in terms of curvature.

Toro's results require more smoothness than one would like, but her conditions do scale correctly, and they are of an appropriately quadratic nature (orthogonality is taken into account).

Siebenmann and Sullivan [SS] have characterized the polyhedra in dimensions ≥ 5 which admit local bilipschitz coordinates from standard Euclidean spaces. People are often surprised that this is nontrivial for polyhedra, but it is. (As pointed out in [SS], examples of Edwards and Cannon [E], [C2], [C3] show that a polyhedron can be a topological manifold without having bilipschitz coordinates. The existence of a quasisymmetric parameterization is not known.)

These are all the positive results that I know about producing bilipschitz or quasiconformal parameterizations from scratch.

One can ask for less and just try to find a bilipschitz embedding of a metric space into some finite dimensional Euclidean space. Some criteria for doing this have been given by Assouad [A1], [A2], [A3], and also in [Se2]. For quasisymmetric embeddings Assouad's theorem provides a complete answer: a metric space M admits a quasisymmetric embedding into some finite-dimensional Euclidean space if and only if it satisfies the doubling condition, to the effect that every ball can be covered by a bounded number of balls of half the radius.

There are several examples, mostly from topology, in which bilipschitz or quasisymmetric parameterizations do not exist for spaces whose geometry is nonetheless reasonably well behaved. (See [Se4], [Se5].) These examples indicate that some of the results about the existence of parameterizations that one might want to be true simply are not true. There remain many situations where one can still hope for positive results (see Section 2 of [Se4]), but one has to face the fact that bilipschitz and quasisymmetric parameterizations simply may not exist in situations where one might want them to.

6 Comparisons with topology

Much of the preceding discussion makes sense simply for homeomorphisms, what about them?

That is, can one tell when a topological space is a topological manifold, so that it admits local homeomorphic coordinates?

This is called the recognition problem in geometric topology, and it is discussed in [C1], [C2], [E], although these references are now somewhat old. More recent references include [Q1], [Q2], [B^∞1], [B^∞2]. The bottom line is that there is a natural class of "generalized manifolds" which are topological manifolds when a certain invariant takes the favorable value, and the latter need not occur, so that there are topological spaces which are very similar to topological manifolds but which are not quite the same. It is not yet known what might be natural geometric structures for these almost-manifolds, but I am optimistic that interesting geometry will emerge from them.

Notice that the question of when a topological space admits an embedding into some finite-dimensional Euclidean space is well understood. (See [HW].) The proof uses the Baire category theorem, which provides a fairly typical example of how topologists can produce a mapping with good properties without having a clue about estimates. (Transversality plays a similarly frustrating role, it can give so much if one does not mind a lack of estimates.)

The recognition problem has a version in the piecewise-linear (PL) category which is instructive. Given a finite polyhedron P – a finite union of simplices sitting in some finite-dimensional Euclidean space – how can we tell if P is a PL manifold, i.e., locally PL homeomorphic to Euclidean space?

This happens if and only if the links of the various simplices in P are PL spheres. That is, if we take a k-simplex σ in P, with $k < n$, where n is the dimension of P, then we should ask that the link λ of σ in P be an $(n-k-1)$-dimensional PL sphere. Recall that this link is the polyhedron which goes around σ in P, so that P looks like the product of σ with the cone on λ near points in the interior of σ.

This characterization of PL manifolds among polyhedra is not so great, but we did win something, we started with a question about an n-dimensional polyhedron and we ended up with a question about polyhedra of dimension $< n$. This can enable one to use induction.

We also lost something. In this dimension reduction we are faced with the problem of recognizing when a polyhedron is a *PL* sphere, and not just a *PL* manifold. In all dimensions except 3 and 4 it is known that a *PL* manifold is a *PL* sphere if it is homotopy equivalent to the sphere; this is the Poincaré conjecture in the *PL* category. (Freedman proved the Poincaré conjecture in the topological category in dimension 4, but it remains open in the *PL* category, which is equivalent to the problem in the smooth category in dimension 4. In dimension 3 the smooth, *PL*, and topological categories are equivalent, and the Poincaré conjecture is unsolved for all of them.)

Assuming that the *PL* Poincaré conjecture is true in all dimensions there is a nicer statement about recognizing when a polyhedron is a *PL* manifold, namely that it suffices to show that the links of all the simplices in the polyhedron are homotopy spheres. That is, we can drop the assumption that they be *PL* manifolds. We can get that property back by induction on the dimension, if we know the Poincaré conjecture in all dimensions.

So in some sense there is a nice recognition theorem in the *PL* category, modulo the remaining parts of the Poincaré conjecture. But there is an important catch here. To show that a polyhedron is a *PL* manifold one has to show that its links are homotopy spheres, that they are simply connected in particular. This may sound like a perfectly reasonable condition to impose, but it is algorithmically unsolvable. There is no algorithm which can determine whether a given finite polyhedron is simply connected or not. This is a well-known consequence of the unsolvability of the word problem for finitely presented groups.

So even in the *PL* category, where everything is finitely complex in principle, we still run into serious difficulty with complexity.

7 Positive results and measure theory

So let us return to the problem of knowing when a space admits a quasisymmetric or bilipschitz coordinates from $\mathbf{R}^n$. For simplicity let us assume that we are working with a subset of some higher-dimensional Euclidean space.

We know that there are examples of nice spaces which do not admit such coordinates even though the geometry is pretty nice, so what kind of good structure can we detect?

I don't want to go into this in much detail, but I would like to at least mention that there are some nice "measure-theoretic" results. If one wishes to find criteria for the existence of bilipschitz coordinates for a set, then it is reasonable to impose the requirement that the given set behave like a Euclidean space of the relevant dimension measure-theoretically, i.e., that it have the correct Hausdorff dimension, locally finite Hausdorff measure in that dimension, etc. This is reasonable in that it is fairly checkable in principle, a matter of covering the set by little balls and asking how many there are. This kind of condition is not completely reasonable for the question of the existence of quasisymmetric coordinates, since the existence of quasisymmetric coordinates would not constrain the Hausdorff dimension the way that bilipschitz coordinates would, but even in that case it is not such an unreasonable requirement to make, so little is known anyway.

Under reasonable measure-theoretic conditions, there are good tools from geometric measure theory [Fd] to check when a set is "rectifiable", which means that it is contained in a countable union of C^1 submanifolds of the correct dimension, except for a set of zero Hausdorff measure. It roughly means the existence of tangents almost everywhere. A set must be rectifiable if it admits bilipschitz coordinates, or if it admits quasisymmetric coordinates and has locally finite Hausdorff measure in the correct dimension. (See [V1] for the second part.) Even if such coordinates do not exist, or if it is impossible to verify, it can be much easier to prove rectifiability.

There is a stronger notion of "uniform rectifiability", in which one might try to find a bilipschitz parameterization for 50% of a given set inside a given ball, with uniform bounds. For this notion there are also a lot of positive results, and many of the results that one would like to be true are true. See the exposition [Se3] for an introduction. Basic references include [D1], [D2], [DS2], [DS3], [DS4].

In uniform rectifiability one makes a trade of topology for estimates. One gives up on getting an actual homeomorphic parameterization of the whole set, and instead settles for a parameterization of only part of it, or a global parameterization which is not a homeomorphism but has some holes and crossings. In exchange for this one demands estimates, and one gets them. It is a statistical phenomenon, what happens if one asks for good behavior most of the time, or on average, instead of everywhere?

Uniform rectifiability has a lot of nice features. In some analytical sense it captures all the good behavior, all the smoothness, that one should have if a given set did admit bilipschitz coordinates. It provides a more flexible theory which captures certain aspects of the geometry of sets that admit bilipschitz parameterizations (or quasisymmetric parameterizations in the presence of suitable measure-theoretic assumptions), it captures very well the amount of smoothness of these sets and the amount of singularity, but it does not capture topological aspects very well.

What if one does want to capture the topology better?

8 Back to topology

What happens if one tries to get estimates in topology?

Topological dimension theory [HW] is interesting in this regard. One gets estimates for some things but not others, it is a nice mixture. (In order to appreciate a bound on the topological dimension it helps to experience a situation where one needs to build mappings through an induction on dimension, e.g., using information about homotopy groups to go from one level to the next. In this case a bound on the topological dimension gives a bound on the number of levels that one has to consider. To see the non-quantitative nature of [HW] notice the role of the Baire category theorem in the proofs.)

For quantitative versions of Alexander duality one has the nice paper of Väisälä [V3].

In building homeomorphisms in topology it seems that one often uses some sort of genericity theorems, Baire category arguments or transversality theorems, which do not lend themselves well to quantitative estimates.

Here is a simple model situation for transversality. Suppose that one has a smooth mapping f from a smooth manifold M of dimension m to a smooth mapping N of dimension n. Almost every value of f is a regular value, and so for almost every $p \in N$ we have that $f^{-1}(p)$ is a smooth embedded submanifold of M of dimension $m - n$. What quantitative estimates might we be able to get? We can get some, on the average volume for instance, but it is hard to get a quantitative statement which is as nice as the qualitative result that $f^{-1}(p)$ is almost surely a smooth submanifold.

So what can one say about complexity in topology? Let us start with the question of cardinality: how many compact topological manifolds are there, up to homeomorphism type? Countably many or uncountably many? The answer is countably many, which is nice, the opposite would be pretty scary. This is proved in [CK]. In order to prove this one needs to know something about homeomorphisms, how to deform them and glue them together, and this is a substantial question. Basic technology for doing this is provided by [Ce], [EK]. The method involves infinite processes which are fairly concrete even if they do not lend themselves easily to quantification. Sullivan [Su] gave another way to do this based on hyperbolic geometry which can give better estimates for the mappings produced (it works for bilipschitz and quasisymmetric mappings while the earlier method did not).

Note that [CK] is nicely down-to-earth, in the sense that it does not rely on the Poincaré conjecture, surgery theory, or anything like that.

So there are only countably many compact topological manifolds, what can we say about estimates? There has been a lot of interest over the years in finding geometric conditions on a collection of manifolds which ensure that there are only finitely many topological types. (See [Ch], [GP], [GPW], [P1], [P2], for instance.) I would like to describe now a theorem of Ferry [Fe3] in this direction.

Fix a positive integer n, the dimension. We want to look at collections $\mathcal{M}$ of compact topological manifolds of dimension n, and we assume that they are given to us as metric spaces, i.e., with specified distance functions. We would like to have modest conditions on these metric spaces which ensure that there are only finitely many topological types.

We first impose geometric "size" conditions on our metric spaces which imply precompactness in the Gromov-Hausdorff sense ([BS], [G1], [G2], [P]). We require that all the metric spaces in the collection $\mathcal{M}$ have uniformly bounded diameter, and that there exist a function $N : (0,1) \to [1,\infty)$ so that if $0 < t < 1$ and $M \in \mathcal{M}$, then M can be covered by at most $N(t)$ balls of radius t. Note that for any compact metric space we can always find such a "covering function" $N(t)$, the point here is to require a uniform choice of covering function $N(t)$ for the spaces in $\mathcal{M}$.

This is the first quantitative assumption on $\mathcal{M}$, the second controls the topology of the elements of $\mathcal{M}$ in a uniform way. We require that there exist a function $\rho : [0, R] \to [0,\infty)$ which is continuous at the origin and satisfies $\rho(0) = 0$, $\rho(t) \geq t$ such that if $M \in \mathcal{M}$, $x \in M$, and $0 < t \leq R$, then the ball in M with center x and radius t can be contracted to a point inside the ball with center x and radius $\rho(t)$. That is, the natural inclusion mapping of the first ball into M should be homotopic to a constant as a mapping into the second ball. Again for a given compact topological manifold we can always find a contractability function $\rho(t)$ like this, the point is to have a uniform choice of contractability function for the whole collection $\mathcal{M}$. We also want a uniform choice of the maximal radius $R > 0$ for this contractability.

Under these conditions Ferry [Fe3] proves that there are only finitely many homeomorphism types of the manifolds $\mathcal{M}$, at least when $n \neq 3$. (When $n = 3$ we run into the Poincaré conjecture.)

This is pretty neat. This is the sort of theorem that one would like to be true. The hypotheses are extremely reasonable and natural. It is a theorem that says that the topologists really know something about constructing homeomorphisms.

Here is another nice theorem about being able to produce homeomorphisms due to Ferry [Fe1] (see also Chapman and Ferry [CF]). Let M be a compact manifold (without boundary, say) of dimension $n \geq 5$ and equipped with a metric. For each $\epsilon > 0$ there is a $\delta > 0$ so that the following is true. Let N be any other compact topological manifold of dimension n, and suppose that $g : M \to N$ is continuous, surjective, and is a *δ-mapping*, which means that $g^{-1}(p)$ has diameter $< \delta$ for each $p \in N$. Then M and N are homeomorphic to each other, and in fact g can be homotoped through ϵ-mappings to a homeomorphism. To understand what this means notice that if $g^{-1}(p)$ has diameter 0 for each $p \in N$, then g is one-to-one, and hence a homeomorphism. Thus Ferry's theorem says that one can make small errors and still be able to find a homeomorphism around. This is very good, the sort of theorem that one would want to be true. Note that the choice of δ depends on M but not on N.

There is a third result of Ferry's [Fe2] which I would like to describe now, which is also about building homeomorphisms. Let M be a compact topological manifold of dimension ≥ 4. Assume that M comes equipped with a metric, and that M is *locally linearly contractable*, which means that there is a $C > 0$ so that any ball in M of radius $r \leq C^{-1} \operatorname{diam} M$ can be contracted to a point in the ball with the same center but radius Cr. In the terminology above this means that there is a linear contractability function $\rho(t)$. For the conclusion we work with $M \times S^1$, where we use any reasonable product metric, but we should rescale S^1 so that it has the same diameter as M. The conclusion is that there is a constant K which depends only on C and the dimension such that given a metric ball B in $M \times S^1$ with radius $t \leq K^{-1} \operatorname{diam} M \times S^1$ there is a topological ball of dimension $n+1$ inside $M \times S^1$ which contains B and which is contained in a ball of radius Kt. In rough terms, we can approximate metric balls in $M \times S^1$ by topological balls.

To understand what this means notice that this kind of linear approximation of metric balls by topological balls always works in compact smooth manifolds (equipped with a reasonable metric). It need not work for general topological manifolds with general metrics. If the space admits local bilipschitz or quasisymmetric coordinates then this kind of property works again. But we don't really know how to tell when a space admits local bilipschitz or quasisymmetric coordinates. We would be very happy if something like the linear contractability condition or linear approximation of metric balls by topological balls were sufficient to ensure the existence of quasisymmetric coordinates, but this is not true, because of examples in [Tu], [V2], [AV]. (See [Se5] for examples where the measure theory is also controlled.) Ferry's theorem is something that we would like to be true, and it is of a similar flavor as statements about the existence of quasisymmetric coordinates which we might like to be true but which are not.

The stabilization in Ferry's theorem (the passage from M to $M \times S^1$) is very natural in the sense that it is a common phenomenon in topology that stabilization is needed and helpful. There are examples in this case to show that it is needed here too.

It is not completely clear to me whether stabilization might have a beneficial effect in problems about the existence of bilipschitz or quasisymmetric parameterization. The examples in [Tu], [V2], [AV] show that stabilization is often bad for the existence quasisymmetric coordinates, but in these examples the spaces are not well behaved measure-theoretically, and I don't know clear results about stabilization when the measure theory behaves well. The examples in [Se5] are well behaved in terms of measure theory, and they clearly do become better spaces after stabilization, but the existence of quasisymmetric coordinates are unclear for them.

The bottom line is that the topologists know a lot about being able to build homeomorphisms in situations where you might want them to be able to do, even with some quantitative control. Control on how far things move, though, not bounds on the moduli of continuity of their homeomorphisms. Their methods do not seem to be geared to this, and there are examples which indicate that it just does not work. (See [Se4], [Se5].)

9 Compromising between topology and estimates

So there is a lot of technology for building homeomorphisms, but not with bounds on their moduli of continuity. "Uniform rectifiability" (as discussed in Section 7) provides tools for constructing bilipschitz parameterizations of large pieces of a given set, and for constructing other types of parameterizations which have good bounds but which also have some kind of topological errors which are controlled only in measure. Can we do something in between, where we have more bounds than topologists normally provide, but more respect for topology than in the measure-theoretic methods?

This is not a well-posed question, and there is plenty of room for interpretation, but one reasonable interpretation of this question is addressed in [DS1], [Se6], [Se7]. Two reasonable interpretations actually, two interpretations which are in a kind of duality.

To motivate the first interpretation let us begin by recalling that for each pair of points x and y in some Euclidean space there is a curve of length $|x - y|$ which connects them, a line segment, and this curve is unique. If we permit curves of length $\leq 2\,|x - y|$, however, then there are plenty of them. We can even choose a nice family of curves, which never overlap but which fill up a region between x and y, a region that looks like the union of two truncated cones with vertices at x and y. We can choose this family of curves so that the arc length measures are well-distributed across space, so that they average out to look like Lebesgue measure, with no unwanted concentrations, just the necessary concentrations at x and y.

These families of curves are useful if one wants to control the values of a function in terms of averages of its gradient in the traditional manner of harmonic analysts (and for the purpose of proving inequalities of Poincaré and Sobolev type). That is, given a function f and a pair of points x and y, we can control $|f(x) - f(y)|$ in terms of the integral of ∇f along any curve that connects x and y. Given a nice family of curves, we can get a good bound for $|f(x) - f(y)|$ in terms of averages of $|\nabla f|$ by doing this for each curve and then averaging over the curves.

So can one find geometric conditions on a space under which nice families of curves like this exist? Curves that are so plentiful and well-distributed that their arc length measures average out to be approximately like the volume measure on the space?

Criteria of this nature are given in [DS1], [Se6], [Se7]. Of course such families of curves exist whenever one has bilipschitz coordinates, because one can then bring the curves over from Euclidean space. It turns out that these curve families will also exist when there are quasisymmetric coordinates and certain measure-theoretic assumptions are satisfied, because of [DS1]. These criteria for the existence of nice families of curves lead to criteria for Poincaré and Sobolev-type inequalities to hold.

A different way to try to respect the topology but still get estimates is the following. Let M be a compact metric space which is also a topological manifold of dimension n. If M admits bilipschitz or quasisymmetric coordinates, then given a metric ball B in M with radius r one can build a mapping from M to the n-sphere $\mathbf{S}^n$ which is constant outside of B, which is topologically nontrivial (homotopically nontrivial, say, or has degree 1 when M is oriented), and which has Lipschitz norm $\leq C\,r^{-1}$ for some constant C. Even when M does not have such good coordinates we can try to find criteria under which such mappings exist, we can move them around in a controlled manner, etc. Criteria of this variety are given in [Se6], related constructions are given in [DS1], [Se7].

These two stories are related. The way that good families of curves are constructed in [DS1], [Se6] is by first building certain mappings with Lipschitz bounds and nontrivial topology, and then obtaining the curves from inside the fibers of the mappings. The co-area theorem from geometric measure theory [Fd] is used to obtain bounds on the distribution of the arc length measures of these curves.

For the purposes of controlling the values of a function in terms of averages of its gradient, one does not necessarily need to go through the story of curves, one can use calculus of differential forms to work directly with special mappings into spheres which enjoy good Lipschitz bounds and which are topologically nontrivial. See [DS1], [Se6], [Se7]. In terms of algebraic topology one can think of using these mappings to produce controlled representatives of a cohomology class (top dimensional and so not so interesting topologically) as well as controlled coboundaries for the difference of two of these representatives.

When thinking about the matter of estimates in topology one should be aware that there is a lot of activity of this general description, if not quite in the vein of the present story. Often people look at the geometry and topology at infinity of a metric space. Here is a specific question of current interest. Let M be a metric space which is homeomorphic to $\mathbf{R}^n$ but which is allowed to have a nonstandard metric. It might look like hyperbolic space for instance. Find good conditions on M which imply that there is a Lipschitz mapping $f : M \to \mathbf{R}^n$ which is proper ($f(x) \to \infty$ as $x \to \infty$) and has nonzero degree (which makes sense for proper maps). Note that M could be much larger than the standard Euclidean space at infinity, in which case f will have to contract distances strongly at infinity. References related to this question include [CGM], [DFW], [FH], [Gr3] (0.3 II'), [Gr4] (especially the beginning of §4 and §$9\frac{1}{4}$). See also [BW], [FeW], [FHP], and [FS] for other matters pertaining to controlled geometry and topology at infinity.

Acknowledgements

I am grateful to A. Carbone, J. Heinonen, R. Rochberg, and K. E. Smith for their comments and suggestions.

REFERENCES

[A1] P. Assouad, *Espaces Métriques, Plongements, Facteurs*, Thèse de Doctorat (January, 1977), Université de Paris XI, 91405 Orsay Cedex, France.

[A2] P. Assouad, *Étude d'une dimension métrique liée à la possibilité de plongement dans* $\mathbf{R}^n$, C. R. Acad. Sci. Paris **288** (1979), 731-734.

[A3] P. Assouad, *Plongements Lipschitziens dans* $\mathbf{R}^n$, Bull. Soc. Math. France **111** (1983), 429-448.

[AV] P. Alestalo and J. Väisälä, *Quasisymmetric embeddings of products of cells into the Euclidean space*, Ann. Acad. Sci. Fenn. Ser. A I Math. **19** (1994), 375-392.

[BW] J. Block and S. Weinberger, *Large scale homology theories and geometry*, preprint.

[BS] M. Bridson and G. Swarup, *On Hausdorff-Gromov convergence and a theorem of Paulin*, L'Enseignement Mathématique **40** (1994), 267-289.

[B^∞1] J. Bryant, S. Ferry, W. Mio, and S. Weinberger, *Topology of homology manifolds*, Bull. Amer. Math. Soc. **28** (1993), 324-328.

[B^∞2] J. Bryant, S. Ferry, W. Mio, and S. Weinberger, *Topology of homology manifolds*, Ann. Math. (2) **143** (1996), 435–467.

[C1] J. Cannon, *The recognition problem: What is a topological manifold?*, Bull. Amer. Math. Soc. **84** (1978), 832-866.

[C2] J. Cannon, *The characterization of topological manifolds of dimension* $n \geq 5$, Proceedings I.C.M. (Helsinki, 1978), 449-454.

[C3] J. Cannon, *Shrinking cell-like decompositions of manifolds: codimension 3*,Ann. Math. (2) **110** (1979), 83-112.

[Ce] A. V. Černavskii, *Local contractibility of the group of homeomorphisms of a manifold*, Math. Sbornik **8** (1969), 287-333.

[CF] T. Chapman and S. Ferry, *Approximating homotopy equivalences by homeomorphisms*, American J. Math. **101** (1979), 583-607.

[Ch] J. Cheeger, *Finiteness theorems for Riemannian manifolds*, American J. Math. **92** (1970), 61-74.

[CK] J. Cheeger and J. Kister, *Counting topological manifolds*, Topology **9** (1970), 149-151.

[CW1] R. Coifman and G. Weiss, *Analyse Harmonique Non-commutative sur Certains Espaces Homogènes*, Lecture Notes in Math. **242** (1971), Springer-Verlag.

[CW2] R. Coifman and G. Weiss, *Extensions of Hardy spaces and their use in analysis*, Bull. Amer. Math. Soc. **83** (1977), 569-645.

[CGM] A. Connes, M. Gromov, and H. Moscovici, *Group cohomology with Lipschitz control and higher signatures*, Geometric and Functional Analysis **3** (1993), 1 - 78.

[D1] G. David, *Morceaux de graphes lipschitziens et intégrales singulières sur un surface*, Rev. Mat. Iberoamericana **4** (1988), 73-114.

[D2] G. David, *Wavelets and Singular Integrals on Curves and Surfaces*, Lecture Notes in Math. **1465** (1991), Springer-Verlag.

[DS1] G. David and S. Semmes, *Strong A_∞-weights, Sobolev inequalities, and quasiconformal mappings*, in *Analysis and Partial Differential Equations*, edited by C. Sadosky, Lecture Notes in Pure and Applied Mathematics **122**, Marcel Dekker 1990.

[DS2] G. David and S. Semmes, *Singular Integrals and Rectifiable Sets in* $\mathbf{R}^n$*: au-delà des graphes lipschitziens*, Astérisque **193**, Société Mathématique de France 1991.

[DS3] G. David and S. Semmes, *Quantitative rectifiability and Lipschitz mappings*, Transactions Amer.Math. Soc. **337** (1993), 855-889.

[DS4] G. David and S. Semmes, *Analysis of and on Uniformly Rectifiable Sets*, Mathematical Surveys and Monographs **38**, 1993, American Mathematical Society.

[DoS] S. Donaldson and D. Sullivan, *Quasiconformal 4-manifolds*, Acta Math. **163** (1989), 181-252.

[DFW] A. Dranishnikov, S. Ferry, and S. Weinberger, *Large Riemannian manifolds which are flexible*, preprint.

[E] R. Edwards, *The topology of manifolds and cell-like maps*, Proceedings I.C.M. (Helsinki, 1978), 111-127.

[EK] R. Edwards and R. Kirby, *Deformations of spaces of embeddings*, Ann. Math. **93** (1971), 63-88.

[FS] B. Farb and R. Schwartz, *The large-scale geometry of Hilbert modular groups*, preprint.

[FH] F. T. Farrell and W. C. Hsiang, *On Novikov's conjecture for non-positively curved manifolds, I*, Ann. Math. (2) **113** (1981), 199-209.

[Fd] H. Federer, *Geometric Measure Theory*, Springer-Verlag, 1969.

[Fe1] S. Ferry, *Homotoping ϵ-maps to homeomorphisms*, American J. Math. **101** (1979), 567-582.

[Fe2] S. Ferry, *Counting simple-homotopy types of topological manifolds*, preprint.

[Fe3] S. Ferry, *Topological finiteness theorems for manifolds in Gromov-Hausdorff space*, Duke Math. J. **74** (1994), 95-106.

[FeW] S. Ferry and S. Weinberger, *A coarse approach to the Novikov conjecture*, preprint.

[FHP] S. Ferry, I. Hambleton, and E. K. Pedersen, *A survey of bounded surgery theory and applications*, preprint.

[GJ] J. Garnett and P. Jones, *The corona theorem for Denjoy domains*, Acta Math. **155** (1985), 29-40.

[Ge] F. W. Gehring, *The L^p integrability of the partial derivatives of a quasiconformal mapping*, Acta Math. **130** (1973), 265-277.

[GP] R. Greene and P. Petersen V, *Little topology, big volume*, Duke. Math. J. **67** (1992), 273-290.

[GPW] K. Grove, P. Petersen V, and J.-Y. Wu, *Geometric finiteness theorems via controlled topology*, Invent. Math **99** (1990), 205-213; erratum, **104** (1991), 221-222.

[Gr1] M.Gromov, *Structures Métriques pour les Variétés Riemanniennes* (J. Lafontaine, P. Pansu, ed.), Cedic/Fernand Nathan, Paris, 1981.

[Gr2] M.Gromov, *Groups of polynomial growth and expanding maps*, Publ. Math. IHES **53** (1981), 183-215.

[Gr3] M.Gromov, *Asymptotic invariants of infinite groups*, Proceedings of the 1991 conference Sussex conference on Geometric Group Theory, Niblo and Koller, editors, London math. Soc. Lecture Notes **182**, 1993, Cambridge Univ. Press.

[Gr4] M. Gromov, *Positive curvature, macroscopic dimension, spectral gaps and higher signatures*, IHES preprint M/95/36, 1995.

[HK] J. Heinonen and P. Koskela, *Definitions of quasiconformality*, Inv. Math. **120** (1995), 61-79.

[HW] W. Hurewicz and H. Wallman, *Dimension Theory*, Princeton Univ. Press, 1941, 1969.

[J] P. Jones, *L^∞ estimates for the $\overline{\partial}$ problem in a half plane*, Acta Math. **150** (1983), 137-152.

[KS2] R. Kirby and L. Siebenmann, *On the triangulation of manifolds and the Hauptvermutung*, Bull. Amer. Math. Soc. **75** (1969), 742-749.

[KS2] R. Kirby and L. Siebenmann, *Foundational Essays on Topological Manifolds, Smoothings, and Triangulations*, Annals of Math. Studies **88** (1977), Princeton University Press.

[Pa] P. Pansu, *Métriques de Carnot-Carathéodory et quasiisométries des espaces symétriques de rang un*, Ann. Math. **129** (1989), 1-60.

[P1] P. Petersen V, *A finiteness theorem for metric spaces*, J. Diff. Geom. **31** (1990), 387-395.

[P2] P. Petersen V, *Gromov-Hausdorff convergence of metric spaces*, Proc. Sym. Pure Math. **54**, Part 3, (1993), 489-504, Amer. Math. Soc.

[Q1] F. Quinn, *Resolutions of manifolds, and the topological characterization of manifolds*, Invent. Math. **72** (1983), 267-284.

[Q2] F. Quinn, *An obstruction to the resolution of homology manifolds*, Michigan Math. J. **301** (1987), 285-292.

[Se1] S. Semmes, *Chord-arc surfaces with small constant II: Good parameterizations*, Adv. Math. **88** (1991), p170 -199.

[Se2] S. Semmes, *Bilipschitz mappings and strong A_∞ weights*, Ann. Acad. Sci. Fenn. A I **18** (1993), 211-248.

[Se3] S. Semmes, *Finding structure in sets with little smoothness*, Proc. I.C.M. (Zürich, 1994), 875-885, Birkhäuser, 1995.

[Se4] S. Semmes, *On the nonexistence of bilipschitz parameterizations and geometric problems about A_∞ weights*, Revista Matemática Iberoamericana **12** (1996), 337–410.

[Se5] S. Semmes, *Good metric spaces without good parameterizations*, , Revista Matemática Iberoamericana **12** (1996), 187–275.

[Se6] S. Semmes, *Finding Curves on General Spaces through Quantitative Topology with Applications for Sobolev and Poincaré Inequalities*, Selecta Math (New Series) **2** (1996), 155–295.

[Se7] S. Semmes, *Some remarks about metric spaces, spherical mappings, functions and their derivatives*, to appear, Publicacions Matemàtiques.

[Si] L. Siebenmann, *Topological Manifolds*, Proc. I.C.M. Nice (1970) Vol.2, 133-163, Gauthier-Villars, Paris, 1971.

[SS] L. Siebenmann and D. Sullivan, *On complexes that are Lipschitz manifolds*, "Geometric Topology", J. Cantrell, ed., Academic Press, 1979, 503-525.

[Su] D. Sullivan, *Hyperbolic geometry and homeomorphisms*, "Geometric Topology", (J. Cantrell, ed.), Academic Press, 1979, 543-555.

[T1] T. Toro, *Surfaces with generalized second fundamental form in L^2 are Lipschitz manifolds*, J. Diff. Geom. **39** (1994), 65-102.

[T2] T. Toro, *Geometric conditions and existence of bilipschitz parameterizations*, Duke Math. J. **77** (1995), 193-227.

[Tu] P. Tukia, *A quasiconformal group not isomorphic to a Möbius group*, Ann. Acad. Sci. Fenn. Ser. A I Math. **6** (1981), 149-160.

[TV1] P. Tukia and J. Väisälä, *Quasisymmetric embeddings of metric spaces*, Ann. Acad. Sci. Fenn. Ser. A I Math. 5 (1980), 97–114.

[TV2] P. Tukia and J. Väisälä, *Lipschitz and quasiconformal approximation and extension*, Ann. Acad. Sci. Fenn. Ser. A I Math. **6** (1981), 303-342.

[TV3] P. Tukia and J. Väisälä, *Quasiconformal extension from dimension n to $n+1$*, Ann. Math. **115** (1982), 331-348.

[V1] J. Väisälä, *Quasisymmetric embeddings in Euclidean spaces*, Trans. Amer. Math. Soc. **264** (1981), 191-204.

[V2] J. Väisälä, *Quasisymmetric maps of products of curves into the plane*, Rev. Roumaine Math. Pures Appl. **33** (1988), 147-156.

[V3] J. Väisälä, *Metric duality in Euclidean spaces*, preprint.

Rice University
Department of Mathematics - MS 136
Wiess School of Natural Sciences
6100 Main Street
Houston, TX 77005-1892

Questions on Quasiconformal Maps in Space

Jussi Väisälä

1 Introduction

The purpose of this article is to list miscellaneous open questions on quasiconformal maps in R^n.

I recall the basic definitions. Let G and G' be domains in R^n, $n \geq 2$, let $f : G \to G'$ be a homeomorphism, and let $K \geq 1$. We say that f is *K-quasiconformal* if

$$M(\Gamma)/K \leq M(f\Gamma) \leq KM(\Gamma)$$

for each path family Γ in G, where $M(\Gamma)$ is the n-modulus of Γ. The smallest K satisfying the first or the second inequality is the *outer dilatation* $K_O(f)$ or the *inner dilatation* $K_I(f)$ of f, respectively. Alternatively, quasiconformality can be characterized analytically: f is quasiconformal with $K_O(f) \leq K$ if and only if f is in the Sobolev space $W^1_{n,\mathrm{loc}}$ and

$$|f'(x)|^n \leq KJ_f(x)$$

for almost every $x \in G$; here $|f'(x)|$ is the max-norm of the derivative $f'(x)$ and $J_f(x)$ is the jacobian. A third characterization for quasiconformality is metrical: For $x \in G$ and $0 < r < d(x, \partial G)$ we set

$$L(x, f, r) = \max \{|fy - fx| : |y - x| = r\},$$
$$l(x, f, r) = \min \{|fy - fx| : |y - x| = r\},$$
$$H(x, f) = \limsup_{r \to 0} \frac{L(x, f, r)}{l(x, f, r)}.$$

Then f is quasiconformal if and only if there is $c \geq 1$ such that (1) $H(x, f) < \infty$ for all $x \in G$ and (2) $H(x, f) \leq c$ a.e. The smallest c for which this is true is the *metric* (or linear) *dilatation* of f, written as $H(f)$.

In the case $n = 2$ we have $K_I(f) = K_O(f) = H(f)$. In the general case, these numbers need not be equal, but they satisfy the following inequalities:

$$K_I(f) \leq K_O(f)^{n-1}, \; K_O(f) \leq K_I(f)^{n-1},$$

$$K_I(f) \le H(f)^{n-1}, \; K_O(f) \le H(f)^{n-1},$$
$$H(f) \le K_I(f), \; H(f) \le K_O(f),$$
$$H(f) \le (K_I(f)K_O(f))^{1/n}.$$

2 Liouville's theorem

This theorem states that if $n \ge 3$ and if f is 1-quasiconformal, then f is a Möbius map. In this form this result was proved by Gehring [Ge1] in 1962; see also [Re]. Gehring's proof was based on deep results on the regularity theory of partial differential equations. A simpler proof was given by Bojarski and Iwaniec [BI] in 1982. However, also this proof is highly analytic. On the other hand, a map $f : G \to G'$ is Möbius if and only if the image of each ball in G is a ball or a half space. It is therefore natural to ask:

Question 1. Is there a geometric proof of Liouville's theorem?

3 The L^p problem

Let $f : G \to G'$ be quasiconformal with $K_O(f) \le K$. From the analytic definition it easily follows that the partial derivatives are locally in L^p for $1 \le p \le n$. A remarkable fact is that they are indeed in L^p also for some $p > n$. For $n = 2$, this was proved by Bojarski [Bo] in 1955, and the general case was established by Gehring [Ge2] in 1973. Indeed, there is a number $p = p(K,n) > n$ such that the partial derivatives are locally in L^p for $1 \le p < p(K,n)$. The radial map for which $|fx| = |x|^{1/K}$ shows that $p(K,n)$ can be at most $nK/(K-1)$. A long-standing conjecture is that one can choose $p(K,n) = nK/(K-1)$. For $n = 2$, this was proved recently by Astala [As]. However, the general case is still open:

Question 2. If f is an n-dimensional quasiconformal map with $K_O(f) \le K$, are the partial derivatives locally in L^p for all $1 \le p < nK/(K-1)$?

4 Convergent sequences

Assume that $f_j : G \to G'$ is a sequence of K-quasiconformal maps converging pointwise to a limit map $f : G \to R^n$. It is known that f is either a constant or a K-quasiconformal map onto G'. Assume that the latter case occurs. It is well known that the convergence is uniform on compact sets and that

$$K_I(f) \leq \limsup_{j\to\infty} K_I(f_j), \;\; K_O(f) \leq \limsup_{j\to\infty} K_O(f_j).$$

From these and from the basic inequalities in Section 1 it follows that

$$H(f) \leq \limsup_{j\to\infty} H(f_j)^{2-2/n}.$$

However, the following question is open:

Question 3. In the situation above, is

$$H(f) \leq \limsup_{j\to\infty} H(f_j)?$$

Added in proof. The question has been answered negatively by T. Iwaniec.

5 Dimension-free results

The definition of the metric dilatation $H(f)$ is dimension-free. In fact, it can be considered for maps between fairly arbitrary metric spaces. On the other hand, the definitions of the dilatations $K_I(f)$ and $K_O(f)$ involve the dimension n, and they cannot be considered in infinite-dimensional spaces. By a *dimension-free result* I therefore mean a result where the condition $H(f) \leq c$ implies some global property of f, which can be expressed in terms of c alone. As an example, I give the following result, proved by Kühnau [Kü] in 1962:

Let G and G' be the spherical rings $1 < |x| < R$ and $1 < |x| < R'$ in R^n with $R \leq R'$. Let $f : G \to G'$ be quasiconformal. Then $\log R'/\log R \leq H(f)$.

To prove this we consider the family Γ of all open paths joining the boundary components of G. A well known modulus formula gives

$$\left(\frac{\log R'}{\log R}\right)^{n-1} = \frac{M(\Gamma)}{M(\Gamma')} \leq K_O(f).$$

Since $K_O(f) \leq H(f)^{n-1}$, this proves the result.

As Kühnau points out, the result is implicitly contained in R. Nevanlinna's paper [Ne1] published in 1955. Nevanlinna's proof was based on the isoperimetric inequality.

I do not know of other dimension-free results in this sense. In particular, the following questions are open:

Question 4. Suppose that $f : R^n \to R^n$ is quasiconformal with $H(f) \leq c$ an that $f(0) = 0$. Let $x, y \in R^n$ with $|x| = |y|$. It is known that $|fy| \leq \lambda(c,n)|fx|$. Is this true with $\lambda = \lambda(c)$?

Question 5. (Schwarz lemma) Let $f : B^n \to B^n$ be quasiconformal with $H(f) \le c$ and $f(0) = 0$. It is known that

$$|fx| \le \varphi(|x|, c, n) < 1$$

for all $x \in B^n$. Is this true with φ depending only on $|x|$ and c?

Remark. Anderson, Vamanamurthy and Vuorinen [AVV] have proved the above results with $\lambda = \lambda(c^{n-1})$, $\varphi = \varphi(|x|, c^{n-1})$.

6 Infinite-dimensional results

Let E be an infinite-dimensional Banach space, for example, a separable Hilbert space. Let G and G' be domains in E and let $f : G \to G'$ be a homeomorphism. To avoid technical difficulties we assume that f is a diffeomorphism. This means that the Fréchet derivative $f'(x)$ exists as a bounded operator at each point $x \in G$ and that the map $f' : G \to L(E, E)$ is continuous. Moreover, f^{-1} has the same properties. The metric dilatation $H_f(x)$ can be defined as before, and we have

$$H_f(x) = |f'(x)||(f^{-1})'(fx)|.$$

We set $H(f) = \sup \{H_f(x) : x \in G\}$.

Question 6. In the situation above, does the condition $H(f) \le c$ imply any global properties of f?

The only result I know in this direction is that if E is a Hilbert space, f is C^4 and $H(f) = 1$, then f is Möbius; see [Ne2]. For example, we may ask whether Kühnau's result on spherical rings in Section 4 is true in every Hilbert space.

I remark that there is a theory of quasiconformality in arbitrary Banach spaces [Vä], but the definition is based on local quasisymmetry or on the quasihyperbolic metric, and it is not known whether maps with $H(f) < \infty$ are quasiconformal in this sense.

7 Quasisymmetry

Quasisymmetry is an offspring of quasiconformality. It is a global property, while quasiconformality is a local property. For maps $f : R^n \to R^n$ they are equivalent. For homeomorphisms between domains in R^n, quasiconformality is equivalent to local quasisymmetry. On the other hand, quasisymmetry is a useful concept in a more general setting. Let X and Y be metric spaces with distance written as $|a - b|$, and let $f : X \to Y$ be injective. Let $\eta : [0,\infty) \to [0,\infty)$ be a homeomorphism. We say that f is *η-quasisymmetric* if $|a-x| \le t|b-x|$ implies $|fa-fx| \le \eta(t)|fb-fx|$ for all x, a, b in X and for all $t \ge 0$. For maps $f : \mathbf{R} \to \mathbf{R}$, this is equivalent to the well-known Beurling-Ahlfors condition.

The relations between quasisymmetry and measure are not easy to prove. It was shown already by Beurling and Ahlfors [BA] that a quasisymmetric map $f : \mathbf{R} \to \mathbf{R}$ may map a set of measure zero onto a set of positive measure. For maps $f : R^n \to R^n$, $n \ge 2$, this is not possible, since the map is quasiconformal. Until very recently, it was not known whether $m(A) = 0$ implies $m(fA) = 0$ for quasisymmetric maps $f : A \to R^n$, $A \subset R^n$, $n \ge 2$. However, a positive answer has now been given by Semmes [Se]. The following question is still open:

Question 7. Let $A \subset R^n$, $n \ge 2$, let $x \in A$ be a point of density in A, and let $f : A \to R^n$ be quasisymmetric. Is fx a point of density of fA?

Observe that this question is non-trivial even if f is bilipschitz. However, a positive answer in this case was given by Buczolich [Bu] in 1992.

Consider a set $A \subset R^n$ and a topological embedding $f : A \to R^n$. In general, there is no extension of f to a homeomorphism $g : R^n \to R^n$. However, there is always an extension to a homeomorphism $F : R^{2n} \to R^{2n}$; cf. [Ru, pp. 74–75]. Moreover, if f is M-bilipschitz, then F can be chosen to be $3M^2$-bilipschitz; see [AV, 5.5]. It is therefore natural to ask:

Question 8. Let $A \subset R^n$ and let $f : A \to R^n$ be η-quasisymmetric. Can f be extended to a K-quasiconformal map $F : R^{2n} \to R^{2n}$ with $K = K(\eta, n)$?

In this form the answer is known to be negative, since there are quasisymmetric maps between sets in $\mathbf{R}$ which are not Hölder continuous; see [TV, 3.16.1]. One must therefore add a condition which guarantees that f is Hölder continuous, for example, that A is connected.

8 The decomposition problem

Suppose that $f : R^2 \to R^2$ is quasiconformal and that $K > 1$. Then f can be written as a composite map of a finite number of K-quasiconformal maps $R^2 \to R^2$; see [Ah, p. 99]. The corresponding problem in dimensions $n \ge 3$ is open:

Question 9. Suppose that $n \ge 3$, that $f : R^n \to R^n$ is quasiconformal and that $K > 1$. Is it possible to write $f = g_1 \circ \ldots \circ g_s$ where each $g_j : R^n \to R^n$ is K-quasiconformal?

References

[Ah] Ahlfors, L.V., *Lectures on quasiconformal mappings*, Van Nostrand, 1966.

[AV] Alestalo, P., Väisälä, J., *Uniform domains of higher order III*, Ann. Acad. Sci. Fenn. Ser. AI Math., to appear.

[AVV] Anderson, G., Vamanamurthy, M.K., Vuorinen, M., *Dimension-free quasiconformal distortion in n-space*, Trans. Amer. Math. Soc. **297** (1986), 687–706.

[As] Astala, K., *Area distortion of quasiconformal mappings*, Acta Math. **173** (1994), 37–60.

[BA] Beurling, A., Ahlfors, L., *The boundary correspondence under quasiconformal mappings*, Acta Math. **96** (1956), 125–142.

[Bo] Bojarski, B.V., *Homeomorphic solutions of a Beltrami system*, Dokl. Akad. Nauk. SSSR **102** (1955), 661–664.

[BI] Bojarski, B.V., Iwaniec, T., *Another approach to Liouville theorem*, Math. Nachr. **107** (1982), 253–262.

[Bu] Buczolich, Z., *Density points and bi-Lipschitz functions in R^m*, Proc. Amer. Math. Soc. **116** (1992), 53–59.

[Ge1] Gehring, F.W., *Rings and quasiconformal mappings in space*, Trans. Amer. Math. Soc. **103** (1962), 353–393.

[Ge2] Gehring, F.W., *The L^p-integrability of the partial derivatives of a quasiconformal mapping*, Acta Math. **130** (1973), 265–277.

[Kü] Kühnau, R., *Elementare Beispiele möglichst konformen Abbildungen im dreidimensionalen Raum*, Wiss. Z. Martin-Luther-Univ. Halle-Wittenberg Math.-Natur. Reihe **11** (1962), 729–732.

[Ne1] Nevanlinna, R., *A remark on differentiable mappings*, Michigan Math. J. **3** (1955/56), 53–57.

[Ne2] Nevanlinna, R., *On differentiable mappings*, Analytic Functions, ed. by L. Ahlfors et al., Princeton University Press, 3–9 (1960).

[Re] Reshetnyak, Ju. G., *On conformal mappings of a space*, Soviet Math. Dokl. **1** (1960), 153–155.

[Ru] Rushing, T.B., *Topological embeddings*, Academic Press, 1973.

[Se] Semmes, S., *On a question of Heinonen concerning the absolute continuity of quasisymmetric mappings*, Rev. Mat. Iberoamericana. To appear.

[TV] Tukia, P., Väisälä, J., *Quasisymmetric embeddings of metric spaces*, Ann. Acad. Sci. Fenn. Ser. A I Math. **5** (1980), 97–114.

[Vä] Väisälä, J., *Free quasiconformality in Banach spaces I*, Ann. Acad. Sci. Fenn. Ser. A I Math. **15** (1990), 355–379.

Matematiikan laitos
Helsingin yliopisto
Helsinki, Finland
jvaisala@cc.helsinki.fi

List of Contributors

Kari Astala
University of Jyväskylä
Department of Mathematics
P.O. Box 35, FIN-40351
Jyväskylä, Finland
astala@math.jyu.fi

Albert Baernstein, II
Washington University
St. Louis, MO 63130
al@math.wustl.edu

A.F. Beardon
Department of Pure Mathematics and Mathematical Statistics
University of Cambridge
16 Mill Lane, Cambridge CB2 1SB
England
A.F.Beardon@pmms.cam.ac.uk

C. Cao
University of Michigan
Ann Arbor, MI 48109
ccao@math.lsa.umich.edu

Peter Duren, Editor
Department of Mathematics
University of Michigan
435 S. State Street, Suite 3014
Ann Arbor, MI 48109-1109
duren@math.umich.edu

Clifford J. Earle
Department of Mathematics
Cornell University
Ithaca, New York 14853
cliff@math.cornell.edu

V.Ya. Gutlyanskii
Institute of Applied Mathematics and Mechanics
NAS of Ukraine
ul. Roze Luxemburg 74
340114, Donetsk, Ukraine

Juha Heinonen, Editor
Department of Mathematics
University of Michigan
435 S. State Street, Suite 3014
Ann Arbor, MI 48109-1109
juha@math.lsa.umich.edu

Tadeusz Iwaniec
Department of Mathematics
215 Carnegie
Syracuse University
Syracuse, NY 13244
tiwaniec@mailbox.syr.edu

Pekka Koskela
Department of Mathematics
University of Jyväskylä
P.O. Box 35, Fin-40351
Jyväskylä, Finland
koskela@math.jyu.fi

Irwin Kra
Department of Mathematics
State University of New York at Stony Brook
Stony Brook, NY 11794
irwin@math.sunysb.edu

Gaven J. Martin
Australian National University & University of Auckland
Canberra & Auckland
Australia & New Zealand
martin@math.auckland.ac.nz

Olli Martio
Department of Mathematics
University of Helsinki
P.O. Box 4 (Yliopistonk. 5)
FIN–00014, Helsinki, Finland

Brad Osgood, Editor
Department of Mathematics
Stanford University
Stanford, CA 94305-2125
brad@math.stanford.edu

Bruce Palka, Editor
Department of Mathematics
The University of Texas at Austin
Austin, TX 78712-1082
palka@math.utexas.edu

Ch. Pommerenke & S. Rohde
Fachbereich Mathematik
TU Berlin
10623 Berlin, Germany
rohde@math.tu-berlin.de

Edgar Reich
Department of Mathematics
127 Vincent Hall
206 Church Street S.E.
University of Minnesota
Minneapolis, MN 55455
reich@math.umn.edu

Seppo Rickman
Department of Mathematics
University of Helsinki
P.O. Box 4 (Yliopistonkatu 5)
FIN–00014, Finland
rickman@cc.helsinki.fi

V.I. Ryazanov
Institute of Applied Mathematics and Mechanics
NAS of Ukraine
ul. Roze Luxemburg 74
340114, Donetsk, Ukraine

Stephen Semmes
Rice University
Department of Mathematics – MS 136
Wiess School of Natural Sciences
6100 Main Street
Houston, TX 77005-1892
semmes@ihes.fr

Jussi Väisälä
Matematiikan laitos
Helsingin yliopisto
Helsinki, Finland
jvaisala@cc.helsinki.fi

Jade P. Vinson
Washington University
St. Louis, MO 63130
jpvinson@artsci.wustl.edu

Matti Vuorinen
Department of Mathematics
University of Helsinki
P.O. Box 4 (Yliopistonk. 5)
FIN-00014, Helsinki, Finland
vuorinen@convex.csc.fi

P.L. Waterman
Northern Illinois University
Dekalb, IL 60115
waterman@math.niu.edu

Li Zhong
Department of Mathematics
Peking University, Beijing
100871, P.R. China

SPRINGER NATURE

GPSR Compliance

The European Union's (EU) General Product Safety Regulation (GPSR) is a set of rules that requires consumer products to be safe and our obligations to ensure this.

If you have any concerns about our products, you can contact us on ProductSafety@springernature.com

In case Publisher is established outside the EU, the EU authorized representative is:

Springer Nature Customer Service Center GmbH
Europaplatz 3
69115 Heidelberg, Germany

The manufacturer's authorised representative in the EU is Springer Nature Customer Service Centre GmbH, Europaplatz 3, 69115 Heidelberg, Germany. If you have any concerns regarding our products, please contact ProductSafety@springernature.com

Printed and bound by CPI Group (UK) Ltd, Croydon, CR0 4YY
29/07/2026
02180528-0008